THE COMMONWEALTH AND INTERNATIONAL LIBRARY

MATHEMATICAL TOPICS

General Editor: C. PLUMPTON

THE PROCESS OF LEARNING MATHEMATICS

THE PROCESS OF LEARNING MATHEMATICS

EDITED BY

L. R. CHAPMAN

PERGAMON PRESS

OXFORD · NEW YORK · TORONTO
SYDNEY · BRAUNSCHWEIG

Pergamon Press Ltd., Headington Hill Hall, Oxford
Pergamon Press Inc., Maxwell House, Fairview Park, Elmsford, New York 10523
Pergamon of Canada Ltd., 207 Queen's Quay West, Toronto 1
Pergamon Press (Aust.) Pty. Ltd., 19a Boundary Street, Rushcutters Bay, N.S.W. 2011, Australia
Vieweg & Sohn GmbH, Burgplatz 1, Braunschweig

First edition 1972
Library of Congress Catalog Card No. 71-178683

Printed in Germany

08 016623 7

Contents

Contributors

MARGARET E. BARON, M.SC., PH.D.	Principal Lecturer in Mathematics, Stockwell College, Institute of Education.
RUTH M. BEARD, M.SC., PH.D.	Director, University Teaching Methods Unit, University of London Institute of Education.
EDITH E. BIGGS	Her Majesty's Inspectorate.
W. BRODIE, B.SC.	Rector, Trinity Academy, Edinburgh.
A. P. K. CALDWELL, B.A.	Senior Lecturer in Mathematics, Coloma College, Institute of Education.
L. R. CHAPMAN, B.SC.	Principal Lecturer in Mathematics, Trent Park College, Institute of Education.

J. F. DEANS, B.SC.	Senior Lecturer in Mathematics, Newlands Park College, Institute of Education, Bristol.
W. A. GIBBY, B.SC., M.A.	Senior Lecturer in Mathematics, Institute of Education, University of London.
C. HOPE, O.B.E., B.SC.	Principal Lecturer in Mathematics, Worcester College of Education, Institute of Education, Birmingham.
C. W. KILMISTER, M.SC., PH.D.	Professor of Mathematics, King's College, University of London.
J. A. MERCER, B.A.	Principal Lecturer in Mathematics, Digby Stuart College, Institute of Education.
W. L. BENEDICT NIXON	Lecturer, Institute of Computer Science, University of London.
J. W. OLIVER, B.SC.	Senior Lecturer in Mathematics, Digby Stuart College, Institute of Education.

A. P. PENFOLD, B.SC., M.A., F.I.M.A.
Senior Lecturer in Mathematics,
Institute of Education,
University of London.

C. PLUMPTON, M.A., PH.D.
Senior Lecturer in Mathematics and Director of Engineering Mathematics,
Queen Mary College,
University of London.

R. R. SKEMP, M.A., PH.D., F.B.PS.S.
Child Study Unit,
University of Manchester.

A. G. VOSPER, M.A.
Principal Lecturer in Mathematics,
St. Mary's College,
Institute of Education.

Foreword

IN THE autumn of 1967 a small sub-committee of the Standing Sub-Committee in Mathematics, University of London Institute of Education, was set up to consider possible arrangements for a two-term course of intercollegiate lectures to prepare students for Section K of the syllabus for Part II (Mathematics) of the newly constituted B.Ed. degree. The title of the section is "The Psychology of Mathematical Thinking and Learning". A course of eighteen lectures was designed and the editor of this book was asked to undertake the organisation of the course.

In structuring the course of lectures, the editor was grateful for help from lecturers both inside and outside the Institute, but a special mention must be made of the valuable suggestions offered by the late Professor D. M. Lee, Professor of Education in the University of London Institute of Education. The present lecture programme is the end product of a collective effort.

The arrangements for the academic year 1969/70 were:

Introduction	Lecture 1.	The Nature of Mathematics
	2.	The Nature of Mathematics—another view
Dealing with these elements	3.	The Role of Intuition
	4.	The Role of Logic
Building with these elements	5.	Generalisation and Structure
Psychology bridge	6.	Attempts to Investigate Mathematical Learning

The Learning Process	7.	The Growth of Logical Thinking
	8.	Schematic Learning
	9.	Intuitive and Reflective Functioning of Intelligence
	10.	Motivation, emotional and inter-personal factors
	11.	Investigational Methods
	12.	Structural Apparatus
	13.	Programmed Learning
	14.	Computer-assisted Instruction
Trends in the schools	15.	Midlands Mathematics Experiment
	16.	Modern Mathematics for Schools Scottish Mathematics Group
	17.	Schools Mathematics Project
	18.	Nuffield Mathematics Project

The material and ideas contained in the lectures were freely selected by the individual lecturers, but the contents of each chapter were designed to provide a starting-point for discussions and seminar work in the colleges.

After listening to several of the lectures, I, as course organiser, realised that a more permanent record was worth while; and the book grew out of this idea. At the beginning the aim was narrow; the book was to form the basis of a course for students reading Section K of Part II of the B.Ed. degree. Later it was thought that the lecture material might appeal to a wider audience. Mindful of this enlarged aim, some lecturers have taken the opportunity to modify or augment their original material. It must be stated, however, that the book retains its earlier purpose. It attempts to provide a series of starting-points and is not necessarily to be read as a statement of the most recent theoretical positions. Each chapter can stand alone and the sections can be read in any order, but a perceptive student should discover the recurring themes in the book. Perhaps the chief value of the work lies in the bringing together of ideas from the "whole spectrum of mathematics".

I wish to express my thanks to the *Daily Telegraph*, the University of London and W. R. Chambers Ltd. for permission to reproduce material.

The production of a multi-author volume is not an easy task, and without the constant help and encouragement of a number of friends and colleagues this book would remain incomplete. I wish to thank Dr. Charles Plumpton of Queen Mary College, University of London, for his ready help and wisdom, also my colleagues Paul Butler and Lionel Lane for their constructive criticisms. My daughter Jill has given many hours of clerical assistance, but my main thanks must go to the contributors and to my wife Margaret for her support and helpful suggestions. To all the above I am very grateful.

Trent Park College L. R. CHAPMAN

The Nature of Mathematics

C. W. KILMISTER

THIS course of lectures shows by its existence the surprising coming together and development of questions and ideas from opposite ends of the spectrum of workers in the field of mathematics. If it were otherwise I would have been afraid to accept the invitation to speak; but as it is I look on my function as being to give some healthy entertainment beforehand so that you can get down to the real work in succeeding lectures. And so I must warn you that I have not set out here with the idea of answering any of the questions you may have in your mind about the nature of mathematics. I do not think that now is the time for that. I have preferred to sketch what has been done and to raise questions which will, I think, only be finally resolved, if at all, by uniting the approaches of logic and highbrow mathematics with practical investigations of this mysterious and immensely difficult process of communicating mathematical truth.

Discussions of the nature of mathematics seem to me often to confuse two questions; one of these, which is really one for psychologists, is the nature of mathematical *proof*. The other, essentially for philosophers, is that of mathematical *truth*. Before I deal with these in turn, I want to put the view that what they have in common is that their solutions—if any final solutions are to be expected—are more likely to be found if we break out of the artificial barrier separating mathematics from the other sciences. I shall return to this later.

Let us consider first the psychological question, which is very difficult to pose, let alone answer. Mathematical proof has something in common with everyday argument; yet the extent to which a mathematical argument can be carried on leaves us stunned when we compare the poverty of ordinary argument.

An analogy is helpful here; we are now familiar with computers as devices for doing large amounts of very simple arithmetic. Yet in fact such a machine has been programmed to find proofs of the theorems in Whitehead and Russell's treatment of the first-order predicate calculus, and found proofs of the first 300 or so in 10 minutes. The question about mathematical proof is, then, whether the analogy is a good one—that is, does a mathematical argument differ from a simple verbal one only in repeating certain simple steps many times as the machine's argument did, or is there some characteristic "mathematical" feature of an argument, so that the analogy fails (perhaps because the 300 theorems mentioned were a particularly simple class). This problem has been considered to some extent by Poincaré at the turn of the century, but otherwise, as far as I am aware, has been neglected. Poincaré considered that the characteristic feature entered mathematics as soon as mathematical induction entered a proof; that is, he felt that ordinary finite steps were only of the same kind as ordinary argument, but as soon as an infinite set of steps were inserted in the argument as a single step something different in kind emerged. But I think, though Poincaré may be right in supposing induction to be one way of being definitely mathematical, he was wrong in supposing it to be the only way; for consider carefully the famous proof that $\sqrt{2}$ is irrational. One assumes that the theorem is false so that two integers p, q exist, without common factors, such that

$$p^2 = 2q^2.$$

Then one appeals to a lemma; the square of an even number is even and that of an odd number is odd: $(2k + 1)^2 = 4k(k + 1) + 1$. So p is even $= 2r$ and so

$$2r^2 = q^2.$$

Hence q is even as well and p and q have common factor 2 contrary to hypothesis. Let us try to look freshly at this; when does the surprise come in? Surely at the end, and the *reductio ad absurdum* is what provides it. Many branches of mathematics have this feature—the theory of finite groups and that of graphs proceed almost entirely by this process of fighting forwards one step at a time by dint of showing that the way back is impossible.

But if, instead of seeking a criterion to mark off the mathematical nature of the argument, we look at the rest of science, we see a somewhat similar pattern. What is the nature of scientific "certainty"? This is the philosopher's "problem of induction". Scientists certainly behave as if scientific laws had a degreee of validity not shared by everyday statements. But this is only one side of the picture; tomorrow these certainties may be cast on to one side, refuted by the weight of experimental evidence. And to *refute* a law not very much experimental evidence is needed; some say only a single contrary result, but that is probably an oversimplification because a single adverse result is not clearly marked out from a wrongly understood one. In the argument which rages still about scientific laws I have the personal belief that Popper is nearest to the truth, when he describes the so-called law as a *conjecture*, and the scientific process is then a search for possible refutations of the conjecture. Now Lakatos has shown, convincingly in my opinion, that this same view applies with equal weight in mathematics. A theorem there can be described as a conjecture; and the proof of the theorem is a search for counter-examples—which, because of the complex nature of mathematical concepts, is usually successful, requiring subsequent modifications either in the enunciation of the theorem or in the definitions of the mathematical objects entering it. If this view is accepted, then the problem of proof is indeed one for the psychologist, but it can now be rephrased; he has to determine why mathematical conjectures give mathematicians such a *feeling* of certainty, a question only to be answered by getting the mathematicians on the couch.

However we may answer the question of proof, it is only a question of intellectual curiosity—of why mathematics is like this. Its answer does not affect the way we do mathematics, and this is perhaps the explanation of its neglect compared with the other question, which I have described briefly as that of truth. By this I mean to subsume all the various investigations into the foundations of mathematics that have taken the attention of so many mathematicians in the past 100 years. So much has been said about these that I must excuse myself for going over what has now become familiar ground; my reason is that I would like to sketch the history in such a way as to make my attitude clear.

In many presentations this history begins with the paradoxes of set theory that began to plague mathematicians from 1895 onwards (when Cantor hit upon the Burali–Forti paradox). I think this is much too late. I would like to draw attention to the contribution of my favourite, Hamilton, in his great memoir on "Conjugate functions or algebraic couples with a preliminary and elementary essay on algebra as the science of pure time" (*Trans. Roy. Irish Acad.* **17**, 293–422, 1837). The motivation here is of interest. J. T. Graves, with some assistance from Hamilton, had published results in 1829 on imaginary logarithms (in effect noting the periods of the complex exponential function) and his results had been unjustly, as it seemed to Hamilton, criticised by Peacock at the British Association 1833 meeting. With the idea of justifying his friend, Hamilton constructed the complex number field as (essentially) the only division algebra of dimension two over the real field, by operating with number-couples in exactly the way that one introduces complex numbers to undergraduates now. This construction was already completed by 1833, but the paper was delayed in publication. It is perhaps characteristic of Hamilton that the immense ingenuity and learning is brought to bear at the wrong point. The question at issue was whether the logarithm of the base (e) should be multivalued as well as those of the variables, i.e. of whether we would have

$$\log x = \log |x| + 2k\pi i$$

or

$$\log x = \frac{\log |x| + 2k\pi i}{1 + 2l\pi i}$$

the denominator being the multivalued logarithm of e. The difference was clearly brought out by Duncan F. Gregory (*Camb.*

Math. Journal **1**, 226, 1838), who settles the matter axiomatically by exhibiting the different properties of the logarithm which had been assumed by the protagonists. But this unfortunate fact, and the equally unfortunate references to "the science of pure time" have indeed caused a complete ignorance of the magnitude of Hamilton's achievement here.

To see this a few words are needed about British mathematics at the time (for continental influences were still small). A major drive towards "symbolic algebra", as it was called, was the success of the analysis, both on the Continent and in Cambridge, with symbolic methods. (We may here neglect Boole, whose *Algebraic Analysis of Logic* appeared in 1847, for he was very much a lone wolf.) The first volume of the Cambridge journal contains papers by Gregory on the D-method for ordinary differential equations (with acknowledgements to Cauchy and Brisson), the corresponding method for equations in finite and mixed differences (including some partial equations), the symbolic solution of partial differential equations of some generality, similarly of simultaneous total equations, and symbolic proofs of the main formulae in the differential and difference calculi as well as some "Notes on Fourier's Heat" in which, for example, the general solution of

$$\frac{\partial^2 v}{\partial x^2} + \frac{\partial^2 v}{\partial y^2} = 0$$

is given in the form

$$v = \cos\left(y\frac{\partial}{\partial x}\right)\phi(x) + \sin\left(y\frac{\partial}{\partial x}\right)\psi(x).$$

Such methods, originally regarded as analogy, soon led to results very difficult to prove independently, and so to the need to show that the supposed analogy was indeed a proof—that is, that the symbols of operation could be detached from the operands and treated like numbers because they satisfied the same axioms as numbers. And Gregory was the Bourbaki of the movement in his splendid (though short) paper "On the real nature of symbolical algebra" (*Trans. Roy. Soc. Edin.* **14**, 208, 1838).

It was Hamilton's genius to see much deeper than the mere axiomatic approach at this level. Having used his couples to justify the complex number field (i.e. the *operator* $i = \sqrt{-1}$) he seems to have divined that he had begun in the middle. The "pure-time" essay he describes as "a much more recent development of an idea against which the author struggled long ... on account of its departing so far from views now commonly received". In it he proposes to do no less than construct successively the ring of integers, the rational field, and the real field, so as to have a sure foundation for the complex numbers. The later steps are quite ordinary; but in constructing the integers he is anticipating Frege and Russell and Whitehead. His construction may be expressed in modern language by saying that he appeals to his intuition of time to provide a set with a total order relation $<$, with the following additional property: that the relation induces, on the set of propositions $\{A < B\}$, an order relation, say $\leqslant$, so that if $(A < B) \leqslant (C < D)$ and if $D < D_1$ and $C_1 < C$, then also

$$(A < B) \leqslant (C < D_1),$$

$$(A < B) \leqslant (C_1 < D).$$

Interpreting $A, B, C, \ldots$ as points of the real line he is then able to construct the step $\vec{AB}$, and by repeating some such arbitrary step the integers arise.

Hamilton here carried out, as well as could be done at the time, the technical steps later perfected by Whitehead and Russell. He was not wishing to take a particular philosophical view. One finds a much more developed view by the end of the century, initiated particularly by Frege (1879) and Peano (1889). I pass over these to something that seems even more to open the door to the twentieth century. Whitehead, in his *Universal Algebra* (1898), says: "Words, spoken or written, mathematical symbols, are alike signs. These can be analysed into (i) suggestive signs (e.g. knot in handkerchief), (ii) expressive signs (e.g. word in language, which *primarily* stands for its meaning), (iii) substitutive signs, which, in thought, take the place of that for which it is substituted. Such are the signs of a mathematical calculus (and of course words sometimes have this as a secondary use, e.g. in poetry or politics)."

Then he goes on to note that in a calculus of substitutive signs one must have *rules*. But in general the result of using arbitrary rules will be a frivolous calculus. There must be a similarity to some *interpretation*.

He then works out in some detail the idea of a substitutive calculus. But he runs into trouble over the use of a calculus which is only partially interpretable. (This difficulty suggests itself to him particularly because of Boole's algebraisation of logic, which differs from modern Boolean algebra in having $x + x = 2x$, an uninterpreted symbol.) Shortly after that Whitehead was to come under the influence of his pupil Russell, and their co-operation is described in Vol. 1 of Russell's autobiography. Essentially they were trying to answer the question of what mathematics was by going back to Hamilton's standpoint (though they do not seem to have known of it) that the integers had to be constructed, and then the rest was easy. What the integers *are* is just the same problem as one has in the infants school, and Whitehead and Russell, taking logic as a supposedly firm beginning, gave essentially the same answer as a modern teacher of five to sevens. She will strive by repeated example of sets of four members to make her charges make the necessary abstraction of "four-ness". Whitehead and Russell would define four as the set of all four-like sets for they cannot describe the mysterious process of making the abstraction precisely. Though, indeed, *four* comes rather later, for the important thing is to define *one* as the set of all unit sets, without of course introducing a circular definition. The achievement of doing this by writing (in a modified notation)

$$1 =_{Df} \hat{x}[\sim(x \in \phi). \&: y \in x. \& z \in x \rightarrow y = z]$$

seems to me the high-water mark of their great book. Let me explain the notation employed here. If $P(x)$ denotes a property of a mathematical variable x, then Whitehead and Russell write, for the set of quantities possessing that property, $\hat{x}P(x)$. Thus 1 is here being defined as a certain set. Let us now look at the property in question. Dots are here used for brackets in an abbreviated notation which Whitehead and Russell introduced, and the two dots indicate the main bracket, which separates off the first part, which we will ignore for the moment, from the second part, which says that any two members of the set x must be equal. This is the characteristic feature of a unit set—that it only has a unique number. However, the arrow symbol had already been endowed by Whitehead and Russell with properties which include the following slightly paradoxical one: that $p \rightarrow q$ is to be a true proposition whenever p is false and q is true. As a result, if x were the empty set it could have no members at all, and so the clause which we have just been discussing would evidently be satisfied in that case. It is therefore necessary to preclude the possibility of the empty set, and this is the reason for the first part of the bracket, where the symbol $\sim$ is introduced to indicate the negation of a proposition. But this achievement has already the seeds of its own downfall for with the same notation one can write down

$$R = \hat{x}[\sim(x \in x)]$$

and so define Russell's set of sets which do not belong to themselves. It is impossible that this set should either belong or not belong to itself, and so we reach a paradox. Whitehead and Russell convinced themselves that the theory of types was essential to avoid this and other paradoxes. Sets of objects of type n form themselves objects of type $(n + 1)$. Then $x \in x$ cannot mean anything; but, on the other hand, a unique 1 is not defined, only a 1 of type 2 or higher. To rescue arithmetic requires the axiom of reducibility, which asserts roughly speaking that, if an object of type n can be defined, then "corresponding" objects of every other type also exist. The simple straightforward clarification of mathematics described by the optimistic young Russell in some of the essays collected in *Mysticism and Logic*—essays astonishingly prophetic as far as mathematical education is concerned—becomes an extraordinarily clever but impossibly complicated and no longer particularly plausible system.

Yet, despite its failure, the Whitehead–Russell school showed convincingly that mathematics and logic are inextricably mixed, and to understand one needs understanding of both; it is only that the prior position of logic was wrong.

How then are we to further this understanding? Certainly we shall need to relate mathematics to the rest of science but, before doing this explicitly, we can talk about Hilbert's approach, which was a truly scientific one. He realised that the problem of mathematical truth could be broken down further; and breaking down problems into simpler ones is the way of progress in all science. Although logic is not prior, it certainly plays a vital role and the mathematical act is that of formulating axioms and deducing conclusions from them. Let us, then, put on one side the question of the truth of the axioms, and study to begin with the axiom systems and their resultant body of theorems. The eventual assault on the castle of truth is to come with the help of greatly increased knowledge of the characteristics of the axiom systems—in particular by using the (proved) consistency and (in some sense) completeness (e.g. in the sense that, for any proposition p there is a proof of p or of $\sim p$ from the axioms) of these axioms. Not that these are the only requirements of an axiom system, though they are most desirable. Quite apart from the fact that axiom systems are devised for particular (already existing) portions of mathematics, so that they must agree with what mathematicians want there, there is the aesthetically pleasing property of independence (i.e. none of the axioms deducible as a theorem from the remainder) and a further most important property which I will return to at the end. Of course the required proof of consistency and completeness has to be carried out in another system, and the Hilbert programme was to use only the simplest finite methods in this "metasystem".

Thirty years of work succeeded in reducing the problem of consistency and completeness of axiom systems for large parts of mathematics to that of elementary arithmetic (i.e. arithmetic of the positive integers). Indeed, once this has been done, it does not seem surprising; analysis shows how the real number field can be constructed by a sophisticated completion construction from the rationals, which in turn come by an ordered-pair construction from the integers, and these in turn arise by another ordered-pair construction from the positive integers. Once the real numbers are given, cartesian methods allows the construction of geometry—and so on. Then Gödel showed that the last step, of proving the consistency and completeness of elementary arithmetic, or of any system containing it, was impossible inside the system. The reason, 30 years after, seems obvious. The proofs and theorems are to be written out using only a finite set of characters (say, the English alphabet and the numerals 0, ..., 9); we can number off these characters and so describe the combinations of them in terms of elementary arithmetic—that is we have a system complex enough to talk about itself. In such a system we can number the sentences; and it is only a technical trick to verify that it is possible to choose N so that the number of the sentence "Sentence number N is unprovable" is itself N. Evidently, either N is provable, in which case the system is inconsistent, or it is unprovable, in which case it ought to be provable since it is evidently true, so the system is incomplete.

This story has been told many times in the past 10 years, by myself as well as others. I only want to repeat it here to draw the moral that, in following up the Whitehead–Russell clue to link logic and mathematics, an important step will be to realise that mathematics really *is* mostly deductions from axiom systems and that some assurance is needed of consistency and completeness of the axiom systems, but we cannot expect elementary proofs of these properties.

Three more remarks about these talking-points of the last 30 years before we leave them to address ourselves to the problems of today. Firstly, of course Brouwer cannot be left out of the picture. For him and his fellow intuitionists—whose influence is out of all proportion to their numbers—the essential feature is mathematics as an *act*, not as a finished thing. New discoveries may make old insoluble problems soluble; and equally may invalidate old proofs and demand improvements. We are forced to say at once that mathematics is indeed *just* like that. Such a view must play an important part in the synthesis we want to make; but if the act of proving is the essential thing, then to say that a proposition is true must be interpreted as saying that the act of proof can be carried out—and to show this it is necessary to do it—and to say it is false must mean that a proof is available of the fact that the proposition is inconsistent with what we know already. As well as these two classes of propositions, there will be the third, of which there is no proof of either the proposition or its absurdity. Because $\sim a$ is absurd it does not follow that we have a proof of a; a new logic is needed, and one in which proofs by *reductio ad absurdum* is no longer possible. And so, as practising mathematicians, we obstinately feel unable to accept the tremendously subtle techniques provided for us by Brouwer. Mathematics for us is *not* so difficult as that.

Secondly, we cannot take refuge in Curry's formalist philosophy as I understand it. He would answer the failure of the Hilbert school—by which I mean that the long-awaited assault on the castle of truth will now never take place—by regarding mathematics as an empirical science, the nature of the experiments being the subjecting of axiom systems to testing for consistency—from the nature of things testing for completeness is not to be expected. Such an approach can only produce more and more restrictive axiom systems, and this is not how mathematics advances. I would give an analogy here; Euler used divergent series freely and could sum $1 - 1 + 1 - 1 \ldots$ to $\frac{1}{2}$ to derive correct answers. This is wrong, because the ability to get the right answer cannot be communicated. So Cauchy (if I let him, for his other sins, bear all the blame here) said—only those series may be used which are such that the sum of n terms may be made to differ from a fixed number by as little as desired by choosing n sufficiently large. The whole theory of convergence—with which we begin to waste students' time as early as the sixth form and carry on all through the university—results. But convergence fails on two accounts; a convergent series may be completely useless as a means of computation, for it could lead to hundreds of hours of machine-time before the approximation was sufficient. On the other hand, I would make a bet that, in the computation of functions by infinite series, only a minority of values have been computed by convergent series—because Cauchy's "differ ... by as little as desired" is hopelessly idealistic. One needs only a series in which *n* can be so chosen that the difference is less than some error—for example, the asymptotic series in which the terms initially decrease, then increase and so diverge, and the error is less than the next term. Not only in numerical computation, too; the perturbation expansions of quantum electrodynamics are thought to be asymptotic series. Certainly they give good numerical values, and yet they seem to diverge.

In the third place, why have these points been so much talked over in the last 30 years? The paradoxes were a great shock because, until the end of the nineteenth century, mathematicians had continued to think of their subject as Kant had described it, giving *a priori* synthetic judgements. Certainty and clarity seemed enshrined there, and first the paradoxes and then Gödel (not to mention Brouwer) ruined this view. Here, though, it is useful to mention the rest of science again. From the seventeenth to the twentieth century science was engaged in a long retreat from the position in which any body of knowledge must have the following structure:†

A set of propositions at the top (axioms) comprising perfectly well-known primitive terms, and infallible truth-value injections: *True*, at the top, so that the truth flows down and floods the whole.

The alternative, to which science slowly and painfully retreated, is to have a number of basic statements at the bottom, comprising only empirical terms, and infallible truth-value injections: *False*, at the bottom which flow upwards and flood the whole.

Now this 400-year painful retreat largely passed mathematics by. (In the eighteenth century, when analysis was in a bad way, mathematics had begun to look like other sciences; but Cauchy saved the day.) But the modern difficulties of mathematics are surely *not* just one more unfortunate incident which still awaits a new Cauchy to allow us all to forget it. Rather they are the indications of a new empiricism which is needed; not the narrow one of Curry, which is

† Here again I follow Lakatos closely.

not going to help in the development of the subject, but something still to be worked out.

So far the consequences of investigations into the foundations of mathematics which we have discussed have been mainly negative. It is time to turn to the more positive aspect of these discoveries and so to enter what I believe to be the important arena for discussion in the second half of this century. Although the general nature of Gödel's discovery is essentially a negative one, restricting the possibilities before us, it is obvious that so much work in the field of logic by so many workers must be bound to produce some positive result. I have been asked not to enter into too much detail, so I propose to give only a single rather detailed example which will show how these positive results have arisen. I start with what will appear to be a digression and we will rejoin the main theme of the argument later on.

In 1958 there appeared a paper by two analysists, Schmieden and Laugwitz, about the foundations of analysis. I think that they looked on their work more as an extension of that of Cauchy and Dedekind than as a development in foundational studies in general. In fact the motivation at the beginning was a criticism of Dedekind's view of the real number systems. Dedekind, as you all know, defined real numbers by dividing the rational numbers into two classes, and identifying this "cut" with the real number which is essentially the intersection of these two classes. Schmieden and Laugwitz point out that from a general intuitive point of view there is a very strong motivation, when the division has been carried out, for supposing that a number is defined by it. There is no motivation for supposing that this number is unique. Dedekind made both of these assumptions. They therefore proposed to initiate a non-Dedekind analysis in which such a cut may define a whole set of numbers.

Some people may prefer to define the real numbers in terms of convergent sequences. In that case any convergent sequence of rationals is defined as a real number. For example, a real number between 0 and 1 means any finite or infinite decimal the terms of the sequence being the successive stages in the writing down of the decimal. But not all sequences are to be counted as different; those that differ by a sequence converging on zero are to be identified. Even in the special case of the decimals, point nine recurring has to be identified with 1. This identification corresponds to Dedekind's assumption of the uniqueness of the real number defined by the cut.

Schmieden and Laugwitz would not make this identification but would regard as different numbers, neither of them zero, the objects defined by the sequences

$$(1, \tfrac{1}{2}, \tfrac{1}{3}, \tfrac{1}{4}, \ldots)$$

and

$$\left(1, \frac{1}{2^2}, \frac{1}{3^2}, \frac{1}{4^2}, \ldots\right),$$

Moreover, any sequence $a = (a_1, a_2, a_3, \ldots)$ of rational numbers for them represents a number which they can work with. In order to add these numbers one adds corresponding members of sequences so that $a + b = (a_1 + b_1, a_2 + b_2, \ldots)$ and in order to multiply, one multiplies corresponding members so $ab = (a_1b_1, a_2b_2, a_3b_3, \ldots)$. Notice that *all* sequences are permitted at this stage, not only the convergent ones.

Of course the rational numbers with which we started can all be added, subtracted multiplied and divided (with the usual proviso about not dividing by zero). That is to say, they form a field. This is no longer the case with the extended system of quantities defined by Schmieden and Laugwitz for there one has divisors of zero, i.e. pairs of numbers, neither zero, such that $ab = 0$. This is obvious since the product of

$$(1, 0, 1, 0, 1, \ldots)$$

and

$$(0, 1, 0, 1, 0, \ldots)$$

will be

$$(0, 0, 0, 0, \ldots)$$

and with the obvious identification this will be zero. However, matters are not so bad as they might be since every number which is not a divisor of zero has a reciprocal; that is to say, we are dealing with what the algebraists call a division ring. All the usual relations between numbers can be used to define corresponding relations between the entities of this ring, for example $a > b$ means ($a_1 > b_1$ and $a_2 > b_2$ and ...), or, if it is more convenient, one can define the weaker relation

$$a > b \leftrightarrow (\exists N)[r > N \rightarrow a_r > b_r].$$

Embedded in this division ring are the integers, the rational numbers, and the real numbers, or at least sets of sequences isomorphic with these, for we can identify the sequence $(x, x, x, x, \ldots)$ with the rational number x, and we can define the convergent sequences with the real numbers. But the ring also has numbers which have no analogue with the numbers of the real or rational field, although they have a surprising analogue with the sort of numbers which our grandfathers spoke about freely. If we open any book, even quite a modern one, on differential geometry, we will find the pages liberally bestrewed with quantities called, quite openly, infinitesimals, and more often than not no real meaning is given to these quantities. They are understood as abbreviations for other quantities which can be related to the real numbers, or else they are simply thought of in a light-hearted way as very small quantities, and the whole procedure is thought of as approximately true, and able to be made "as true as we like" by choosing these quantities sufficiently small. When we get back to the nineteenth century, however, such quantities occur freely, for example, in books on algebraic geometry. Salmon's *Higher Plane Curves* often contained arguments about proceeding from a point on a curve to the neighbouring point, and then to the next point. Differences between the coordinates of the point one begins with and a neighbouring point are the same kind of quantities as the differential geometers use. All these kinds of shaky arguments can be put right at one blow if a proper meaning can be given to these peculiar numbers. The ring defined by Schmieden and Laugwitz contains such numbers. For example there is the number $\Omega = (1, 2, 3, \ldots)$ which evidently, from the definition of greater than, is greater than all the ordinary integers. The reciprocals of such *infinite* numbers will be the *infinitesimal* numbers in the ring.

The motivation for Schmieden and Laugwitz' work was indeed the desire to do something about the improper functions which had been defined, notably by Dirac in the 1920s but originally at least as early as Heaviside at the end of the nineteenth century. Dirac defined his δ-function to be a function which was zero everywhere away from the origin and whose integral over any interval containing the origin was 1. Such a function must have an infinite value at the origin and Schmieden and Laugwitz were able to construct a whole set of δ-type functions of which one example is

$$\delta(x) = \frac{1}{\pi} \frac{\Omega}{1 + \Omega^2 x^2}.$$

Previously the scandal generated by the δ-function had been explained away by the analysists by extending the idea of *function*, either by means of the theory of distributions or by improper functions. Schmieden and Laugwitz layed the scandal low by extending the idea of the number field in which they were working.

We begin by wondering at this point what has happened, for we have always been taught that the real number field is the largest field which we can expect to deal with and which is ordered in the usual way. It is worth looking a little more closely at the achievements, firstly, of the Greek geometers, especially Eudoxus, in constructing this field in the first place; secondly of Dedekind and the late nineteenth-century analysists in rediscovering it. They began, to put it in modern terms, with a field, the rational field, which was totally ordered by a relation $x < y$. Not only were any two elements of the field comparable by means of this ordered relation, but the ordering also satisfied the axiom of Archimedes, that is, that a sufficient number of copies of one element would together be bigger than another element. The astonishing achievement was to construct another field, also totally ordered and also satisfying the axiom of Archimedes, and containing the rationals as proper sub-field. In fact they achieved the best possible result. The real number field is the largest totally ordered Archimedean field. Now when we look at Schmieden and Laugwitz' work, as far as we have described it up to now, the extended number system is not a field but a division ring. It is not totally ordered and this ordering is not Archmidean. For example, one cannot, by taking any number of copies of $1 = (1, 1, 1, \ldots)$, exceed Ω. The Dedekind construction, however, consists not merely of taking sequences of rationals but of identifying certain different sequences and talking only about the whole classes of sequences. Evidently, unless analysis is to be quite unworkable, some similar identification needs to be carried out here. Schmieden and Laugwitz realised this and attempted to make some such identification in the later part of their paper, but this was unsuccessful in achieving the desired result. They tried the rather obvious trick of identifying two sequences that differed in only a finite number of points. To put it another way, which is useful later, two sequences a, b were identified if the set of values r for which $a_r = b_r$ is the complement of a finite set. But, rather puzzlingly, what seemed to be needed was rather the identification of sequences whenever the set of values r belonged to a certain family of sets—*not* that particularly family defined by taking the sets to be all complements of finite sets. The situation was only cleared up by a thoroughgoing application of the logical ideas which had come to light in discussing axiom systems.

To see the connection between this new analysis and logic we need to go back a little in time to the work on the theory of *models* of an axiom system. The idea of a model as a set of objects fulfilling the axioms is a familiar one, and seems completely clear. Let us look a little more closely at the axioms that Peano chose for elementary arithmetic:

1. Every number n has a successor n'.
2. There is a number, 0, which is not a successor.
3. If $m = n$ then $m' = n'$.
4. The axiom of mathematical induction, that if 0 has a property f, and if whenever n has f so has n', then all numbers have the property f.

It is proved in many books that these axioms characterise the integers in this sense: if one constructs two models of them, then the objects in one can be put in 1 : 1 relation with those in the other and the relation carries all true statements about objects in one into the corresponding true statements about the corresponding objects in the other. Such a property of the axiom system—that it characterises the objects it describes up to an isomorphism—is called being *categorical.* Once one has realised that consistency and completeness are hopeless ideals, categoricity becomes something of evident importance.

Yet despite the simplicity of the proof just mentioned (which consists of merely building up the two models by means of the successor operation from the zero object and verifying the isomorphism as you go along), the result is completely at variance with a theorem of Skolem's in 1934, that no axiom system could be categorical. Skolem's original application of this was to set theory; there were a collection of axioms meant to describe sets with varying degrees of infinity—enumerable sets like that of all the integers contrasted with the set of reals between 0 and 1 for example, and possibly more infinite ones. Yet, Skolem showed, this axiom system, and indeed every one, had a model satisfying it with only an enumerable number of objects. (The importance of the enumerable sets is tied up with the finite alphabet used to write the expressions of finite length which are propositions in the theory; there are an enumerable number of possible propositions.) But I am wandering: in the case of Peano's axioms we can use Skolem's result to predict that there is a model of the axioms which is *more* complex than the usual integers.

What has gone wrong? The proof that Peano's axioms are categorical includes a hidden assumption—that the only models to be considered are such that, if an object A occurs in one as a set of objects $x, y, z, \ldots$, then the corresponding A' in the other model must also be a set of objects $x', y', z' \ldots$ If one limits oneself to such models, they are indeed all isomorphic. But Skolem proved that, as well as these standard models, there will always be non-standard models, which serve to satisfy the axioms but in which the set-belonging relationship (which was not mentioned in the axioms) goes wrong. The study of a non-standard model of Peano's axioms is that of non-standard arithmetic.

It is easy to begin constructing a non-standard arithmetic. If we use *natural numbers* to mean the positive integers in the (usual) standard model of Peano's axioms, we can define the integers to be any sequences of natural numbers.

If

$$N = (a_1, a_2, a_3, \ldots)$$

we can define the successor

$$N' = (a'_1, a'_2, a'_3, \ldots)$$

whilst 0 is defined as (0, 0, 0, ...). The first three of Peano's axioms are easily seen to be satisfied. The real trick is in the fourth one. To express this adequately one has to define what it means for N to have a property; the obvious thing—which does not help—is to say that N has the property F if and only if $a_1, a_2, a_3, \ldots$ all have the property f. This way only leads to standard models dressed up in a more complicated guise. Instead one can make a more complicated model by saying that N has the property F if the set $(p, q, r, \ldots)$ such that $(a_p, a_q, a_r, \ldots)$ are just those a_i having the property f is one of a particular family of sets. Then it is just a matter of technical cunning to choose this family of sets in such a way that we have a non-standard arithmetic.

Without going into details you will not be surprised to hear that the same technical trick is what is needed to put right the troubles in Schmieden and Laugwitz' paper. Two sequences are to be identified there when the suffixes of their equal members belong to one of this family of sets. When such an identification is made the division-ring collapses to a field which is again totally ordered. That is, the achievement of Eudoxus and Dedekind is almost repeated—but of course not quite, since we know that we already have in the real numbers the largest totally ordered Archimedean field. Here the axiom of Archimedes fails, as it must do in any field which is to contain infinite and infinitesimal quantities. The corresponding result may be considered to be the theorem that every non-infinite member of the field can be written in the form $a + \varepsilon$ where a is an ordinary real number and ε is an infinitesimal member. In a certain sense—somewhat more sophisticated than with ordinary arithmetic and analysis—one can say that non-standard analysis is just the analysis one reaches if one starts with non-standard arithmetic. In my view this has proved one of the major positive results from research into foundations. The consequences of this widening of analysis are likely to be even wider in applied mathematics and theoretical physics than in pure mathematics. But that lies still in the future.

Bibliography

Two elementary introductions to the "classical position", i.e. as of about 1930 or so:

H. MESCHOWSKI, *Evolution of Mathematical Thought* (translated J. H. GAYE), Holden-Day.

C. W. KILMISTER, *Language Logic & Mathematics*, English Universities Press.

More detailed view of the "classical position":

A. A. FRAENKEL and Y. BAR-HILLEL, *Foundations of Set Theory* (omitting the more technical parts).

An updated and very readable account:

The Philosophy of Mathematics (edited J. HINTIKKA), Oxford.

More technical work on non-standard analysis, but by no means easy:

M. MACHOVER and J. HIRSCHFELD, *Lectures on N S A*, Springer.

A. ROBINSON, *Non-Standard Analysis*, North Holland.

Chapter 1 and 2 and the appendices (which form nearly half the book) of G. KREISEL and J. L. KRIVINE, *Elements of Mathematical Logic*.

The Nature of Mathematics—Another View

Margaret E. Baron

Introduction

What is mathematics? A large and varied body of thought which has grown up from the earliest times purports to answer this question. But upon examination, the opinions which range from those of Pythagoras to the theories of the most recent schools of mathematical philosophy reveal the sad fact that it is easier to be clever than clear.

Kasner, E. and Newman, J., *Mathematics and the Imagination* (2nd ed., London, 1949).

Most people who use mathematics as a tool in science, industry or commerce find themselves, from time to time, puzzled by doubts as to the nature of the material they are using; such doubts are quickly dismissed if the mathematical techniques bring the required results and if these results make sense in terms of the material world in which we live and work. For teachers, questions concerning the nature of mathematics are likely to arise more frequently and attempts to resolve them must inevitably be more sustained in duration. In what terms does one justify the introduction of a new syllabus or a new topic? What is the relation between pure and applied mathematics? Which is more important, the components of a syllabus or the power of mathematical thinking which is developed in the pupils? If we introduce modern mathematics into our schools is it because we are interested in new content and more powerful applications, or is it because we are interested in new methods and new ways of thinking about mathematics? What, in fact, are the aims and purposes of mathematics teaching? Unless we give some careful thought to such questions our teaching is likely to remain at a mediocre level and our day-to-day procedures will be based on tradition, convention and imitation. We will never have the courage to initiate reforms or to break away from the practices of our predecessors.

My own interest in the nature of mathematical thinking developed largely from attempts to resolve questions of this kind which arose over 20 years of learning and teaching mathematics. In the classroom much of the mathematics which we teach is *applied* mathematics in the sense that it arises directly from everyday life activities connected with pushing, shoving, tearing, weighing, measuring and so forth. Even when we reach a level where the work appears to be more abstract and theoretical in nature there always seems to be a boy or girl around who demands an immediate justification in terms of *use* and *application.* How sad it is to have to come down from the clouds and answer questions such as, "What use is this to *us*?", or, "Why do we have to learn this?", and even, "What use is this to *anybody*?" None the less, a teacher who is concerned to retain the goodwill of his classes must always go armed with a wide and varied collection of applications and illustrations of any new piece of theory—if he fails to do so there will be many children who "can't see the point".

Has pure mathematics as such any kind of existence of its own independent of the applications which it finds or the power over material affairs which it conveys?

It was in an attempt to solve some of these problems that, more than 12 years ago, I departed from the main stream of mathematics and embarked on a course of reading and study in the history and philosophy of science. Until then, although I had occupied myself with mathematical activities fairly continuously, I had never, so far as I know, contributed anything new to mathematics; that is to say, I had never invented or discovered anything which someone had not found before me. I felt that, in order to get a better understanding of the nature of mathematics it was necessary to look carefully at

(i) the nature of mathematical invention, i.e. how new discoveries are made,

(ii) the role of mathematics in scientific theories.

For me the central problem hinges on the relation between mathematics and natural science and I feel that only by getting as close as possible to the discovery process can one hope to throw light on the way in which ideas pass from one field to another. Contemporary studies are, of course, immensely valuable, but the records are usually unavailable and the process incomplete. It follows that to throw light on the nature of invention in science and in mathematics it is necessary to use historical material. During the last 10 years I have been engaged in studies in the history of mathematics. This has meant, amongst other things, studying the day-to-day growth of new concepts and new ideas in the notebooks, published and unpublished, of Huygens, Fermat, Leibniz and Newton and the very many lesser men who contributed in one way or another to the development of the infinitesimal calculus.

The study of the history of mathematics does not, of course, answer the question, "What is mathematics?" It is, however, both suggestive and illuminating and contributes, above all, a sense of perspective, which is always valuable in any discussion of the nature of an intellectual discipline at a given point in time. Through such studies, one notes particularly how each generation criticises the unconscious assumptions made by its parents. At this point it may be as well to put firmly on record that, so far as I am concerned, I regard Pythagorean number theory, Euclidean geometry, Cavalierian indivisibles, Newtonian fluxions, Boolean algebra, ..., all as *mathematics*. I do not wish to restrict the terms of this discussion to the particular mathematical structures which have been erected in the twentieth century.

The Word Mathematics

Inquiries as to the nature of mathematics often become confused because the word *mathematics* can be used in two distinct and different senses, i.e. the *methods* used to discover certain truths and the *truths* which are discovered.

To understand and appreciate the nature of mathematics at a given time it is necessary to take into consideration not only the formal structure in which the proofs are presented but also the network of conceptual thinking in which it is embedded and the kind of applications which it finds. Historically, of course, mathematics seems to have grown up largely as a result of (i) social needs, as shown in everyday life, commerce, science and technology, (ii) the intellectual need to connect together existing mathematics into a single logical framework or proof structure. I am not disposed to admit any arguments which suggest that either of these is prior, or more significant as an influence than the other, although many left-wing historians give primacy to social factors in explaining all developments in science and mathematics.

Egyptian mathematics was undoubtedly largely practical in nature and it seems reasonable to link together the empirical formulae developed for mensuration purposes with land surveys and the annual flooding of the Nile. Even so, there is plenty of evidence that some of the work done vastly exceeded any social, agricultural or commercial needs: this is particularly the case with the Egyptian "unit fractions" and the "aha calculations" (equivalent to linear and quadratic equations).† Even at this early stage (1800 B.C.), intellectual curiosity and the desire to develop a set of consistent and comprehensible techniques provided, at least in part, the motivation for mathematical development.

The Greeks were mainly concerned to develop a unified proof structure and logical framework in terms of which all mathematical theorems could be expressed. Plato attached importance to the study of mathematics because, as he says in the *Republic*, "The study of mathematics develops and sets into operation a mental organism more valuable than a thousand eyes, because through it alone can truth be apprehended". Aristotle's *Logic* was based on a codification of the rules and methods of reasoning already established on a strictly rigorous basis by the great Greek mathematicians before

† VAN DER WAERDEN, B. L., *Science Awakening*, 2nd English ed., trans. DRESDEN, A. (New York, 1963).

him. Notwithstanding, there are many examples of problems and solutions depending upon "kinematic" and "mechanical" methods which forced their way in, despite complete incompatibility with the established proof structure. In the seventeenth century (A.D.), special problems abounded, many thrown up by speculation as to the nature of the universe and the forces which kept the planets in their courses. The new methods which grew up did so partly to provide solutions to the particular problems presented and partly to provide universal methods through which all such problems could be solved—with a consequent saving of time and labour.

In the eighteenth and nineteenth centuries progress towards tidying up the new mathematics and securing its foundations went on alongside the almost feverish efforts to chart the heavens and explore the oceans by means of the new and powerful tools. In the present century many new fields of mathematics have developed almost overnight as a result of problems thrown up in engineering, science and technology; at the same time, probably the most sustained and determined attack of all time has been made on the foundations and structure of pure mathematics with the aim and purpose of strengthening and unifying the logical framework which holds it together.

It follows from all this that, in considering the nature of mathematics, we must concern ourselves with the developing whole—not only with the formal accepted logical framework, but also with the methods of work and the thought processes of mathematicians, the connecting threads of suggestion and analogy, the language and notation through which communication is established. In talking *about* mathematics as much attention must be paid to the psychology of mathematical invention as to the logic of mathematical proof.

In the world today mathematics is a universal language used as a powerful tool by banks and insurance companies, surveyors and architects, builders and engineers, scientists and technologists. It is, at the same time, a discipline, logical and rigorous. Advanced studies in mathematics command the highest respect because of their abstract nature and total incomprehensibility to the general reader. In school and in college, creative and imaginative thinking in mathematics should be rated at least as highly as deductive and logical thinking. What is it that distinguishes mathematical thinking from other kinds of thinking? What, indeed, is mathematics?

Is Mathematics Useful?

Since the central problem, "What is mathematics?", is essentially extremely difficult it seems strategic to chip, or gently peck at it a little from the side, in order to try to let in a little light. We might, for instance, try to answer what appears at first sight a comparatively trivial question, "Is mathematics useful?" Even this might profitably be modified, "Is *all* mathematics useful?" "Is *some* mathematics useful?" "Is *some* mathematics useless?" "Is *all* mathematics useless?"

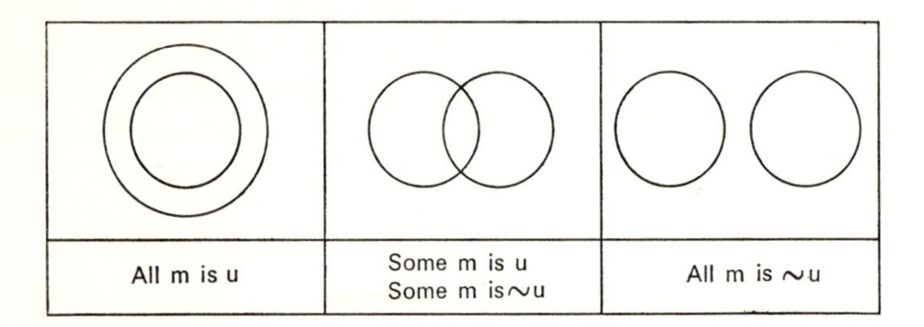

The usefulness, or otherwise, of mathematics, is a matter on which great mathematicians have, over the years, expressed themselves freely. Let us take one or two characteristic points of view. We might start by taking the well-known story told of Euclid who, when asked by someone, "What shall I get by learning geometry?" immediately called his slave, saying, "Give him threepence, since he must gain from what he learns".

Robert Recorde (a sixteenth-century mathematician who probably did more to make the English numerate than anyone before or after) takes the opposite point of view. In *The Grounde of Artes* (London, 1540), a textbook which went through at least eighteen editions in the sixteenth century and a dozen more in the seventeenth, Recorde makes an impassioned "*Declaration of the Profit of Arithmeticke*". His rhyme on the value of "numbring" runs thus,

> If number be lacking it maketh men dumbe,
> So that to most questions they must answer *Mum*.

He continues: "Wherefore as without numbring a man can do almost nothing, so with the helpe of it you may attaine to all things".

A mathematician nearer our own time, G. H. Hardy, takes the contrary position. He writes: "I have never done anything *useful*. No discovery of mine has made, or is likely to make, directly, or indirectly, for good or ill, the least difference to the amenity of the world".†

Recorde, of course, was talking about arithmetic. Hardy was a pure mathematician. Even so, his statement was not entirely true as he did make a contribution to genetics—Hardy's Law, of central importance in the study of blood groups and the treatment of haemolytic disease in the newborn. It also turns out that some of his work on prime numbers can be used in the theory of pyrometry (the investigation of the temperatures of furnaces). The fact remains that the mathematics Hardy was concerned with was entirely abstract and any applications found for it were entirely fortuitous and irrelevant.

When we say that a piece of mathematics is useful we normally mean that applications have been found for it so that it can be used, for instance, in finding the distances of a ship at sea, or telling us something about the solar system, or that it can be helpful in the advancement of science by elucidating some of the very complex theories of physicists, chemists and biologists. When a set of equations, or mathematical formulae, is used in this way, we say that we are fitting a mathematical model. One of the most difficult tasks of the mathematician working in industry today is to identify a problem, fit a mathematical model and solve the problem by deploying his mathematical resources. What is of interest to us is that the same model can frequently be used to resolve a variety of different physical

† *A Mathematician's Apology* (Cambridge, 1948).

situations. The usefulness of a given piece of mathematics is clearly a matter of degree, depending on the number of different situations in which it can operate as a model.

Pure and Applied Mathematics

The idea of usefulness in mathematics leads us to the distinction between *pure* and *applied* mathematics. This distinction was first drawn in the nineteenth century and, in the school and university sense, in England, *applied* mathematics was usually taken to mean theoretical physics, i.e. the system of concepts, techniques and laws, put together by Newton in the *Principia* (1687) and which, throughout the eighteenth and nineteenth centuries, was extended to include elastic and fluid media, the theory of optics and electricity and magnetism. Today, the main concepts of classical physics have been reformulated in terms of relativity theory. Even so, because the scope and influence of mathematics has been vastly enlarged in the present century, it is common to use the term *mathematical applications* to cover the whole area in which mathematics is applied to sets of objects, or elements in the external world. Through the manipulation of symbols in a fitted mathematical model, we are enabled to deduce some of the properties of electrical circuits, radio waves, or even the combining of genes in reproductive processes.

In school and college mathematics courses in England a great many subjects, such as algebra, geometry, calculus, analysis, are grouped together under the general title, *pure* mathematics, and innumerable textbooks and examination syllabuses bear this name. Is this material really pure mathematics? Good teachers (those we can understand) often begin with a problem and, when we have solved it, go on to generalise our method and develop some abstract theory. The ideas we have about the mathematics we learn are built up over many years of study and become associated in our minds with all the applications and illustrations we have encountered on the way. It is always easier to explain what we can do with a piece of mathematics than to say what it is. Elementary school algebra of the traditional kind is certainly *about number* and consequently helps us in our counting operations. We can use school geometry to measure and to map the space around us. Shall we call arithmetic, algebra and geometry *pure* mathematics or mathematical applications?

The word *geometry* denotes *earth measurement* and few would deny that, with the ancient Egyptians, this is exactly what was involved. In Greek mathematics, however, and particularly in the *Elements* of Euclid, the *Conics* of Apollonius and the treatises of Archimedes, we find geometry utilised as a language through which. the structure of mathematics could be organised and symbolised. For over 2000 years Euclid's *Elements* represented a model of a deductive system based on axioms, postulates and common notions (or rules of reasoning) and the only recognised valid form in which mathematical proofs could be presented. Notwithstanding, and despite the axiomatic form in which Greek mathematics came down to us, the initial definitions, peculiar as they are, seemed to identify recognisable, though idealised, elements existing in the real world. In consequence, theorems deducible within the system appeared to represent true statements, confirmable within the limits of human error, by measurement. It follows that, at least until the nineteenth century, Euclidean geometry was regarded as a valid account of the physical space in which we live and there are those who, to this day, cannot regard it in any other light.

Is Mathematics an Empirical Science?

This leads us inevitably to another question, "Is mathematics an empirical science?" Those who maintain that mathematics is a true description of something external to us must indubitably answer this question in the affirmative. The development of mathematics will then be regarded as a *discovery process*. So much current use is made of the phrase "discovering mathematics" that it seems desirable to ask what it is that is actually being discovered. Because these matters are important to teachers, particularly, it may be as well to clarify the issues so far as we are able.

We say that Columbus discovered America meaning that he took a ship and sailed across the Atlantic; when he returned he told us of the existence of a vast continent on the other side. In fact, America was there—Columbus *told us* that it was there. Now, let us try an example from mathematics; if we say, "Isaac Newton *discovered* the binomial theorem", do we mean the same thing? Let us try a few more examples: "Leibniz discovered the infinitesimal calculus"; "Hamilton discovered quaternions"; "Cayley discovered matrices". More generally, do mathematical truths exist in some sort of absolute sense independently of human minds, or are they man-made inventions?

Quoting some nineteenth-century views we have, from Edward Everett, the American scholar and statesman,

> In the pure mathematics we contemplate absolute truths which existed in the divine mind before the morning stars sang together and which will continue to exist there when the last of their radiant host shall have fallen from heaven.

G. H. Hardy supports this point of view. He says

> I believe that mathematical reality lies outside of us, and that our function is *to discover*, or *observe* it, and that the theorems which we prove, and which we describe grandiloquently as our 'creations' are simply notes on our observations.

Twentieth-century philosophers of science look at the matter quite differently. P. W. Bridgman says

> It is the merest truism, evident at once to unsophisticated observation, that mathematics is a human invention.

From E. Kasner and J. Newman (*Mathematics and the Imagination*) we have

> We have overcome the notion that mathematical truths have an existence independent and apart from our own minds. It is even strange to us that such a notion could ever have existed.

What is Pure Mathematics About?

If pure mathematics is not concerned with the external world we must surely enquire what it is concerned with? What is the subject matter, or content, of pure mathematics? The most quoted statement here is that of Bertrand Russell:

> Mathematics may be defined as the subject in which we never know what we are talking about, nor whether what we say is true.

Now, if it is the case that we do not know what we are talking about it would, at first sight, seem that what we are saying is worthless but Russell makes the point clearer when he says

> If our hypothesis is about *anything* and not about some one or more particular things then our deductions constitute mathematics.

These ideas are, of course, related to the notion of *mathematical abstraction*, a subject on which a great deal has been written but nothing very clear seems to have been said. There are, it is certain, different levels of abstraction. Whilst many of us felt that the elementary algebra we learnt in school was extremely *abstract*, looking back on it we realise that the elements were always *numbers* and the operations were restricted to a limited range of arithmetic processes. Later, the idea of an algebra was extended to include any freely constructed mathematical system defined in terms of specified operational rules. Here, for instance, is an algebra of two elements,

o	a	b
a	a	b
b	b	a

It tells us nothing about anything in particular but defines a mathematical system whose properties can be explored. One of the finest examples of abstraction in mathematics is group theory. Newman describes it as follows:

> The theory of Groups is a branch of mathematics in which one does something to something and then compares the result with the result of doing the same thing to something else, or something else to the same thing.

New concepts of algebra arose about 1830 when George Peacock laid down the general principle that "if mathematical relations hold in special systems under restricted conditions one may define new systems wherein the same relations hold without these restrictions". A system extended in this way Peacock called *symbolical algebra.* In 1837 Hamilton introduced the formal definition of complex numbers as number pairs, thus,

$$(a, b) + (c, d) = (a + c, b + d),$$

$$(a, b)(c, d) = (ac - bd, ad + bc).$$

Hamilton again, probably motivated by physical considerations, succeeded (about 1843) in developing the *algebra of quaternions* in which a single element was composed of four real numbers. The definitions of addition and multiplication were as follows:

$$(a, b, c, d) + (e, f, g, h) = (a + e, b + f, c + g, d + h),$$

$$(a, b, c, d)(e, f, g, h) = (ae - bf - cg - dh, af + be + ch - dg,$$

$$ag + ce + df - bh, ah + bg + de - cf).$$

Both operations are associative but only addition is commutative. The algebra of quaternions was the first example of a non-commutative algebra. Subsequently, ideas linked with those of Hamilton were developed by Grassmann who, starting from a more general standpoint with ordered n-tuples, developed a variety of algebras (published 1844). Further significant advances were made by George Boole (*The Laws of Thought*, 1854) who developed a new algebra, partly as an algebra of propositions and partly as an algebra of sets. In 1857, Cayley investigated the properties of matrices and produced another example of non-commutative multiplication. Although many other types of algebra emerged in the nineteenth century those in which multiplication is non-associative are of more recent origin. It is even possible for an algebraic system to exist in which multiplication is neither commutative nor associative.†

† I am indebted to C. J. SCRIBA for bringing this material together in *The Concept of Number*, Mannheim, 1968 (chap. V).

Similar developments in the field of geometry led to the establishment of non-Euclidean systems by Bolyai and Lobachevski; a new concept of pure geometry emerged, i.e. geometry dealing with no particular subject-matter and asserting nothing about physical space. The pure geometer is indifferent as to whether the axioms find any counterparts in the world of experience—he is only concerned to deduce the consequences. The task of deciding which of the available geometries applies in a given situation is a matter for the scientists. Geometry, considered as a pure science of ideal space, is an exercise in logic comparable to a game played with formal rules.

What is the Connection between Mathematics and the Physical Universe?

If we accept tentatively the view that mathematics is a free creation of the human mind, that it is *invented* rather than *discovered*, it becomes a matter of some urgency to explain its apparently close relationship with the external world. Historically, it is clear that some of the best inspirations in mathematics have come through the application of physical concepts. We have already noted that Euclidean geometry had its origin in the empirical measurements of the Egyptians. With the Greeks, and until the development of algebraic geometry in the seventeenth century, concepts of function were largely determined by the extent of known curves—mainly those which arose through cutting sections of a cone, i.e. the circle, ellipse, parabola and hyperbola. Those were the curves which later Kepler used in plotting the paths of the planets. Concepts of motion have played, and continue to play, a dominant role in developing new ideas of function and certainly were very much present with Newton when he developed his fluxional calculus. There are many examples of new branches of mathematics, developed in the twentieth century and springing directly from technological problems. It appears then that most mathematics, in its early stages, may originate in ideas and concepts associated with physical form and shape—things moving, touching, flowing; the subject then begins to live a peculiar life of its own, creative and self-generating, dominated often by purely aesthetic considerations. Even so, we have examples of apparently abstract theories such as differential geometry and the theory of groups which, after a decade in the one case and a century in the other, both turn out to be useful in physics. How free, one wonders, is this creation of the human mind, since most of its ideas, in their inception, seem to be physical in origin and since, however abstract the theory ultimately developed, it always seems to find application somewhere in the universe.

Making Pictures

> A mathematician, like a painter or a poet, is a maker of patterns. If his patterns are more permanent than theirs, it is because they are made with ideas.†

To weave patterns with ideas which are essentially abstract is difficult and all mathematicians require to give these abstract ideas some form of concrete representation. *Pictures* and *symbols* are central to all mathematical thinking. When we make pictures in mathematics, the pictures are not intended to depict the actual material, whether abstract or concrete, but to *represent*, by some kind of spatial imagery, the elements, operations, relations or functions involved. Pictures, either mental or visual, are frequently used in thinking about a mathematical problem so that the relations between the points, lines and spaces in the figures may suggest relations between the abstract elements and sets of elements in the problem.

It may be helpful to make a picture, or diagram, illustrating the nature of mathematical thinking. The most that such a picture can do, in this case, is to provide a framework around which ideas can be grouped. There is always a danger in any kind of visual imagery; it may turn out to limit the development of thinking because the lines in the figure set up boundaries and limitations which may not, in fact, exist.

† *A Mathematician's Apology*, *op cit*.

The Structure of Mathematical Thinking

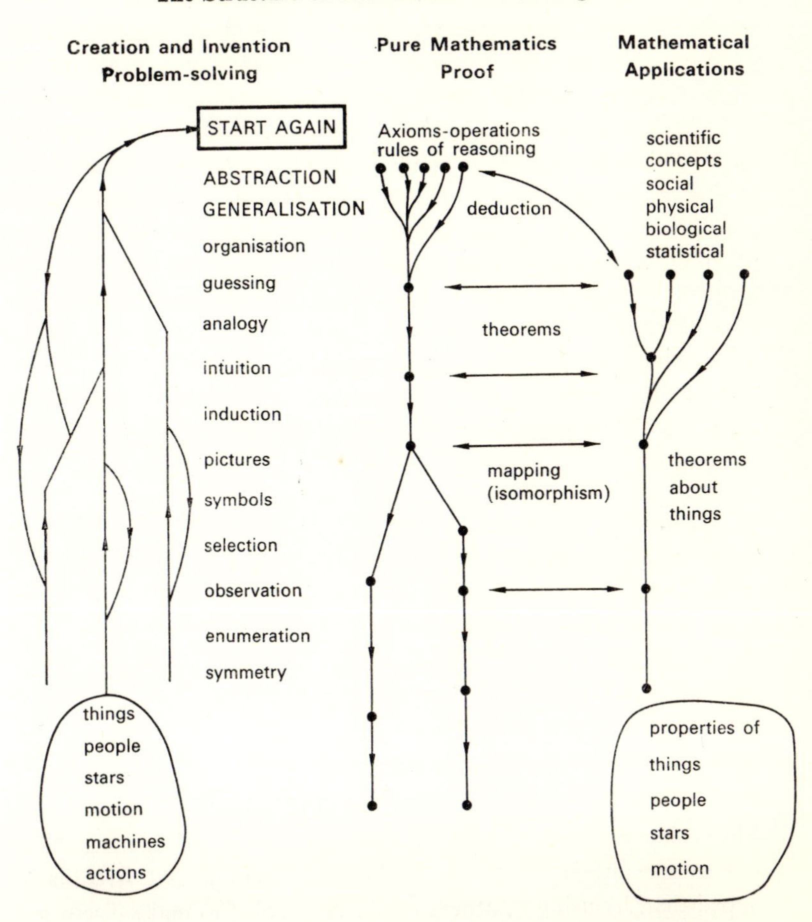

Most of the terms introduced into the above diagram will be discussed elsewhere in this book. It is only possible, at this stage, to make a single broad point relating to the general structure and it is this:

1. Mathematics in the making is experimental and inductive and knows no laws.
2. Mathematics presented as a rigorous proof structure is systematic, deductive and abstract.

Because of limitations of time and space it is not, unfortunately, possible to allow for discussion of mathematical thinking in progress. A good deal of material has been produced recently concerning the role of problem-solving and the development of creative thinking in the teaching and learning of mathematics.† Polya‡ provides a detailed treatment of many aspects of the creative and constructive approach which is indicated on the left of the diagram. For most of us the solution of unclassified problems involves, at the outset, a good deal of fiddling about. Judgements of relevance must be made and the task clearly formulated in relation to the information available. Pictures and diagrams often play a part in stimulating thought and some kind of notation must inevitably be introduced so that the steps in the argument can be formulated. Ultimately, of course, no single special problem can provide the groundwork for a mathematical theory and the best problems are those which lend themselves to development and generalisation and suggest further problems. The grouping of problems of a similar nature leads to the search for more sophisticated tools.

Most new mathematical work is published in systematic logico-deductive form (shown in the centre of the diagram) and we rarely have enough information to see, or to judge, how the problem was first conceived and how, in the end, it was resolved. For example, the whole of Greek mathematics came down to us in this form and it was not until 1906 that the discovery of an Archimedean treatise entitled *The Method* enabled us to see how Archimedes first tackled the problems which he subsequently proved on a rigorous basis. It was specifically written as a *discovery* method and Archimedes hoped that, by using it, others would be enabled to make discoveries as he had done. Another example of the publication of a discovery

† *The Development of Mathematical Activity in Children: the Place of the Problem in this Development:* A Report prepared for the Sub-Committee on Mathematical Instruction of the British National Committee for Mathematics by the Research and Development Panel of the Association of Teachers of Mathematics (A.T.M., 1966).

‡ Polya, G., *How To Solve It* (Princeton, 1945).

method in the history of mathematics is the *Arithmetica Infinitorum* (1655) of John Wallis. By publishing his abortive attempts to square the circle, Wallis succeeded in interesting the young Newton who rapidly developed the general binomial expansion and, by means of it, a series for $\pi/4$. Incidentally, it is worth noting that Wallis was severely criticised by Fermat for departing from the rigorous proof structure of the ancients.

In teaching, we frequently introduce the problem first, so that, particularly with young children, we are mainly concerned with the creative and constructive side of mathematical thinking. In more advanced mathematics the ability to understand and appreciate a rigorous deductive proof is of great importance.

Whilst the logico-deductive systems which constitute the essence of formal mathematics can be analysed and checked, it is not possible in the same way to assess the factors leading to successful creative work in mathematics. Historical studies, if they are close and detailed enough, throw a good deal of light on the qualities associated with successful achievement. Among these qualities must certainly be noted persistence and determination, ability to profit by the work of others and care and attention to detail. Studies in the psychology of learning, where these exist and are soundly based, can also be employed, particularly where these relate directly to the development of creative thinking in mathematics. As for the great mathematicians of today, few have concerned themselves to tell us how their successes were achieved. Of the limited amount of material available here, Poincaré has been more helpful than most. He writes

> In fact, what is mathematical creation? It does not consist in making new combinations with mathematical entities already known. Any one could do that, but the combinations so made would be infinite in number and most of them absolutely without interest. To create consists precisely in not making useless combinations and in making these which are useful and which are only a small minority. Invention is discernment, choice.†

† Given from Poincaré, H., "Mathematical Creation" in *The World of Mathematics*, IV (London, 1960), pp. 2041–50.

Poincaré admits that often the emergence of a new idea is associated with a sudden insight emerging unexpectedly and unheralded. He says:

> Often when one works at a hard question, nothing good is accomplished at the first attack. Then one takes a rest, longer or shorter, and sits down anew to the work. During the first half-hour, as before, nothing is found, and then all of a sudden the decisive idea presents itself to the mind.

He goes on to describe the nature of the second period of conscious work, the arrangement of the results in the form of an orgnised demonstration, or proof:

> It is necessary to put into shape the results of this inspiration, to deduce from them the immediate consequences, to arrange them, to word the demonstrations, but above all is verification necessary. I have spoken of the feeling of absolute certitude accompanying the inspiration; in the case cited this feeling was no deceiver, nor is it usually. But do not think this is a rule without exception; often this feeling deceives us without being any the less vivid and we can only find it out when we seek to put on foot the demonstration.

The Importance of Symbols

The use of symbolic notation in mathematics is so commonplace that the role of symbols is often by-passed in discussion. It is only when one finds oneself deterred from entering a chosen field by the rash of new symbols on each page that one pauses to ask, "Is all this really necessary?" It is, however, the case that only by means of specialised symbols is it possible to represent and preserve long trains of reasoning. In this way, mathematics can be made available on a large scale to those with vastly inferior powers than the actual inventors. Many problems can be solved by sixth-form pupils after a first course in the calculus which were extremely difficult for Leibniz, the inventor of the original notation. Symbols provide for an economy of thought through which many of the elements of mathematics can be stacked away in a well-organised form so that verification is possible. Established proofs can be, so to speak, lifted and applied to relevant problems, often by mere beginners. Engineers, physicists, astronomers, business men and many others are all concerned with *mathematical applications*; these applications can only be effective if the calculations are summarised and structured by means of good notations. Often, indeed, effective symbols, once introduced, seem to live a life of their own and, by their very existence, suggest relations and operations which, without them, would never have arisen. The invention of algebraic geometry by Descartes and Fermat in the seventeenth century is a case in point. For the first time mathematicians were able to move outside the range of known curves: algebraic functions could be created by a stroke of the pen. The use of symbols in the construction of the axiomatic systems which concentrate the central core of *inherited* mathematics at a given time is, of course, central.

Axiomatics

Any discussion of the axiomatic structures which lie at the very kernel of abstract pure mathematics today is complex and difficult since these structures are concerned with the foundations of mathematics itself. It is only possible here to indicate the nature of some of the problems involved.

The idea of an axiomatic structure in mathematics is not, of course, a twentieth-century invention. Euclid's *Elements* and Newton's *Principia* both represented axiom structures. Indeed, it is important to realise that the whole of Newtonian mechanics was written by Newton in order to provide a firm basis in theory for his System of the World. Criticism of the axiom structures of Euclid and of Newton has, over the years, been directed, at least in part, at their incompleteness, in the sense that assumptions crept in from outside which were not stated in the initial axioms or postulates. Mainly, however, the attack has been directed at the apparently meaningless definitions given at the outset, e.g. "A point is that which has no parts" (Euclid), and "The quantity of matter is the measure of the same, arising from its density and bulk conjointly" (Newton). In the axiom structures of mathematics and natural science today everything essential is contained in the axioms themselves so that one begins with *assertions*; these assertions are the axioms or postulates of the theory. They are formulated in terms of certain basic, or primitive, concepts for which no definitions are provided. Once the primitive terms and postulates have been specified the propositions can be deduced by rules of reasoning (the principles of logic).

An example of an axiom system of this sort is Peano's axiom system from which the arithmetic of the natural numbers can be derived. Peano's system contains three primitive terms, "0", "number", "successor", and five axioms, or postulates.

A 1 0 is a number.

A 2 The successor of a number is a number.

A 3 No two numbers have the same successor.

A 4 0 is not the successor of any number.

A 5 If P is a property such that, (i) 0 has the property P and (ii) whenever n has the property P, the successor of n has the property P, then every number has the property P (The Principle of Mathematical Induction). In set notation, we can write

A 1 $0 \in N$,

A 2 $n \in N \Rightarrow n' \in N$,

A 3 $a' = b' \Rightarrow a = b$,

A 4 $\forall a \in N, \quad a' \neq 0$,

A 5 $(n \in P \Rightarrow n' \in P) \Rightarrow \forall n, \ n \in P$.

From these properties we can define the natural numbers, thus, 1 is the successor of 0, 2 is the successor of 1, and so on; the definition of addition follows and hence the construction of elementary arithmetic. In the construction of more advanced theories such as the integers, rational numbers, and so forth, it is possible to build on the theory of the natural numbers by the use of ordered pairs.

Finally, although there is nothing debatable about the essential components of an axiom system, i.e. primitive, or undefined terms, axioms, definition of further terms from the axioms, propositions derived by logical deduction from the axioms, there remain certain unresolved problems and those I would like to leave with you.

(i) What is the origin of the primitive, or undefined terms? Are they purely intuitive? Are they derived in some way from experience? Can they be expressed in terms of purely logical concepts? Is mathematics pure logic?

(ii) What is the origin of the logical principles used in reasoning? Are these principles abstractions from experience of the external world? Are they independent of the world we live in? Is this is the case, are they to be regarded as essential properties of the human mind? Are they absolute and immutable in terms of development and change?

All these are big questions and you may find them interesting enough to follow up by further reading and discussion. In this lecture I have tried to provide a broad framework of inquiry around which reading and learning can take place. Many aspects of the philosophy, psychology and logic of mathematics will be dealt with in greater detail in subsequent lectures.

Bibliography and Reading List

The World of Mathematics (ed. J. R. NEWMAN), 4 vols., London, 1960, contains many extracts from works which are relevant. This book gives many references to other material and can provide a starting-point for further studies.

In particular, I would like to commend the following:

HARDY, G. H., *A Mathematician's Apology*, Vol. IV, pp. 2027–38.

POINCARÉ, H., *Mathematical Creation*, Vol. IV, pp. 2041–50.

POLYA, G., *How to Solve it*, Vol. III, pp. 1980–92.

RUSSELL, B., *Mathematics and the Metaphysicians*, Vol. III, pp. 1576–90.

VON MISES, R., *Mathematical Postulates and Human Understanding*, Vol. III, pp. 1723–54.

The Role of Intuition

J. W. OLIVER

IN SOME miscellaneous works of Buckle, writing in 1872, we come across the following phrase: "That peculiar property of genius which, for want of a better word, we call intuition." Yet again, Dickens writes of one of his characters, "Mr. Boffin ... had a deep respect for his wife's intuitive wisdom". Two references to the same word which seem to place it on two planes of a very different level. But, on whichever plane we choose, the importance of intuition is emphasised.

To no less an extent is the importance of this particular quality of the human mind emphasised today in the field of mathematics. Many mathematicians stress the value of intuitive thinking. Many teachers of mathematics have its development as one of their prime objectives. Bruner says: "Intuitive thinking, the training of hunches, is a much neglected and essential feature of productive thinking ... the shrewd guess, the fertile hypothesis, the courageous leap to a tentative conclusion, these are the most valuable coin of the thinker at work."

Seeing the word used by different writers and in different media, one comes to realise that it has many subtle shades of meaning but let us take for the purposes of this discussion the broad meaning of the word as it is used today. The *Oxford English Dictionary* states that intuition consists of the immediate apprehension, without the intervention of any reasoning process, of knowledge or mental perception.

We might formulate the same thing in a slightly different way merely to emphasise it and to bring it more firmly into the field of Mathematics, by saying that the nature of intuition is the mental act of teaching formulations, or conclusions about previously open situations, without going through a step-by-step analysis. Working definitions of this sort enable us to begin talking about the subject, but they tell us very little about the real nature of it.

What is the nature of intuitive thinking? How does it work? When does it happen? Under what circumstances is it most likely to occur? Why does it suddenly appear? Very little is written on the subject, and when it is, for example in Bruner's writings, he says again and again, at present no research on the matter is available

In fact research in this field is notable by its absence, therefore it is not possible to quote authoritative findings on the subject. In one way this is, of course, a disadvantage, but in another it is not because it leaves the subject highly alive. And, although it may not be specifically referred to in much written work, there is a considerable amount of writing which throws some light on it. At least, this allows the beginning of speculation, kindles the interest and eventually leads to a fascination with it.

Using the definitions quoted earlier, I would like to talk of the subject in terms of what I call "an intuitive leap" and we might start by trying to find examples within our own experience of where this occurs. One example that I observed recently was at a picnic where several adults and a number of children were present. Reaching the end of the picnic we were left with three apples and four children wanting some apple. The question arose "how shall we divide these apples equally among the four children?" Before I had even begun to start dividing by four or operating with fractions, one of the housewives said, "we'll cut a quarter out of each apple, give these to one of the children and give the three remaining children the apples with the pieces cut out of them".

I asked her how she had worked this out. At first she could make no attempt to answer this question and said, "I really have no idea". Then she said, "I suppose I just saw the apples cut up". She had very little formal mathematical background and I'm sure if she had come across a similar question phrased in formal terms in a school textbook, she would have had a real struggle to come to some conclusion.

Some evidence of this assumption was given on a later occasion when six sausages were to be divided among seven children. She was completely lost for an answer and all the adults and children were wrestling with this problem in all sorts of weird and wonderful ways, cutting off bits here and adding bits there. This time I saw immediately—cut a seventh from each and do the same thing as she had done before. Naturally, because my training has led me in this direction, I almost simultaneously took the next step, which was question and answer packaged together, and divided n objects equally among $n + 1$ children by cutting an $(n + 1)$th from each of the n and completing the problem as before. This was done in an instant and could almost be called intuition, at least there was no consciousness of "the intervention of any reasoning process". But if one did call it intuition, then it would be intuition of a different order, because I spend my working hours looking at structure and pattern, and had merely imposed the structure the housewife had produced on two different situations. It does lead, however, to another interesting question and that is how much does an awareness of structure aid the intuitive leap. There is no doubt in the minds of teachers of Mathematics of the importance of learning to see or find structure, and this in its own right, and also because it is at the very root of "transfer of training", which has come to be rather discredited in its old form when children were forced to spend hours learning Latin because it was thought to train the mind to deal with life situations. As a slight diversion let me say that an awareness of structure enables the person to transfer an operation in one system very readily to another system thus making a transfer of a very real sort.

To return to my example, I reminded the housewife of her previous solution concerning the apples. She said she was sorry, but it was no good, she would have to have the objects lined up in front of her and be able to handle them. This leads me to a further speculation, that she had probably had considerable experience in the kitchen of dividing up small numbers of objects between people and her intuition had come from this varied experience, although it was not there in her consciousness at the time of her response to the initial problem. And this leads me to say that I do not believe the intuitive leap comes just from nowhere, I think it comes from varied experiences which are in the past, which somewhere touch on the problem and are stored in the unconscious ready to trigger the intuitive leap into consciousness without the necessity of these experiences themselves being dragged up on to the conscious level to be manipulated into a form which will produce the correct solution to a problem. The last statement arises from the very simple example quoted, but I think it is equally valid for more complex examples. The sort of formal understanding of mathematics which is so often engendered in school learning, the emphasis on explicit formulations, on ability to reproduce formulae, to fit them into a problem or to apply the right one to a given problem would seem far less valuable than a deep intuitive understanding of the subject. This does not mean that to be able to verbalise a piece of mathematics in a clear and ordered way is not important, of course it is. But an attempt to train children in this ability before their intuitive understanding has been touched seems doomed to failure.

And the question might well be asked, "Is this sort of emphasis actually inimical to the later development of a good intuitive understanding?" Examinations in particular encourage this ability to make explicit formulations and may bring such a pressure to bear on the teacher that he is forced into the sort of procedure where what is actually happening is that the pupils are being trained to pass the examination with very little emphasis being placed on the real mathematical factors involved or the understanding of them.

Today there are many attempts being made to find a way of testing a student's mathematical ability which do not have the effect described above. To give but two examples, there is the sort of written examination where the child is allowed as much time as he likes and as many textbooks as he wishes to consult, and in some Colleges of Education a student's own investigation into a problem, tackled and developed in his own way with all the time needed, is replacing at least the greater part of what used to be the traditional type of written examination.

In school mathematics it is the field of Geometry, perhaps especially, where a good intuitive feel is lacking; where the pupils, at least some, become skilled in checking the validity of an argument or in remembering a proof, but not able to begin to discover proofs for themselves. This is, of course, to a great extent changing, and it would seem clear that the development of Motion Geometry will aid this—the children manipulating cardboard shapes themselves, or at least witnessing a proof by visual demonstration, which must by its very nature lean more on the intuitive aspect of understanding. It could be argued whether a visual demonstration constitutes a proof, but perhaps we have too rigid an idea of what a proof is; or alternatively one might say that the more rigorous proof is a last, and not the first, step in a long chain of mathematical events leading to an ever deepening understanding.

In the absence of systematically gathered knowledge about intuition it is interesting and fruitful to do the sort of thing we have done so far, that is to ask questions; to make observations; to ask questions about those observations and to try to come at last to some tentative conclusion.

The questions divide broadly under two headings:

(a) What is intuitive thinking?
(b) What experiences affect its growth?

Let us look a little more closely at the first type of question. In order to see it more clearly, it is helpful to contrast intuitive thinking with analytical thinking which proceeds step by step, the thinker being able to report adequately each step to another person, and this is where a great difference lies. This can be seen very well in some examples which will be quoted later. Analytical thinking proceeds with full conscious awareness. It will probably involve deductive reasoning and a clear plan of attack. The intuitive thinker, however, arrives at an answer with very little awareness of how he has reached it. His answer may be right or wrong. He has usually had considerable experience of the area of knowledge involved and his awareness of structure seems to play a very large role. This seems to make it possible for the thinker to jump from point to point in a series of intuitive leaps. One might draw the simple analogy of the person who has a "sense of direction", again an unconscious non-explicit awareness. This enables him to go from point A to point B in a large city like London without having first mapped out all the single streets he must travel. Possibly he knows several landmarks; this is experience of the area in general. In his imagination he probably overlays the map of London with a square grid whose lines go north and south and east and west; this could be compared to a knowledge of structure. Arriving at a junction of roads he will know intuitively whether to turn, say, sharp left or whether to avoid this and take the road which bears gently to the left. The intuitive thinker can rarely provide an adequate account of how he obtained his answer.

At this point I would like to describe to you a conversation with one of my students tackling the problem of sketching the graphs of two of the inverse hyperbolic functions. During this work I actively encouraged her to make intuitive leaps if she could. She satisfies most of the points just raised, a fair background knowledge, a knowledge of the structure of an inverse function and the structure of how a curve is related to its axes, which involves knowing that if they move it will move relatively to them. What is interesting is that, of the two leaps she makes, one gives a correct solution, the other just misses it. What is fascinating is the way that she quite involuntarily shows that although she can later say what she did, although at first she can say nothing at all, she still cannot supply an adequate account because she just has no idea of why she did it.

The student had sketched the graph of $y = \cosh x$ which we had talked about. She had the graph, shown in Fig. 1, in front of her.

We had also talked a little about inverse hyperbolic functions, saying that one of these might be $y = \cosh^{-1} x$ which we could reverse and put in the form of $x = \cosh y$. I then said to her, "now have you any idea at all, without bothering to think too much, what sort of graph you would get for $y = \cosh^{-1} x$." Without a pause she sketched the graph (shown in Fig. 2). I merely pointed to the intersection of the curve with the x-axis and she put in the number 1. The graph was perfectly correct.

I commented on the fact that her answer had been instantaneous and asked her if she could think of anything at all that had come into her mind before she had "plonked" the sketch down on the piece of paper in front of her. She paused, began an attempt at a

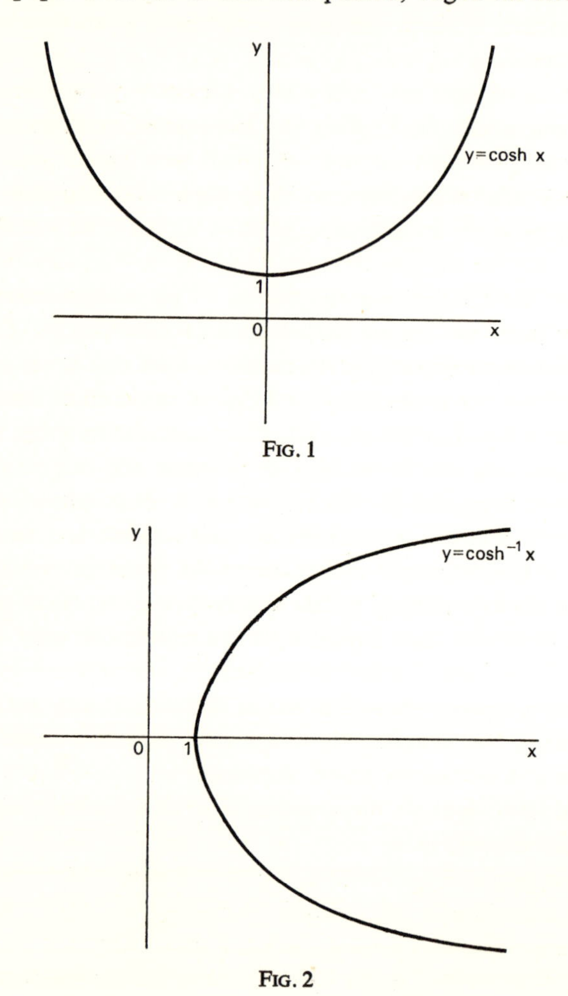

FIG. 1

FIG. 2

supposition of some sort of comparison, gave up and simply said, "I don't know, I just guessed".

I pressed the point, "do you remember any mathematical thought coming before you made the guess?" She laughed and said, as if she were admitting some misdemeanour, "no I don't". We talked further and I discovered that she had been pretty certain that she was correct and, furthermore, that she probably would have given no answer if she had not been.

She went on to say that "if anybody guesses anything they think they are right, don't they?" This was her immediate response and, I think, some comment on how she had been led to look upon guessing in mathematics. She did later agree that someone might give a tentative guess answer to a question without being sure of its correctness, but the feeling was that it would be prefaced by some remark like "this is only a guess, but ...". This student had shown clearly that there was no immediate conscious awareness of where her answer to this problem had come from. I decided to try to go a little deeper into this to see if we could get at what might have gone on, so I asked her if she could justify her sureness in some way or if in retrospect she could give any explanation whatsoever for her answer. She said she thought she had looked at the graph as if she was behind it and sideways on so that the y-axis had become the x-axis.

It is important to remember that before she could go even this far with an explanation, which nowhere approaches an analytical process of thought, some time had elapsed since she had first literally slapped the sketch of $y = \cosh^{-1} x$ down on paper.

I commented on the speed with which she had actually performed these actions in her imagination in order to make this aspect conscious to her and she seemed quite surprised.

We went on to look at the graph of $y = \sinh^{-1} x$ in exactly the same way as before (Fig. 3).

Again her response was immediate. She sketched her attempt at $y = \sinh^{-1} x$ practically before I had finished speaking (Fig. 4).

Again I asked her how certain she was that her sketch was correct. She said "the same as the other one". We saw that this sketch had also come immediately and she volunteered the comment that she was more certain of the correctness of this one because the other one had been correct. (An interesting sidelight on her quick gain in confidence to guess because of a positive experience.)

Again we found she had done the same imaginary actions as before, i.e. she had gone behind the paper and looked sideways on.

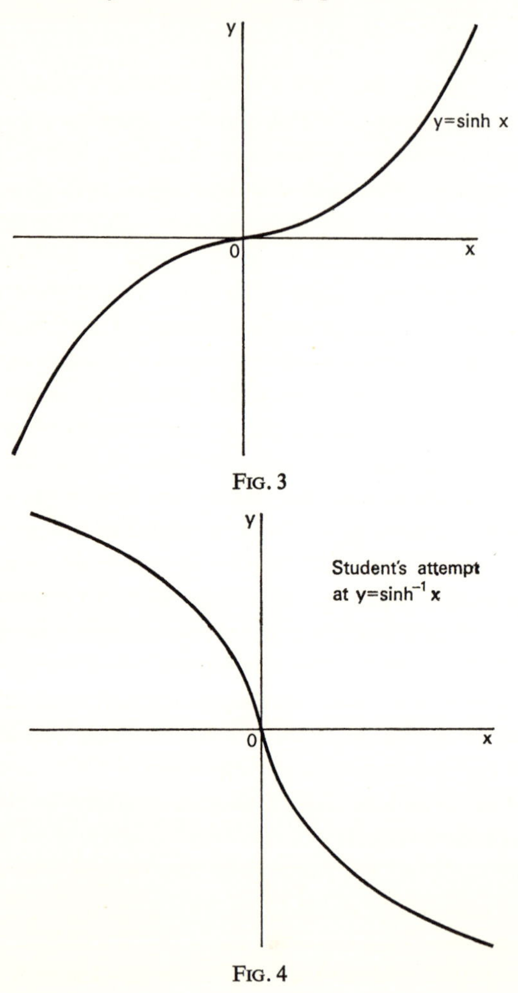

FIG. 3

FIG. 4

I now asked her to convince me of the correctness of her answer by "a more step by step process" and helped her to do it by her own method. We saw that she was trying to swap the x- and the y-axes. She saw that she had in fact turned negative x into positive y. I then asked her if her curve was in the right position and she said, "no, its back to front. I don't know whether you can see." I told her I could and asked her if she could correct her first attempt. She did so (Fig. 5).

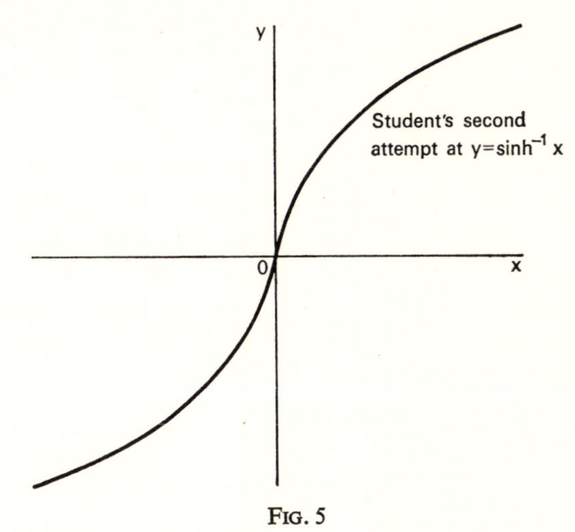

FIG. 5

And then (which is interesting), she turned to me and said "is this right now?" I gave her back the question and asked her what she thought. There was a long thoughtful pause and eventually she said, "I don't know". I had become the authority for her once more and she was now treading very carefully and wanting my judgement. She had fallen silent. After a time, I asked her if there was any flaw in any of the steps she had taken. I then went through it with her again step by step showing that the outcome of these steps was her final sketch. Her response was quite amazing. She said, "but why should I look at it from the back?"

"You tell me", I replied.

"I don't know", she said, "I had no reason at all for it!"

This was an absolutely classic example of the thinker being quite unable to give an explanation of why she had arrived at the answer and in fact ending up quite puzzled at her own actions, in so far as they had become conscious. Since she was so puzzled, we tried briefly together to find a reason for her "looking at it from the back". We saw that by going behind before rotating, she got the positives and negatives in the right position. "Yes!", she said. And then it came again; she was really genuinely puzzled. "I wonder why I went behind though?", she said. "I had no reason for doing it. I hadn't thought of a reason until you gave one to me!"

She had become so interested in this that I decided to spend time trying to look a little deeper. I asked her if she could think of anything at all in the past where she might have been encouraged to do this sort of thing. After a very long pause she came up with nothing. I tried a quick possibility and talked to her about reflection in an axis, and how some people flip things over to do this. (An equivalent to her "going behind".) I wondered whether she might use her method of looking at the picture from the back when working on reflection, but this obviously was not it and we just had to accept the fact that for the moment we did not know the answer to her question.

This student had given unmistakable evidence of her ability to make intuitive leaps so we now went on to see if any lead could be found for her tendency to think intuitively in this way. She had no memory of any explicit attempt on the part of any of her teachers, either Primary or Secondary, to encourage guessing. On the other hand, she had not been discouraged and would not have felt afraid to make guesses in school. However, she was very sure that she would not, at that time, have felt it positively a good thing to do. She felt that the "more relaxed atmosphere and smaller groups" of her $2\frac{1}{2}$ years of College Mathematics sessions had encouraged her to guess and the fact that guesses were taken seriously and followed through whatever they were was a definite encouragement and therefore important and especially important in teaching children. It seems that here was some small evidence that, because this student had not been actively discouraged from guessing, her tendency to think intuitively had not been killed and later when she had come into an atmosphere where guessing had been taken seriously, this tendency had begun to flourish. Further, what had applied to her might apply generally in teaching in the development of this kind of thinking.

Let us not assume that in the actual process of thinking the intuitive approach and the analytical are separated rigidly; that there can only be one or the other. In fact it would seem that they are complementary. Of those engaged in original mathematical research the intuitive thinker may often arrive at solutions which he would not achieve at all working analytically. But then the only means by which he can check his solutions are analytic.

Bruner says this, "More important, the intuitive thinker may even invent or discover problems that the analyst would not". I would like to put this point much more strongly and say that intuition must invariably be the first step to Mathematics of a creative nature, that creative Mathematics is inevitably linked with the intuitive leap. The analytical approach is needed to validate the new discoveries in a rigorous manner.

When we think of the schools in the light of what has just been said, the pupils are so often presented with the finished product which has been already formalised and it is this accent on formalism in much school learning which has devalued intuition. So the children are not doing Mathematics so much as learning of the finished products of those who have done it. If this is to be changed it seems inevitably linked with an attempt to develop the experience of intuitive understanding. Formal methods of proof can come afterwards or indeed they may well come naturally out of this and retain more life and reality for the school pupil. Certainly at the present time there are many attempts going on to move away from this static formal approach. Of these you will read more in later chapters.

Quoting again from Bruner, he says, "Precise definition (of intuitive thinking) in terms of observable behaviour is not readily within our reach at the present time. It is not easy to recognise a particular problem solving episode as intuitive or to identify intuitive ability." He goes on to say that research cannot be delayed until such a time as pure and unambiguous definition is possible along with precise techniques for identifying intuition when it occurs. It suffices as a start to ask can we agree one person's style being more intuitive than otherwise; can we identify some episodes as more intuitive than others?

My feeling is that perhaps the very nature of intuition precludes the sort of precise definition which is called for here. It is as if there is a need to define or recognise in a formal rigorous way something which is too ephemeral to catch in this particular net. It seems clear that an intuitive leap can be recognised as such and the recognition agreed upon, and if too precise a definition is called for, we run the risk of defining it out of existence and leaving ourselves, like Witgenstein and the logical empiricists, with nothing to talk about. But even if greater precision of a helpful nature can be achieved, I think it is very right to say research cannot be delayed for this event. I feel certain that intuitive ability can be recognised and here I am not referring to great mathematical ability but rather the ability of even a very young school child to approach mathematics intuitively.

Before going further it seems important to describe some conversations with students, involving "problem-solving episodes", which illustrate the last few statements. The first is with a group grappling with Wertheimer's famous altar-window problem. Here I think it can be seen clearly when an intuitive leap takes place although in this case some help is needed to trigger off the leap.

The problem was described to the students by saying that in a certain church there was a circular window which the authorities wanted to embellish. They asked a painter to paint around the window by first drawing tangents on either side, going vertically up as far as the top of the window and down as far as the bottom, then drawing a semicircle at the top on the ends of the tangents "the same as that one". As I spoke, I drew the diagram for them and pointed to the top half of the window. Continuing, I said, "A semicircle was then drawn at the bottom on the ends of the tangents the same as that one", pointing to the bottom half of the window (Fig. 6). Finally, supposing you know the diameter of the original circle, $2a$ if you like, "can you say what the area to be painted is?" There was a very, very long pause so I suggested they tried to talk about what thoughts they were having. There was another long pause then one of the group said "I would try to make even shapes like this". She drew in the lines shown (Fig. 7).

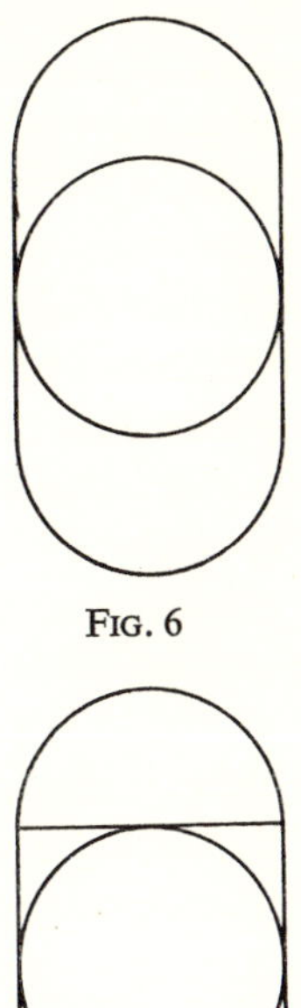

FIG. 6

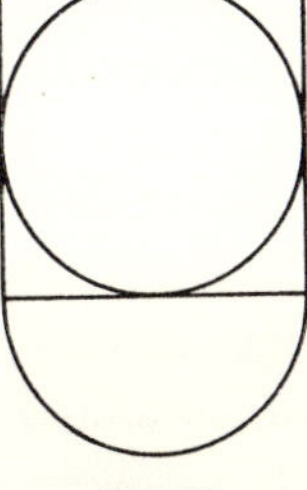

FIG. 7

Then comments came from various students such as "then you've got the area of the circle again"; "the straights bits are $2a$"! They then proceeded to work out the area of the circle which led to another pause while they wondered about the "little bits left over". Eventually they decided to find the area of the square and subtract from it the area of the circle to give them the area of these bits.

Here they seemed inclined to stop so I threw a small spanner into the works by saying "What about the fact that you can't evaluate π exactly?" One student thought it was a circular shape anyway, so it was not possible to get an exact answer. This seemed to be the general feeling so I led the situation even more firmly towards the possibility of an intuitive leap by getting the group to consider whether it was possible to do the question without using π. I went so far as to redraw the figure leaving out the lines which had been put in previously by one of the students. Then, feeling that the lines may in fact help a leap, I encouraged them to look at whichever drawing they felt helped most.

There followed quite a long discussion going backwards and forwards with several ideas. The same student as before said even more firmly that it was not possible to find the exact area because it was circular and π had to be used and it was not possible to get away from this. (This in itself was very interesting, the way that the idea of the circle seemed to be blocking any other possibility.) Suddenly in the middle of an animated discussion one of the students said quietly, "It's the area of that square there". She said it so quietly that I had to draw the attention of the other students to this and ask for it to be repeated. On questioning, this student said the answer had just come. On further questioning she could remember thinking "the outside arcs make up the circle in the middle".

It seems to me that this was eminently recognisable as an intuitive leap. A little time had gone by now and the rest of the students had had time to consider this answer. I asked them whether they agreed with it. They all did. Interestingly the same girl who had stuck on π before said "I didn't see it straight away because I was too convinced you could not do it without using π". We spent a little time talking about attitudes towards guessing when these students had been at school. The one who had actually made the leap said, "at school you weren't supposed to give guesses, you were supposed to have worked everything out". In general the students had not been actively encouraged to guess but neither had they been discouraged. The student who had had most bother with the question said, "If you guessed wrongly you usually felt bad about it". None of this, of course, is substantial enough to base any conclusions upon, but perhaps it gives an indication that an intuitive approach can be helped by the way the pupil is taught. It is also interesting that I had phrased the question in such a way that a definite diameter had been given for the circle, a fact which may well have led this group straight into an analytical approach. An intuitive leap may have come earlier had this not been done.

The second conversation is with one student who, I think, shows clearly a high degree of intuitive ability. As the discussion develops she shows very definitely that, as Bruner puts it, her style is intuitive. She is that sort of person. She first tackles the same problem as the others. It was described to her in the same way except that this time no mention of any diameter was made. It was just a window to be painted around. There was a short pause, she was about to speak and drew back, then she quickly drew in the two lines that had been drawn in before by the previous group and said, "The area of that square". No analytical procedure whatsoever had been used here.

We went on to talk about a situation which had arisen during a lecture-work session with a small group of students of which she was

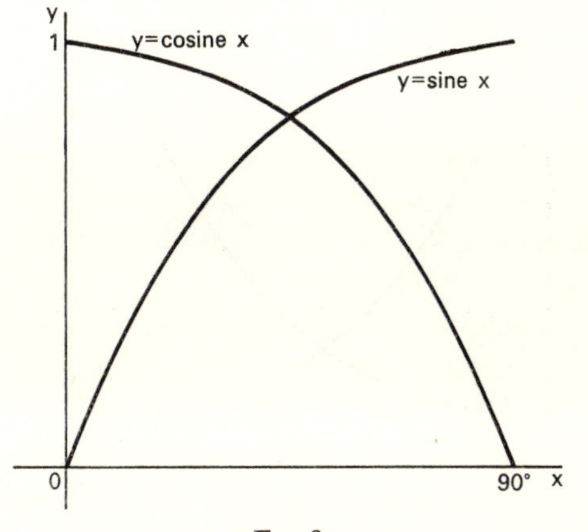

FIG. 8

one. They had all sketched the graphs of y = sine x and y = cosine x for $90° \geqq x \geqq 0°$ (Fig. 8). I had held one of the finished products (Fig. 8) up in front of the group of students and said, "does anything come to your mind when you look at this?" This student had said immediately "complementary angles". We later saw that she could amplify this brief answer and verbalise what she already was thinking, i.e. that the sine and the cosine of complementary angles are equal.

When we talked later about this I showed her the picture and asked her if she could remember what she had said. She started to give me a long analytical explanation. I stopped her and explained that although we were talking about something which had happened earlier could she remember what she had said then. She said, yes she did remember, but she "remembered at the time that I just jumped to the conclusion. I didn't really think about it." This was said rather as if she hadn't given me this answer again now because she ought to be thinking about it and trying to justify it.

We went on talking and she said she had been certain she was correct (which was tantamount to saying the graph itself showed the fact that the sine and cosine of complementary angles are equal). She added that she could not at the time justify her answer, neither

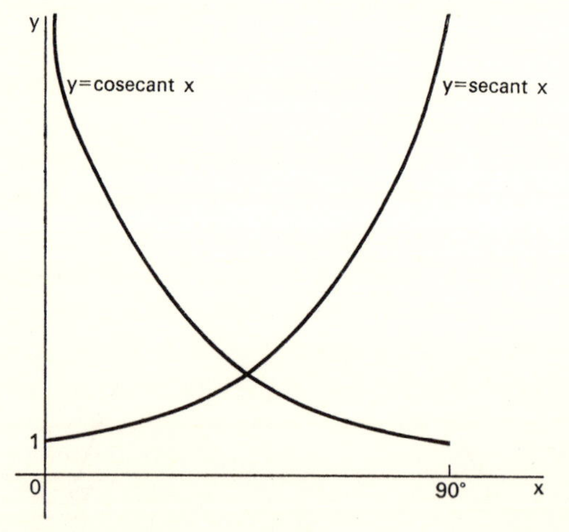

FIG. 9

could she remember being aware of having any reason in her mind for giving the answer, "it had just come". We went on to try to justify it and I asked her to look at the graph in order to do this. She noticed that one curve was the reflection of the other but at first said they were reflected in the point of intersection. Then we got as far as seeing they were reflected in the line $x = 45°$. I asked her if this helped at all with the justification. She said yes, and started to try to do it, in a short time she had become verbally very mixed up and said, "Oh, I can't explain it".

We then went on and she sketched the graph of $y = \text{secant}\, x$ for $90° \geqq x \geqq 0°$. I asked her to look at this, and see if she could put in straight away the graph of cosecant x. She did so (Fig. 9). There was a slight hesitancy about where the curve would be when $x = 90°$ but I merely had to ask her what the value would be there and she immediately said, "1". Again I asked if any thought had gone before her answer and she said, "Not really, it was a similar instinct that I had with cosine and sine".

This time she had not been certain straight away of her answer being correct. She said, "I *knew* it had to be a similar shape and to somehow get the mirror image". This would explain her hesitancy described above as a slight lack of fluency in actually drawing a mirror image and not as a lack of knowing intuitively the result. Here I think was a classic example of intuitive ability and certainly highly recognisable as such.

We looked a bit at her school background and she said she was always doing this sort of thing—"assuming" as she herself called it. She was not discouraged from doing this as long as she could make some attempt to back it up with some sort of justification. Although she was not encouraged to guess there was an atmosphere in the classroom of guesses being accepted as long as they were looked at closely afterwards. I asked her about her work at the college and specially that part of it which was investigating mathematical problems of her own choice in her own way. I asked if this work had encouraged her to make guesses. "Well I've always been the type of persons who guesses anyway", she said. So although she liked "investigation", she didn't think they altered her attitude at all because she already worked in this way. So indeed this student was conscious herself that her general approach to mathematics was an intuitive one.

Finally I think it worth describing conversations with two schoolboys about the altar-window problem. One reason being that they are still at school and therefore nearer to their earlier mathematical experiences. It is still possible I feel to recognise an intuitive leap when it comes but in both cases help is required to produce "the leap". These particular discussions illustrate more the connection between background knowledge and experiences and the possibility of an intuitive approach developing. The two boys are brothers, the elder being 14 years 3 months. The problem was described to him as before, making sure he understood it. An interesting point here is that his very first comment was the question, "What's the diameter of the circle?" I told him this was not known and we could make it anything we liked. We decided on 10 feet. He paused for a while, obviously thinking hard. Then said, "I know", and proceeded to solve the problem analytically very clearly, precisely and quickly using his knowledge of the formula for the area of a circle. I think this sort of work was still fresh in his mind from school and he tackled it with great confidence.

I then tried to lead him to another approach by saying that it was interesting that he had told me the two end pieces were together equal to the area of the circle (he had drawn in lines to make a square around the window), and would that give him a clue "to something perhaps even shorter than the way you are doing it?" His response was immediate, and almost before I had stopped talking he said, "The area of the square". We found that by saying "even shorter" this had prompted him to throw away his formal approach and he had "just put those two in there". In other words he had moved the two semicircles on the ends of the diagram into the circular space in the middle.

This similarity of his first attempt with the way one of the students in an earlier discussion had tackled the problem was striking. It seems to me that here was the intuitive approach just waiting to be used. All it needed was a little encouragement, and just this one suggestion of a variation of approach had brought it out.

The question arises how important is it to vary the approach to the same problem and try it in as many different ways as possible. This leads very definitely away from the tendency in much mathematical education to standardise the approach and to give instruction in a standard method. We found that at school he was not encouraged to make guesses, but "to look over what you know and see how it might fit in". He was not discouraged from making guesses; on the other hand the response of the teacher might be "Yes, you are half-way to it; or no, you are confusing yourself; or no, you are getting side-tracked; or yes, that's right, can you explain a bit clearer?" All answers which seem to suggest a single right approach. As he put it himself, "we are not discouraged, we are free to make a guess but not aimless guesses". On looking closely at this situation it would appear that an inhibiting factor to the intuitive leap *is* present.

The younger brother was 12½ years old. His first comment was "What do you mean, what area?" I answered, "how much surface he has got to cover with paint". He laughed and said he didn't know what I meant. I then asked him if he would know what I meant if I said there was a rectangle 3 feet by 2 feet to be painted. Immediately he replied "that would be six, six square feet". This would suggest a limited background knowledge of the idea of area.

Then he said he didn't know the size of the window. I told him he could make it any size he liked. This seemed to satisfy him. After a considerable pause he said "I'd say a hundred square feet". I asked him what size window he was making and he replied that it was quite a big one. I suggested there was no means of knowing whether he was right or wrong, to which he replied well perhaps it could be worked out by "sort of proportions". When I suggested it might help to make the window a certain size he thought this would make it more difficult and laughingly added "because I haven't got really any idea". I suggested he had a good look at the drawing and that he could put in any extra lines he wanted. After a long pause he started drawing in squares (Fig. 10). He knew about symmetry and so positioned his squares symmetrically about an axis of symmetry and decided he need only bother with the top half of the drawing. He said he could work out the number of squares, and the bits at the edges were approximately half squares so he could do it, but only roughly.

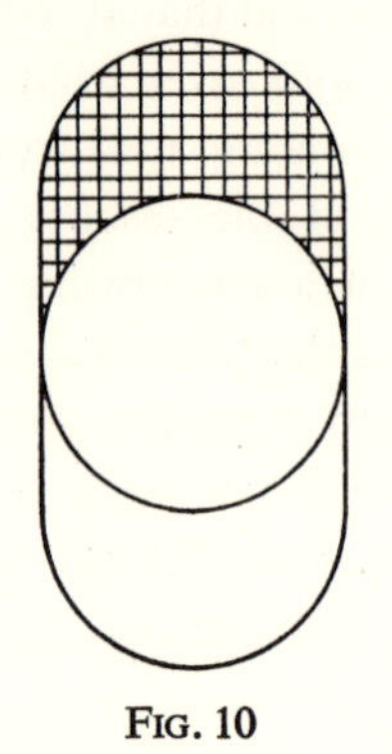

FIG. 10

We found out that he could not remember tackling area in this way at school so this in itself was quite an intuitive leap of a very basic nature. We agreed that he could now work out roughly the area and then we left this idea.

We next saw that he had at some time in the past done some work on the area of a circle and had a very vague memory of a formula. We didn't pursue this any further.

I then suggested I gave him a tip and drew in two lines to form the square. He seemed to become more confident. He became very absorbed and proceeded to try several different avenues discarding them as they came to a dead end, but in a very short time he said with a great air of satisfaction "A square".

What this discussion illustrates clearly is that, without a certain background, it was impossible for him to get anywhere at all with this problem by himself, but that, with even the limited background experience he had in this field, with a little help, he was able to make the intuitive leap necessary for him to solve the problem.

In a less obvious way than with the student described earlier one might say here was a case of somebody with intuitive ability and, remembering the very first guess, 100 square feet, somebody with a naturally intuitive style.

I think that some areas of mathematics are more likely to give rise to an intuitive approach or should I say that they are more easily recognisable to us as teachers as being that sort of situation. Certainly since becoming interested in this field I find myself in the classroom suddenly recognising that we are discussing something which might contain such a possibility and I find myself asking the sort of question, as with the sine and cosine graphs, which might bring it out. This is something we can all try. The altar-window problem obviously has a built in intuitive short cut waiting for those, who can, to take it. There are many approaches which research might use to find out about intuition, but when Bruner talks of "Procedures or instruments ... to characterise and measure intuitive thinking", I think this and those like it are doomed to failure for the reason that I mentioned before. It seems much more likely that getting lots of people to talk about what they are doing as they are actually doing mathematics is something which might bear fruit.

What is likely to influence the development of the intuitive approach? What is the role of the teacher in this respect? If the teacher himself thinks intuitively and is seen to do so by his pupils the power of this approach is seen. The openness to making mistakes is seen, the very important fact that one does not *have* to be right first time is seen. Then perhaps a certain freedom is engendered which loosens something in the pupil, allowing the intuition to leap. On the negative side, unless a student is very strongly orientated intuitively, it seems unlikely that he would develop confidence in this approach if he never saw it used by his tutors.

The teacher who is willing to make intuitive leaps in the classroom and then to analyse these with the class may be a catalyst to this approach which could be laying dormant in some of the students. The teacher who poses an open-ended situation to the students and is then willing to examine it with them in the insecurity of not knowing the answer, I feel convinced, will be even more effective. The student can always reach behind the words and feel whether the teacher already knows the answer and is trying subtly to lead the way or whether he is just as much in the dark, but perhaps has had more experience of finding his way through the darkness before.

Is it possible that varying the experience to the utmost in any given field will increase the possibility of intuitive leaps in that field? Varied experience would appear to lead us back along the path to structure which I feel sure is important in developing an intuitive approach. Would the teaching of heuristic procedures have an effect, i.e. teaching what is in essence a non-rigorous method of solving problems. This is in direct contrast to the use of an algorithm which can be learned by rote and is certain of success if the problem is solvable.

Should this type of procedure be verbalised? Is the verbalising of such techniques, which is equivalent to providing an algorithm for a heuristic approach, actually inimical to the intuitive approach since the student is led to reduce the process to an analytical one. I think to encourage something which is undoubtedly unconscious is better done implicitly allowing the ideas to be assimilated at a level which is perhaps not so highly conscious. But perhaps widening the door to the unconscious in this way could leave the unconscious even more ready to be activated. This is after all similar to what happens in depth psychology.

Bringing the discussion to a more mundane level should the student be encouraged to make guesses? This may certainly bring a freedom to the personality, the lack of which may inhibit the intuitive leap. The heavy repression of guessing in the classroom may certainly inhibit the intuitive approach and develop a desire for certainty before any step is taken, throwing the student back on entirely analytic procedures, using known facts, and consciously learned and remembered procedures. Is it not better to guess than to be struck dumb when the answer is not immediately known. Bruner says, "It is our feeling that perhaps a student would be given considerable advantage in his thinking, generally, if he learned that there were alternatives, that could be chosen that lay somewhere between truth and complete silence". The development of self-confidence and courage in the student would certainly appear to encourage intuitive thinking. This requires a willingness to make mistakes and one who is insecure or who rather is afraid to enter the realm of insecurity must be unwilling to make such mistakes. Where judgement of mistakes enters to a high degree then the risk will not be taken.

This takes us back to the classroom and grading and the giving of marks for the correct answer and appears to point away from the traditional examination and away from rewards and punishments and more firmly to the open-ended situation and the development of interest in the activity for the sake of the activity and not to the reward of a *good*, in inverted commas, performance in that activity.

Perhaps, in conclusion, one can say nothing more strongly than that to encourage the pupil to be himself in the classroom, and indeed in life itself, and not to attempt to make himself into the image which he thinks the authority, whatever or whoever he be, requires of him is the most important work a teacher can try to do in order to develop intuition and creative activity in mathematics.

Suggestions for Further Reading

Members of A.T.M., *Notes on Mathematics in Primary Schools*, C.U.P., Cambridge, 1967.

BRUNER, J. S., *The Process of Education*, Harvard University Press, Massachusetts, 1962.

DIENES, Z. P., *Building up Mathematics*, Hutchinson, 1961.

POLYA, G., *Mathematics and Plausible Reasoning*, Princeton University Press, Princeton, 1954.

POLYA, G., *Mathematical Discovery*, John Wiley & Sons, New York, 1962.

WERTHEIMER, M., *Productive Thinking*, Social Science Paperbacks in association with Tavistock Publications, London, 1961.

The Role of Logic

J. A. MERCER

Introduction

The part played by logic in mathematics is so obvious and so extensive that it has led some people to assert that mathematics is simply a branch of logic. It is not the intention here to accept or deny this philosophical standpoint, but rather to investigate some of the logical procedures used by mathematicians in an attempt to clarify these procedures and make them more understandable. Secondly, we shall examine different methods of proof employed in mathematics and arising from this, the development of axiomatic systems, which in turn provides an opportunity to see how some errors of reasoning may sometimes be discovered and even avoided.

It is often thought that, apart from the "natural" use of logic in pursuing a piece of mathematical work, e.g. in proving a theorem or a conjecture, and perhaps in discovering errors of reasoning, the function of logic as such is to consolidate, give a good foundation to mathematics as already discovered (almost a sort of philosophical luxury); but that it is not in any sense a tool used by mathematicians in the actual process of discovering new mathematics. So a final object will be to examine the connection, if any, between logic and mathematics in the development of new areas in mathematics.

I. Logical Procedures

The amount of actual knowledge of logic required in this study will be kept to a minimum. Much of it will be what has previously been called "natural" logic, for instance:

$$\text{If } A = X \text{ and } B = X, \text{ then } A = B$$

which may be recognised as asserting that equality satisfies the transitive law.

The reason for talking about "natural" logic is that even if a person has never heard of the transitive law, the above statement of it is often known and used quite happily. This is not to say that we are born with a built-in capacity for applying this law, but that it is more often than not caught rather than taught.

This particular law can serve as the introduction to the first part of our study—the use (often implicitly rather than explicitly applied) of certain logical procedures in mathematical deductions.

EXAMPLE 1. Piaget talks of his observations of small children. Let us consider one such child aged about 6 years. The child has a lot of building blocks of various shapes and sizes and decides to build a tower on his table. Having done this, he builds another one alongside it. Then he builds a third one, this time on a chair. Then the teacher comes up and asks him which is the tallest tower. The child easily decides about the two on the table, for one reaches higher than the other. They both reach a greater height than the one on the chair, but the child is not too sure whether they are both taller than this one. So he opens his arms and "measures" the taller one on the table, moves to the tower on the chair (happily assuming that he has kept his arms the same distance apart) and compares the two heights and so makes his decision. He has used the principle that

$$\text{if } A = X \text{ and } B = X, \text{ then } A = B$$

where A is the height of the tower on the table, X is the distance between his hands (assumed constant) and B is the height of the tower on the chair.

EXAMPLE 2. A student of modern algebra wishes to prove something he was taught at school, viz.

$$(-a)(-b) = ab.$$

Assuming the associative law for real numbers under addition, let

$$x = ab + \{a(-b) + (-a)(-b)\}$$

then

$$\begin{aligned} x &= ab + (-b)\{a + (-a)\} \\ &= ab + (-b) . 0 \\ &= ab + 0 \\ &= ab. \end{aligned}$$

Also

$$\begin{aligned} x &= \{ab + (-a)\, b\} + (-a)(-b) \\ &= b\{a + (-a)\} + (-a)(-b) \\ &= b . 0 + (-a)(-b) \\ &= 0 + (-a)(-b) \\ &= (-a)(-b) \end{aligned}$$

so

$$ab = (-a)(-b).$$

Another form of "natural" reasoning is what logicians call the Law of the Hypothetical Syllogism, viz.

If $a \Rightarrow b$ and $b \Rightarrow c$ then $a \Rightarrow c$.

As with "=", we now have $\Rightarrow$ obeying the transitive law. The great majority of theorems employ this idea, usually in an extended form.

EXAMPLE

$$\begin{aligned} & x^2 + 3x + 2 = 0 \\ & \Rightarrow (x + 2)(x + 1) = 0 \\ & \Rightarrow x + 2 = 0 \quad \text{or} \quad x + 1 = 0 \\ & \Rightarrow x = -2 \quad \text{or} \quad -1. \end{aligned}$$

N.B. This is a repeated application of the above law, with each application in a condensed form. It could be written as:

$$\begin{aligned} & x^2 + 3x + 2 = 0 \\ & \Rightarrow (x + 2)(x + 1) = 0 \end{aligned}$$

and
$$(x + 2)(x + 1) = 0$$
$$\Rightarrow x + 2 = 0 \quad \text{or} \quad x + 1 = 0$$
and
$$x + 2 = 0 \quad \text{or} \quad x + 1 = 0$$
$$\Rightarrow x = -2 \quad \text{or} \quad -1 \qquad \text{respectively.}$$
So
$$x^2 + 3x + 2 = 0$$
$$\Rightarrow x = -2 \quad \text{or} \quad -1$$

which, with a usual notation, is of the form

$$\{(a \Rightarrow b) \wedge (b \Rightarrow c) \wedge (c \Rightarrow d)\} \Rightarrow (a \Rightarrow d).$$

Another of these "natural" laws says in effect that you cannot have something true and false at the same time,

i.e. a and $\sim a$ cannot both be true,

i.e. $\sim(a \wedge \sim a)$ ... Law of Contradiction.

EXAMPLE. To prove that $\sqrt{2}$ is not a rational number.

Assume $\sqrt{2}$ is a rational number, i.e.

$\sqrt{2} = p/q$ where p and q have no common factor other than 1. (a)

Now
$$\sqrt{2} = p/q$$
$$\Rightarrow 2 = p^2/q^2$$
$$\Rightarrow 2q^2 = p^2$$
$$\Rightarrow p^2 \text{ is an even number}$$
$$\Rightarrow p \text{ is an even number.}$$
Then
$$2q^2 = 4r^2$$
$$\Rightarrow q^2 = 2r^2$$
$$\Rightarrow q^2 \text{ is an even number}$$
$$\Rightarrow q \text{ is an even number.}$$

Then p and q are both even and hence have a common factor. So $2 = p/q$ where p, q are both divisible by 2.

This contradicts (a)

So $\sqrt{2} \neq p/q$, i.e. $\sqrt{2}$ is not a rational number. This example uses a little more than the principle $\sim(a \wedge \sim a)$. It assumes not only that $a \wedge \sim a$ is always false, but that either a is true or $\sim a$ is true, i.e.

$a \vee \sim a$ ... Law of the Excluded Middle.

So the same example serves to illustrate both these principles.

N.B. The amount of "natural" logic acquired (without formal training) depends on the individual—and the way he has been taught.

II. Methods of Proof

The reader is assumed to have a knowledge of the basic logical constants and perhaps the use of truth tables, or other means of determining the equivalence of certain propositions. In particular, we shall make use of the equivalence of a proposition and its contrapositive, and for completeness the four related propositions are given below:

Given implication: $p \Rightarrow q$
Contrapositive: $\sim q \Rightarrow \sim p$
} equivalent.

Converse: $q \Rightarrow p$
Inverse: $\sim p \Rightarrow \sim q$
} equivalent.

If one requires to prove a theorem of the form $p \Rightarrow q$, it is often easier to prove the equivalent proposition $\sim q \Rightarrow \sim p$.

Quantifiers

There is a well-known difference in mathematics, and one often causing confusion to school children, between:

Identities, e.g., $\forall x\colon x \in R \quad ((1 + x)^2 = 1 + 2x + x^2)$

and

$$\text{Equations, e.g., } \exists x\colon x \in R \quad ((1 + x)^2 = 4).$$

The use of quantifiers is implicit in almost all mathematical theorems and statements.

EXAMPLE 1. For all real x, $\sin^2 x + \cos^2 x = 1$.

$$\forall x : x \in R \quad (\sin^2 x + \cos^2 x = 1).$$

EXAMPLE 2. It is impossible to trisect an arbitrary angle with ruler and compasses alone.

$\exists x : x$ is an angle (it is impossible to trisect x with ruler and compasses alone).

EXAMPLE 3. If a, b, c are real numbers, the equation

$$ax^2 + bx + c = 0$$

has two real distinct roots if $b^2 - 4ac > 0$.

$\forall a, \forall b, \forall c\colon a, b, c \in R; \exists x \quad ((b^2 - 4ac > 0) \Rightarrow (ax^2 + bx + c = 0$ has two real, distinct roots)).

Methods of Proof

Direct proof. A very familiar type that proceeds from propositions already accepted (axioms or theorems) via a chain of syllogisms to the desired conclusion.

EXAMPLE 1. See the previous example proving that

$$(-a)(-b) = ab.$$

EXAMPLE 2. The type of examination question in calculus that starts: "Prove from first principles that the differential coefficient of x^2 is $2x$".

Indirect proof (or *reductio ad absurdum*). If p is to be proved, assume $\sim p$ is true and hence derive a contradiction. Then the assumption $\sim p$ must be false. So $\sim \sim p$ is true, i.e. p is true.

EXAMPLE 1. See the proof that $\sqrt{2}$ is irrational.

EXAMPLE 2. Euclid's famous theorem: "The number of primes is infinite."

Assume number primes is finite.
Then there is a largest prime, k.
Primes are $a, b, c, \ldots, k$.
Let $P = a \,.\, b \,.\, c \,.\, k$.
Then $P + 1$ has no prime factors, i.e. is a prime number for $P + 1 = (a \,.\, b \,.\, c \ldots k + 1)$.
But $P + 1 > k$.
Contradiction.
So the number primes is infinite.

N.B. Another application of this is in Disproof by Contradiction. The only difference being that instead of trying to prove p, we attempt to disprove it, i.e. to prove $\sim p$. So assume p is true and look for a contradiction.

Proof using Contrapositive

If P is of the form $a \Rightarrow b$, then prove $\sim b \Rightarrow \sim a$.

EXAMPLE. If two lines are cut by a transversal so that a pair of interior alternate angles are equal, the lines are parallel.

The contrapositive is "If two lines cut by a transversal are not parallel, then the interior alternate angles are not equal".

Proof. If the two lines are not parallel, then they meet in some point O.

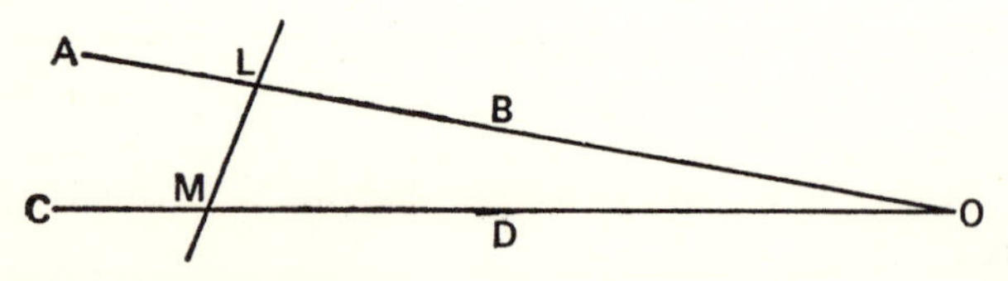

Then $\angle ALM = \angle DML + \angle BOD$
and $\angle ALM > \angle DML$.
So the alternate angles are not equal.

Proof of Existence

Simply exhibit an example.

EXAMPLE 1. Every equation of form $ax + b = 0$ $(a \neq 0)$ has a solution in the real number field, i.e.

$$\forall a, \forall b: \quad a, b \in R, \quad \exists x(ax + b = 0 \Rightarrow x \in R).$$

Proof. $x = -b/a$ is a solution.

EXAMPLE 2. There exist nil-potent 2×2 matrices (i.e. $A^p = 0$ and $A^{p-1} \neq 0$: A is nil-potent of index P).

Proof:

$$A = \begin{pmatrix} 0 & 3 \\ 0 & 0 \end{pmatrix}: \quad A^2 = \begin{pmatrix} 0 & 0 \\ 0 & 0 \end{pmatrix}.$$

Hence A is nil-potent of index 2.

Disproof by counter-example. Used when proposition conjectured is of universal form. Simply exhibit a counter-example.

EXAMPLE 1. The sum of any two odd numbers is odd.

Disproof. This proposition is false, since $3 + 5 = 8$, which is even.

EXAMPLE 2. Any three vectors in the vector space $V_3(R)$ are linearly independent.

Disproof. This proposition is false, since $(1, 2, 1)$, $(-2, -4, -2)$, $(3, 0, 5)$ are linearly dependent for $-2\,(1, 2, 1) + 1\,(-2, -4, -2) + 0\,(3, 0, 5) = (0, 0, 0)$.

III. Axiomatic Systems

In the very first method of proof, use was made of propositions which were accepted as sufficient reasons to carry the deduction through. These propositions could be theorems already proved or axioms.

We shall now consider the general structure of axiomatic systems.

The system that was perhaps the best known of all for about 2000 years is that in Euclid's *Elements*. In Book 1 of the *Elements* Euclid lists

23 definitions,
5 postulates,
5 common notions or axioms.

Definitions

1. A point is that which has no part.
2. A line is length without breadth.
3. The extremities of a line are points.
4. A straight line is a line which lies evenly with the points on itself.
5. A surface is that which has only length and breadth.
6. The extremities of a surface are lines.
7. A plane surface is a surface which lies evenly with the straight lines on itself.
8. A plane angle is the inclination to one another of two lines in a plane if the lines meet and do not lie in a straight line.
9. When the lines containing the angle are straight lines, the angle is called a rectilinear angle.
10. When a straight line erected on a straight line makes the adjacent angles equal to one another, each of the equal angles is called a right angle, and the straight line standing on the other is called a perpendicular to that on which it stands.
11. An obtuse angle is an angle greater than a right angle.
12. An acute angle is an angle less than a right angle.
13. A boundary is that which is an extremity of anything.
14. A figure is that which is contained by any boundary or boundaries.
15. A circle is a plane figure contained by one line such that all the straight lines falling upon it from one particular point among those lying within the figure are equal.

16. The particular point (of Definition 15) is called the centre of the circle.
17. A diameter of a circle is any straight line drawn through the centre and terminated in both directions by the circumference of the circle. Such a straight line also bisects the circle.
18. A semicircle is the figure contained by a diameter and the circumference cut off by it. The centre of the semicircle is the same as that of the circle.
19. Rectilinear figures are those which are contained by straight lines, trilateral figures being those contained by three, quadrilateral those contained by four and multilateral those contained by more than four straight lines.
20. Of the trilateral figures, an equilateral triangle is one which has its three sides equal, an isosceles triangle has two of its sides equal, and a scalene triangle has its three sides unequal.
21. Furthermore, of the trilateral figures, a right-angled triangle is one which has a right angle, an obtuse-angled triangle has an obtuse angle, and an acute-angled triangle has its three angles acute.
22. Of the quadrilateral figures, a square is one which is both equilateral and right-angled; an oblong is right-angled but not equilateral; a rhombus is equilateral but not right-angled; and a rhomboid has its opposite sides and angles equal to one another but is neither equilateral nor right-angled. Quadrilaterals other than these are called trapezia.
23. Parallel straight lines are straight lines which, being in the same plane and being produced indefinitely in both directions do not meet one another in either direction.

Postulates

Let the following be postulated:

1. A straight line can be drawn from any point to any point.
2. A finite straight line can be produced continuously in a straight line.

3. A circle may be described with any centre and distance.
4. All right angles are equal to one another.
5. If a straight line falling on two straight lines makes the interior angles on the same side together less than two right angles, the two straight lines, if produced indefinitely, meet on that side on which the angles are together less than two right angles.

Common Notions

1. Things which are equal to the same thing are also equal to one another.
2. If equals be added to equals, the wholes are equal.
3. If equals be subtracted from equals, the remainders are equal.
4. Things which coincide with one another are equal to one another.
5. The whole is greater than the part.

In his definitions, what Euclid was trying to do was to define every technical or logical term used. This sounds perfectly reasonable at first sight because an attempt is being made to fit all 465 propositions (theorems and constructions) on a good and solid base. Unfortunately, as anyone will know who has tried to find the meaning of a word in a dictionary, you reach a point where you get back to the word you started with, e.g.

Line: shortest distance between two points,

Point: intersection of two lines.

If you really intend to get the full meaning, you can do this with every word: "shortest", "distance", etc. But this gets you absolutely nowhere because under each definition more new words are encountered, and so on.

So it does appear that *some* words, like for example, "point", "line", "exist", "and", "surface", etc., have to be accepted as "known" and then other words defined in terms of these. Such undefined words come under the general heading of *primitive terms* which serve as a basis for defining certain other words and terms used.

To Euclid, postulates and common notions were distinct ideas. A postulate referred to a statement dealing with the subject under discussion, and accepted as true without proof. For as with definitions, we need to start with some "facts" or no start could ever be made. A common notion or axiom was again a statement accepted without proof, but not a technical statement. It was something accepted by anyone at all—a general truth.

Nowadays no distinction is made between these two types of basic statement for reasons which will appear later. But it does appear that Euclid considered both postulates and axioms as *true* statements, with the postulates referring to the properties of the space in which we live:

e.g. P 2 seems to suggest that the universe is infinite, i.e. a straight line goes on and on and on without ever reaching an end. But this might not be so, as will be indicated later.

Because this system seems to refer to objects in and properties of physical space, it is not quite an abstract study—such as, for example, the formal definition of a group where we have "elements", a "set", an "operation" and axioms referring only to these quite abstract entities. Euclid's Elements, if viewed in this way, would be called an example of a "material axiomatic system" where the axioms are assumed to be true statements.

As mentioned above, the formal idea of a group is much more abstract; elements are elements, they belong to a set, there is a binary operation on this set. There are four axioms—basic laws accepted without proof. But compared with the axioms and postulates of Euclidean geometry, they cannot exactly be called or considered as "true" or "false", in the sense that they obviously and clearly refer to certain objects—even though they *can* refer to well-known objects. The terminology seems deliberately intended to dissuade one from making any obvious connection.

There has been a clear development from material axiomatic systems to what are called formal axiomatic systems. In these systems, everything is usually phrased in an abstract manner and the proofs of theorems are carried through very formally, with what appears to be no reliance on intuition.

EXAMPLE. The 1967 Teachers' Certificate examination, Paper III. 9 (i) (London):

H is a non-empty subset of the elements of a multiplicative group G whose identity element is e. Given that H is such that if $a \in H$ and $b \in H$, then $ab^{-1} \in H$, prove that:

(a) $e \in H$,

(b) if $b \in H$, then $b^{-1} \in H$,

(c) if $a \in H$ and $b \in H$ then $ab \in H$.

Proof. $a \in H, b \in H \Rightarrow ab^{-1} \in H$. (i)

(a) Put $b = a$ in (i).

$$a \in H, a \in H \Rightarrow aa^{-1} \in H$$
$$\Rightarrow e \in H \qquad \text{(identity, inverse).}$$

(b) Put $a = e$ in (i).

$$e \in H, b \in H \Rightarrow eb^{-1} \in H$$
$$\Rightarrow b^{-1} \in H \qquad \text{(identity).}$$

(c) From (b), if $b \in H$, then $b^{-1} \in H$.

So $a \in H$ and $b \in H \Rightarrow a \in H$ and $b^{-1} \in H$

$$\Rightarrow a(b^{-1})^{-1} \in H \qquad \text{(from (i))}$$
$$\Rightarrow ab \in H.$$

Here there need be no reliance at all on intuition or a knowledge of the material world.

The structure of a formal axiomatic system consists of

1. Primitive terms—e.g. element, set, exist.
2. Defined terms—using the above.
3. Axioms (or postulates)—propositions assumed without proof.
4. Theorems—propositions proved from axioms and previous theorems.

5. Statements at the end of a theorem asserting that the theorem T is logically implied by the postulates P (this is not always expressed), e.g. Hence the theorem,

Q.E.D.,

◇,

■.

It becomes obvious why there is now no distinction between postulates and axioms. The postulates (axioms) use terms that do not refer to anything—they are completely abstract, so no distinction is possible. Notice that apart from no. 5, the axioms and theorems are neither true nor false—because the primitive terms are undefined. So perhaps we can understand Bertrand Russell's dictum "mathematics may be defined as the subject in which we never know what we are talking about, nor whether what we are saying is true".

But if we look at no. 5 we can say something that is true or false, viz. Q.E.D., ◇, ■, etc. And if this is not put explicitly we can add

6. The postulates P imply the theorems T.

This is true or false. So, perhaps "mathematics is the science that draws necessary truths" (B. Peirce).

We can, if we wish, interpret the symbols in some way so as to make all the postulates true statements (propositions), i.e. we have an interpretation or model of the branch of pure mathematics, or a branch of applied mathematics. The same piece of pure mathematics can have several models. This is a material axiomatic system, e.g. Group can be interpreted as integers under addition

or

symmetries of a plane figure under "followed by", etc.

So perhaps mathematics is "the giving of the same name to different things" (Keyser).

Example of how we can have one formal system with several applications (models).

Example. Consider a set K of arbitrary elements $a, b, c, \ldots$, and a binary relation R. aRb means a is R-related to b, $a = b$ means a, b are identical, $a \neq b$, they are distinct.

Four postulates:

P. 1 If $a \neq b$, then either aRb or bRa.
P. 2 If aRb then $a \neq b$.
P. 3 If aRb and bRc then aRc.
P. 4 K consists of exactly 4 elements.

Various theorems follow, which we have not time for.

Application 1. Let elements of K be four men:

a man
his father
his father's father
his father's father's father.

Let R mean "is an ancestor of".

Application 2. Let elements of K be four distinct points on a horizontal straight line and let R mean "is to the left of".

Application 3. Let the elements of K be 1, 2, 3, 4 and let R mean "is less than".

(For further development see Eves and Newsom, pp. 170–3)

N.B. In drawing up a formal axiomatic system it is nearly always true that there is some underlying concept in mind, i.e. psychologically it first takes shape in a material way—there is some application or concept in mind. If there were not, it would be difficult to proceed.

For example, take some undefined words—*aba*, *daba*—dreamed up from somewhere. It is very difficult to know what to do with these unless there is some concept in mind to which they refer.

So one might have, for example, Ancient Egyptians setting up boundaries to their fields:

Practical situation (concept)	Material axiomatic counterpart	Formal counterpart
corner	point	*aba*
boundary	line	*daba*
field	plane	*abdab*

IV. Errors

Some things can be overlooked because of over-familiarity with the subject-matter. For instance, Euclid probably did not realise that his axiomatic system was incomplete in the sense that not all of his theorems can be strictly proved from his axioms (and postulates).

For example, his very first proposition in Book I, Proposition I.1, assumes that circles intersect. But there is no axiom stating the continuity of lines and curves. This is a subtle omission because it seems so obvious—and no real harm is done, because we "know" we can construct an equilateral triangle. We are just assuming that the two arcs do not somehow slip through one another, and in fact are misled by the figure. Less familiarity with the subject and/or greater care in justifying every assumption might have avoided this.

Also in I.16, "Exterior angle is greater than either interior angle", he assumes that a straight line can be extended infinitely. But it could come back on itself, and then the proof would be void; e.g. if a person who lived on another planet and understood straight lines to be great circles, then points at opposite poles have an infinite number of straight lines joining them. In addition to this he would not consider a straight line to be of infinite length.

But there are cases where the figure misleads us with harmful results.

EXAMPLE. To prove that every triangle is isosceles

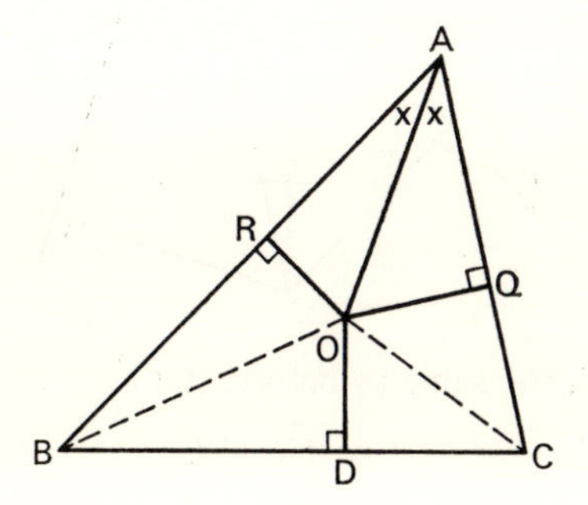

To prove $AB = AC$.

Draw the bisector of A to meet perpendicular bisector of BC in O. Drop perpendiculars to other two sides. Join OB, OC.

Proof. $\Delta\, ODB \equiv \Delta\, ODC$ (S.A.S.).

So

$$OB = OC$$

$$\Delta\, ARO \equiv \Delta\, AQO \quad \text{(A.A.S.)}.$$

So

$$AR = AQ,$$

$$OR = OQ,$$

$$\Delta\, ORB \equiv \Delta\, OQC \quad \text{(right angle, Hyp., S.)}.$$

So

$$RB = QC.$$

But

$$AR = AQ,$$

and so

$$RB + AR = QC + AQ,$$

i.e.

$$AB = AC.$$

The error here is in assuming that O lies within the triangle. If it was in fact drawn accurately it would be:

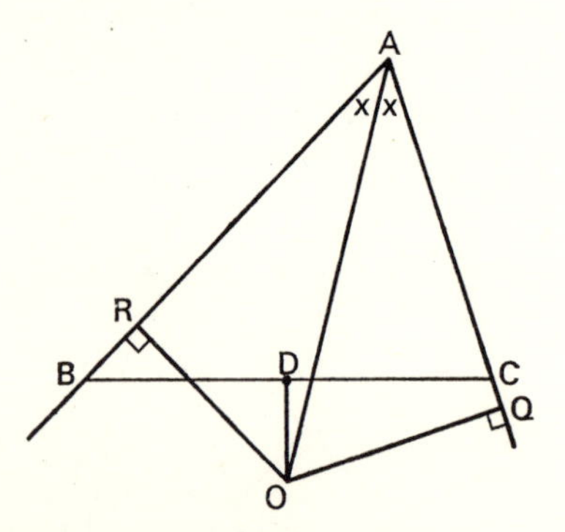

Can proceed much the same as before, but

$$AB = AR + RB,$$

$$AC = AQ - QC.$$

So

$$AB \neq AC \quad \text{in general}.$$

Here the unstated assumption that O lies within the triangle needs first to be exposed before it can be judged to be in error.

V. The Role of Logic in the Development of Mathematics

In order to investigate any possible link between formal logic and mathematics, we shall consider a sort of "case history". The one chosen is one of, if not *the* most, important in the history of mathematics—the famous fifth postulate of Euclid's *Elements*.

Even a cursory glance at the list of postulates makes P 5 appear as of an altogether different nature from P 1 to P 4, or at least from P 1 to P 3. (P 4 also differs from P 1 to P 3 in the sense that the latter are postulates of construction, whereas P 4 is in the form of a proposition.) But most people would accept P 1 to P 4 as being acceptable as postulates, realising that there have to be some unproved statements to start with and these four are simple and straightforward enough. But P 5 is neither simple nor straightforward and looks very much as if it ought to be proved somehow. This impression is strengthened when it is seen that several much simpler sounding propositions are proved as theorems,

e.g. Any two sides of a triangle are together greater than the third.

In addition to this, the converse is actually proved as a theorem and the postulate itself is not used until I.29.

It turns out to be a mark of the genius of Euclid that he saw it was necessary to include this as a postulate, for without it, the theory of parallels just could not be effected.

None the less, it is understandable that from the very first it was attacked. There were three ways in which people disputed it:

(i) Attempts were made to prove it from P 1 to P 4 and I.1–28.

(ii) Some tried to put it in different words, so that it looked less cumbersome.

(iii) A few people actually questioned its truth.

The attempts at proving it all failed at one point. Somewhere in the proof an assumption was made that was equivalent to P 5. The most famous equivalent statement is probably Playfair's Axiom:

> Through any point, not on a straight line, we can draw only one line parallel to the given line.

Another equivalent statement is that of Gauss:

> There is no upper limit to the area of a triangle.

Geminus was one who questioned the truth of it. He drew a comparison between, say, a hyperbola and its asymptote. These converge but never actually meet. So perhaps the truth of P 5 was so questionable that it needed to be proved. This he proceeded to do (with the usual unwitting assumption in his "proof").

Jumping many years, we come to the Italian mathematician, Saccheri, who in 1733 published a book called *Euclides ab omni naevo vindicatus*. This was a book intended, as its title suggests, to wash Euclid's image clean of all the attacks that had been made on him. It is clear, therefore, that he set out to prove the truth of P 5. But in doing so, he investigated very fully the consequences of its denial.

He began by considering an isosceles bi-rectangle *ABCD*:

Using only P 1–P 4 and I.1–28 he easily shows $\hat{C} = \hat{D}$. On the assumption of P 5, these are right angles. But if P 5 is held to be false, these angles can be

either acute

or

obtuse.

He investigated both these cases and produced many theorems that now form part of non-Euclidean geometry.

He disproved (so he thought) both these hypotheses by obtaining contradictions (disproofs by contradiction). Unfortunately he made unjustifiable assumptions in both cases. In point of fact, no contradictions should occur in any of the cases considered, which arises from the independence of P 5.

Even though Saccheri's work failed in its aim, it was of great importance because it was the best attempt so far on investigating P 5, and the fact that his final argument "disproving" the acute angle hypothesis was so weak led people to wonder whether a perfectly consistent geometry might not be found by denying P 5 and asserting the acute angle hypothesis—which in Playfair's Axiom meant that an infinite number of lines can be drawn through a point parallel to a given line.

Equivalent hypotheses

Playfair	*Saccheri*
just one line	right angle
no line	obtuse angle
infinite number lines	acute angle

Bolyai (Hungary) and Lobachevsky (Russia) are the two accredited as the founders of non-Euclidean geometry, though Gauss also played a notable part.

They worked on the acute-angle hypothesis (more than one line) and failing to find any contradictions, published their work. (Around 1830, separately and independently.)

Like Saccheri they found a contradiction in the obtuse-angle hypothesis. It arose from the assumption of the infinite extent of a straight line, made by all so far.

The first to discover the geometry of the obtuse-angle hypothesis was Riemann (Germany) in a probationary lecture in 1854. To do this he had to amend P 1, P 2 slightly, and thus produced a perfectly consistent geometry.

What these men were beginning to do was to look upon the axioms of Euclid not so much as being true or false, but as purely formal statements—axioms from which certain theorems could be derived, irrespective of the truth or falsity of these axioms. In other words, they were playing a vital role in moving from material axiomatic systems to the formal axiomatic systems as we know them today. If they had insisted on viewing axioms as true or false (as did Saccheri) they would not now be recognised as the founders of non-Euclidean geometries—and perhaps our understanding of formal axiomatic systems would not be as advanced as it is today.

References

Insights into Modern Mathematics, chaps. IV, IX (National Council of Teachers of Mathematics).

The World of Mathematics, Book 3, Parts XI and XII, XIII, edited by JAMES R. NEWMAN (Allen & Unwin).

An Introduction to the Foundations and Fundamental Concepts of Mathematics, EVES and NEWSOM, chaps. 1–3, 6, 9 (Holt, Rinehart & Winston).

Introduction to Logic, SUPPES, P. (Van Nostrand).

Fallacies in Mathematics, MAXWELL, E. A. (C.U.P).

Generalisation and Structure

C. Plumpton

1. Introduction

It has been my duty and pleasure during the last 20 years to teach mathematics to first-year university students. Most of the students are reading engineering and are taught in classes of ever-increasing size. In the eyes of university pure mathematicians I teach applied mathematics, since at university level, calculus, differential equations, problem solving and so on, are all designated as applied mathematics. At the most, 5 per cent of those who do mathematics at A-level, go on to read mathematics at Honours standard. The remainder who enter higher education become technologists, physicists, chemists, economists, geographers, or enter colleges of education to train as teachers of mathematics. My remarks are concerned with these latter groups rather than with Special Honours mathematicians, but the ideas developed in this chapter are relevant to A-level teaching, since most students of mathematics are products of a sixth form.

In such a context, this chapter consists essentially of some reflections on the teaching of mathematics and the thought processes involved. In particular, emphasis is placed on the ideas of generalisation, structure, approximations and the evolution of some particular aspect of applied mathematics as defined above, from the viewpoint of a teacher interested in ideas, rather than in the verbiage peripheral to "modern" pure mathematics.

Consider first the implications of the following three "postulates":

Mathematics is an experimental subject. For example, the processes of long division, of integration, the solution of differential equations and inductive reasoning are all experimental operations. Do some long division yourself and check that you carry out a series of trials until you come to part of the correct answer and then proceed with further trials.

With integration one examines whether an integral (a) is of standard form, (b) is a slight modification of a standard form, or (c) can be done by substitution or by parts, or by reduction formulae or some other standard general technique. Finally, if these trials all fail with a definite integral, we can usually resort to numerical integration. The process of solution of differential equations is simply that of trying out a number of tests and techniques one after the other.

In illustration consider

(a) the evaluation of the integral $I = \int x \tan^{-1} x \, dx$,

(b) the solution of the differential equation
$dy/dx \sin x + 2y \cos x = \cos x$.

(a) A survey of the table of standard integrals or, more sensibly, a search, conscious or unconscious, among the standard forms we store in our memories (even the cleverest mathematician should know some of the "vocabulary" of his subject) fails to produce an answer. Nor is the integrand easily twisted into a standard form. The next technique is to apply the "grammar" of integration and first try a substitution. What substitution? To clear the troublesome $\tan^{-1} x$, the substitution $x = \tan\theta$ suggests itself but leads to $\int \theta \sec^2\theta \tan\theta \, d\theta$ which seems no better than our starting-point. The next suggestion of our "grammar of integration" is integration by parts. To use this method we must write I in the form $\int u(dv/dx) \, dx$. Then either $dv/dx = \tan^{-1} x$ or $dv/dx = x$. The former appears complicated; then let us try $v = \frac{1}{2}x^2$ and write

$$I = \int \tan^{-1} x \frac{d}{dx}\left(\frac{1}{2}x^2\right) dx$$

$$= \frac{1}{2}x^2 \tan^{-1} x - \int \frac{1}{2}x^2 \frac{d}{dx}(\tan^{-1} x) \, dx$$

$$= \frac{1}{2}x^2 \tan^{-1} x - \frac{1}{2}\int \frac{x^2}{1 + x^2} dx.$$

Can we evaluate this last integral (of a rational function of x)? If we use a standard technique, writing

$$\int \frac{x^2}{1+x^2}\,dx = \int\left(1 - \frac{1}{1+x^2}\right)dx = x - \tan^{-1} x,$$

the problem is now solved, the result being

$$I = \tfrac{1}{2}(x^2 \tan^{-1} x - x + \tan^{-1} x).$$

Naturally you check the result by differentiation. All experiments must be checked if possible:

(b) To solve the differential equation we ask ourselves the questions: "What general types of differential equations can we solve?" In particular, "Is this differential equation separable or homogeneous or linear or ...?" (In fact, the first stage of our experimental method is *classification* of the problem. In example (a) above we solved the integral by classifying it as a type.) When we examine these possibilities one after the other we find that the differential equation is linear† with the integrating factor sin x and can be written

$$\frac{d}{dx}(y \sin^2 x) = \sin x \cos x$$

with solution (integral)

$$y = \tfrac{1}{2} + C \operatorname{cosec}^2 x.$$

Pure mathematicians use experimental techniques extensively. For example, they use the concept of disproof by counter-example or by contradiction frequently. What is this but experimental reasoning?

To illustrate the use of the counter-example technique we show that the condition $a_n \to 0$ (which is *necessary* for the convergence of the series $\sum_1^\infty a_n$) is *not sufficient* for the convergence of $\sum_1^\infty a_n$. We *try* as our counter-example

$$s_k = \sum_1^k n^{-1/2}.$$

† It is separable also—I missed this joint when writing this chapter.

Clearly the nth term $n^{-1/2}$ tends to zero as $n \to \infty$, but, since every term on the right-hand side except the last exceeds $n^{-1/2}$,

$$s_k > k \,.\, k^{-1/2} = k^{1/2} \quad \text{for} \quad k > 1.$$

Therefore $s_k \to \infty$ as $k \to \infty$ and so the series diverges although the nth term tends to zero. This counter-example technique is essentially experimental. We could have tried

$$\sum_1^\infty n^{-1/3} \quad \text{or} \quad \sum_2^\infty (\ln n)^{-1}$$

with equal success.

To illustrate the idea of proof by contradiction we exhibit the uniqueness of the resolution of a vector in a plane.

If $\boldsymbol{p}_1$ and $\boldsymbol{p}_2$ are any two non-collinear non-zero vectors, then any vector $\boldsymbol{a}$ lying in their plane can be expressed in the form

$$\boldsymbol{a} = \alpha_1 \boldsymbol{p}_1 + \alpha_2 \boldsymbol{p}_2, \tag{1}$$

where α_1, α_2 are unique scalars.

Let the vectors $\boldsymbol{a}, \boldsymbol{p}_1, \boldsymbol{p}_2$ be applied at O, Fig. 1, and let $\overrightarrow{OA}$ denote the vector $\boldsymbol{a}$. Through A we draw a line AA_1 parallel to the vector $\boldsymbol{p}_2$ to cut the line of action of the vector $\boldsymbol{p}_1$ in the point A_1. Similarly through A we draw AA_2 parallel to the vector $\boldsymbol{p}_1$ to cut the line of action of the vector $\boldsymbol{p}_2$ in the point A_2. Then, by the parallelogram law,

$$\boldsymbol{a} = \overrightarrow{OA} = \overrightarrow{OA_1} + \overrightarrow{OA_2}.$$

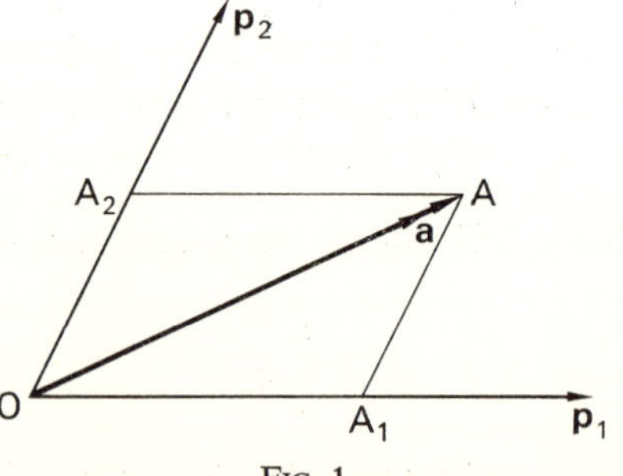

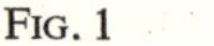

FIG. 1

But since the vectors $\overrightarrow{OA_1}$ and $\boldsymbol{p}_1$ lie along the same straight line, the vector $\overrightarrow{OA_1}$ is a scalar multiple of $\boldsymbol{p}_1$, say $\alpha_1\boldsymbol{p}_1$. Similarly $\overrightarrow{OA_2} = \alpha_2\boldsymbol{p}_2$. Equation (1) follows at once.

We show that α_1, α_2 are unique by assuming that the vector $\boldsymbol{a}$ can also be expressed as

$$\boldsymbol{a} = \alpha_1'\boldsymbol{p}_1 + \alpha_2'\boldsymbol{p}_2, \qquad (2)$$

where at least one of the inequalities $\alpha_1 \neq \alpha_1'$, $\alpha_2 \neq \alpha_2'$ is satisfied, and then obtaining a contradiction. To prove a uniqueness theorem, the usual procedure is to assume that the theorem is untrue and from this assumption to derive a contradiction. It follows that the theorem must, in fact, be true. [In older books on Euclidean geometry this is referred to as a *reductio ad absurdum*.]

Without loss of generality we suppose that $\alpha_1 \neq \alpha_1'$. Subtraction of eqn. (2) from eqn. (1) gives

$$(\alpha_1 - \alpha_1')\boldsymbol{p}_1 + (\alpha_2 - \alpha_2')\boldsymbol{p}_2 = 0$$

or

$$\boldsymbol{p}_1 = \frac{(\alpha_2' - \alpha_2)}{(\alpha_1 - \alpha_1')}\boldsymbol{p}_2$$

and this relation implies that the vectors $\boldsymbol{p}_1$ and $\boldsymbol{p}_2$ are collinear or both zero vectors contrary to our original hypothesis. It follows that $\alpha_1 = \alpha_1'$. Similarly $\alpha_2 = \alpha_2'$ and so the scalars α_1, α_2 are unique.

The expression of the vector $\boldsymbol{a}$ in the form (1) is called its resolution with respect to the *basis* $\boldsymbol{p}_1, \boldsymbol{p}_2$. We have shown that the resolution of $\boldsymbol{a}$ with respect to the basis $\boldsymbol{p}_1, \boldsymbol{p}_2$ is unique.

The process of thinking is essentially that of asking questions. When I am faced with a mathematical problem, I do not look at it in blank horror and say "I cannot do this', instead I say: 'What type of problem is this? Where have I seen this type of thing before? What is it about? What methods are available? How can I use all the information given?" One of the most difficult things in teaching students is to make them appreciate that it is absolutely essential to tackle problems by asking oneself questions. Of course, it is important to ask the right questions!

This process of asking oneself questions was illustrated above where an integral and a differential equation were solved by experimental approaches. We now consider an additional example giving our discussion in a manner which seems unfamiliar when printed but closely resembles the thought processes involved.

(a) Solve the equation:

$$f(x) = x^3 + x^2 + 5 = 0.$$

Q. What formula can I use?

A. None, or at least I cannot remember the results of a method (Cardan's) which once confused me. I must deal with this one from first principles or give up altogether (as poor students do).

Q. What is the degree of this equation?

A. Three—it is a cubic.

Q. How many roots has it?

A. Three.

Q. What type (real or complex)?

A. Since complex roots (of equations with real coefficients) occur in pairs it must have one or three real roots.

Q. How can I determine how many real roots?

A. Use Descartes' rule if I remember it but why not sketch a graph? [There are roots where $y = x^3$ meets $y = -(x^2 + 5)$.] Or consider

$$f'(x) = (3x^2 + 2x)$$

$$\Rightarrow f(x) \text{ has a maximum } 5\tfrac{4}{27} \text{ when } x = -\tfrac{2}{3},$$

$$f(x) \text{ has a minimum } 5 \text{ when } x = 0.$$

The graph of $f(x)$ is, in fact, as illustrated in Fig. 2 and so there is one real root only and it is just less than -2. (After reading section 2 you will appreciate that this method is very suitable for generalisation.) Check by working out $f(-2) = 1$, $f(-2{\cdot}2) \approx -0{\cdot}8$. In fact, this suggests by linear interpolation or common sense (since the graph is to a first approximation a straight line between $x = -2$ and $x = -2{\cdot}2$) that the real root is approximately $-2{\cdot}1$.

Q. How can I get the root more accurately?

A. By some iterative (step-by-step) method preferably Newton–Raphson. However, accurate tabulation does in fact show that $f(-2{\cdot}11) > 0$, $f(-2{\cdot}115) > 0$, $f(-2{\cdot}12) < 0$ and so the root lies between $-2{\cdot}115$ and $-2{\cdot}12$. Therefore this root is $-2{\cdot}12$ correct to two decimal places. (Accurate calculation gives $-2{\cdot}116$ correct to three decimal places.)

Q. How do we calculate the complex roots?

A. We could divide $(x^3 + x^2 + 5)$ by $(x + 2{\cdot}116)$ and solve the quadratic. But why not use the symmetric functions of the roots? Thus, if the roots of $f(x) = 0$ are $\alpha \pm i\beta$, $-2{\cdot}116$, then

$$(x + 2{\cdot}116)(x - \alpha - i\beta)(x - \alpha + i\beta)$$

$$\approx x^3 + (2{\cdot}116 - 2\alpha)x^2 - (4{\cdot}232 \times \alpha - \alpha^2 - \beta^2)x + 2{\cdot}116 \times (\alpha^2 + \beta^2)$$

$$\approx x^3 + x^2 + 5.$$

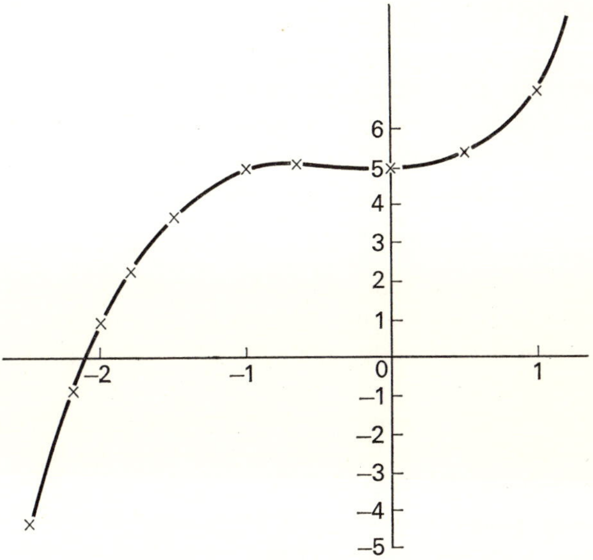

FIG. 2

Equating coefficients gives:

$$x^2: 2{\cdot}116 - 2\alpha = 1, \quad \therefore \alpha = 0{\cdot}558.$$

$$x^0: (\alpha^2 + \beta^2) \times 2{\cdot}116 = 5, \quad \therefore \beta^2 = 2{\cdot}052, \quad \beta \approx 1{\cdot}43.$$

The complex roots are therefore $0{\cdot}56 \pm \mathrm{i}1{\cdot}43$.

A *check* (always useful and usually necessary in numerical work) is provided by the coefficient of x:

$$\alpha^2 + \beta^2 - 4{\cdot}232 \times \alpha = 0.$$

Although most people think inductively, nevertheless, after discovering some fact they tend to shroud it in the language of deductive thought. This can be extremely deceiving, particularly so in a subject such as formal analysis, where a series of axioms followed by theorems and formal proofs can be very soul-destroying and lead to much questioning by students who find that they lose the point of the whole argument in elaborate and austere proofs.

A simple illustration of this point is provided by the proof of the modulus inequality

$$|z_1 + z_2| \leqq |z_1| + |z_2|. \tag{3}$$

By definition the algebraic operation of addition is performed with complex numbers in exactly the same way as with real numbers provided that the symbol i is treated as a number. Therefore, if $z_1 = x_1 + \mathrm{i}y_1$, and $z_2 = x_2 + \mathrm{i}y_2$, then

$$z_1 + z_2 = (x_1 + x_2) + \mathrm{i}(y_1 + y_2).$$

The representation of addition on the Argand diagram leads to a "parallelogram law" for the addition of two complex numbers, Fig. 3. For, if $z_1 = x_1 + \mathrm{i}y_1$ and $z_2 = x_2 + \mathrm{i}y_2$ are represented by P, Q respectively, the point A representing $z_1 + z_2$ has coordinates $x_1 + x_2$, $y_1 + y_2$ and hence lies at the fourth vertex of the parallelogram which has OPQ as three vertices. From the parallelogram law it is immediately obvious that inequality (3) holds since the fundamental properties of a triangle require that

$$OA \leqq OP + PA.$$

[One side is less than or equal to the sum of the other two sides.] But the analytical proof of inequality (3) is as follows:

$$\{|z_1| + |z_2|\}^2 - |z_1 + z_2|^2$$
$$= 2[\sqrt{\{(x_1^2 + y_1^2)(x_2^2 + y_2^2)\}} - (x_1x_2 + y_1y_2)]. \tag{4}$$

But, by Cauchy's inequality

$$(x_1^2 + y_1^2)(x_2^2 + y_2^2) - (x_1x_2 + y_1y_2)^2$$
$$= x_1^2y_2^2 - 2x_1x_2y_1y_2 + x_2^2y_1^2 = (x_1y_2 - x_2y_1)^2 > 0$$

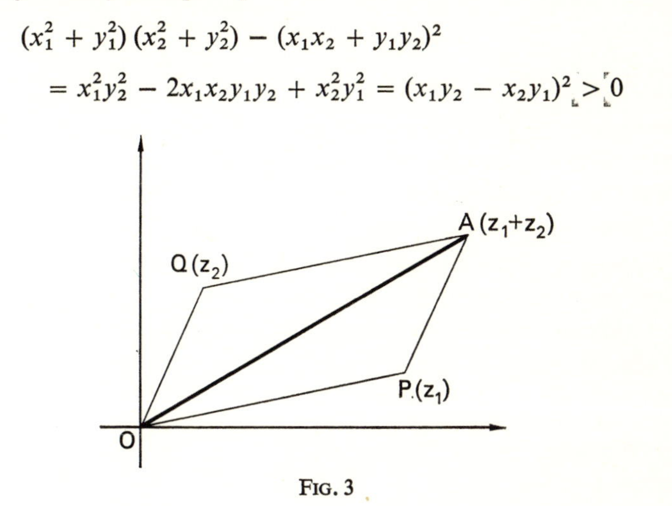

FIG. 3

unless $y_1/x_1 = y_2/x_2$. It follows, on taking positive square roots, that the right-hand side of eqn. (4) is positive unless z_1, z_2 have the same argument. Therefore

$$\{|z_1| + |z_2|\}^2 \geqq |z_1 + z_2|^2$$

and, taking positive square roots, the modulus inequality (3) follows. This proof is not very extensive but without a figure it is not very illuminating. Later, p. 99, this result is generalised.

As a second and even more simple illustration we take the mean value theorem of differentiation, the key to much of the differential calculus.

If a function $f(x)$ has a derivative at all points within the interval $a < x < b$, its graph is a smooth curve starting at height $f(a)$ and finishing at height $f(b)$. It is clear from a figure that the tangent to the curve is parallel to the chord at at least one point inside the interval (see Fig. 4). In fact

$$f'(\xi) = \frac{f(b) - f(a)}{b - a}, \tag{5}$$

where $a < \xi < b$. This is the *mean value theorem.*

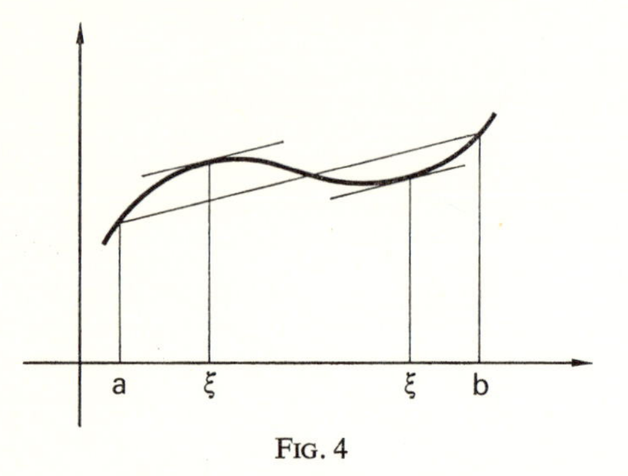

FIG. 4

If $f(a) = f(b) = 0$ this result means that there is at least one value $x = \xi$ $(a < \xi < b)$ for which the derivative vanishes; or we may say that at least one zero of $f'(x)$ separates two zeros of $f(x)$. In this form the theorem is known as *Rolle's theorem.* This result is one of such importance in calculus that a proof which simply appeals to a figure is not strictly sufficient. Usually in books on analysis a proof of Rolle's theorem is given first and the mean value theorem is then deduced.

However, many books on analysis do not give the above simple and intuitive demonstration of the probable validity of this fundamental theorem.

Finally consider the expression for the expansion of a triple vector product, viz.

$$\boldsymbol{a} \times (\boldsymbol{b} \times \boldsymbol{c}) = (\boldsymbol{a} \cdot \boldsymbol{c})\, \boldsymbol{b} - (\boldsymbol{a} \cdot \boldsymbol{b})\boldsymbol{c}. \tag{6}$$

There are many proofs of this identity, for example:

Choose the coordinate axes so that Ox is directed along $\boldsymbol{a}$, and the plane of Oxy is the plane of $\boldsymbol{a}$ and $\boldsymbol{b}$. Then we can write

$$\boldsymbol{a} = a_1\boldsymbol{i}, \quad \boldsymbol{b} = b_1\boldsymbol{i} + b_2\boldsymbol{j}, \quad \boldsymbol{c} = c_1\boldsymbol{i} + c_2\boldsymbol{j} + c_3\boldsymbol{k}.$$

$$\therefore \boldsymbol{b} \times \boldsymbol{c} = b_2c_3\boldsymbol{i} - b_1c_3\boldsymbol{j} + (b_1c_2 - b_2c_1)\boldsymbol{k},$$

$$\boldsymbol{a} \times (\boldsymbol{b} \times \boldsymbol{c}) = (a_1b_2c_1 - a_1b_1c_2)\boldsymbol{j} - a_1b_1c_3\boldsymbol{k}.$$

But

$$(\boldsymbol{a} \,.\, \boldsymbol{c})\,\boldsymbol{b} - (\boldsymbol{a} \,.\, \boldsymbol{b})\,\boldsymbol{c} = a_1c_1\boldsymbol{b} - a_1b_1\boldsymbol{c} = (a_1b_2c_1 - a_1b_1c_2)\boldsymbol{j} - a_1b_1c_3\boldsymbol{k}.$$

However, a relation between vectors is independent of any particular frame of reference and hence, since $\boldsymbol{a} \times (\boldsymbol{b} \times \boldsymbol{c})$ and $(\boldsymbol{a}.\boldsymbol{c})\boldsymbol{b} - (\boldsymbol{a}.\boldsymbol{b})\boldsymbol{c}$ have the same elements referred to the frame chosen above, eqn. (6) follows at once.

This "proves" the result, but how did anyone ever find it in the first place?

Remembering that $\boldsymbol{b} \times \boldsymbol{c}$ is perpendicular to $\boldsymbol{b}$ and to $\boldsymbol{c}$, so that $\boldsymbol{a} \times (\boldsymbol{b} \times \boldsymbol{c})$, which is perpendicular to $\boldsymbol{a}$ and to $\boldsymbol{b} \times \boldsymbol{c}$, therefore lies in the plane of $\boldsymbol{b}$ and $\boldsymbol{c}$, our results (p. 90) on the basis representation of a vector indicate that

$$\boldsymbol{a} \times (\boldsymbol{b} \times \boldsymbol{c}) = l\boldsymbol{b} + m\boldsymbol{c},$$

where l, m are scalars. *Then* the above proof of eqn. (6) *can be discovered by trial.*

2. Generalisation

The *Concise Oxford Dictionary* defines generalisation as "the forming of a general notation or proposition obtained by induction". In other words, in general, we take a simple proposition and extend it to a more elaborate theorem of which the *simple proposition is a special case*. Frequently, mathematical induction is used in the generalisation process. Further, the simple result itself plays a key role in the generalisation proof.

Dienes has written freely and penetratingly about the twin processes of generalisation and abstraction.† He defines generalisation as "the

† Z. P. Dienes, "On abstraction and generalisation", *Harvard Educational Review* (1961); *An Experimental Study of Mathematics Learning*, Hutchinson, 1963.

discovery that a general rule extends beyond the first few known cases"; while he states that abstraction is, "the awareness that the rule applies in a number of other situations". He has attempted to classify various levels of generalisation in the following manner:

1. extension of the rule from a finite number of cases to an infinite number;
2. the generalisation from one infinite class to another infinite class;
3. what Dienes terms a "mathematical generalisation or the formation of an isomorphism between one class and a sub-class of another class".

The following example will make the point clearer. A child given the "function"

$$y = 2x,$$

where x is a whole number, will find the value of y given a value of x. This will be repeated for a finite number of values of x until suddenly the extension is made that given *any* value of x, then the corresponding value of y can be found. This is the first and easiest stage.

If the coefficient of x is changed to 3, or 4, or 5, or to any whole number, the generalisation is made between the infinite number of cases when, say, $y = 2x$ and $y = 3x$, with $x \in \{1, 2, 3, 4, 5, \ldots\}$. The process is developed further with the realisation that the coefficient of x can be generalised and we can write

$$y = ax.$$

The third stage in his classification may be illustrated:

given $[y] = [a]\,[x]$, where $[y]\,[a]$ and $[x]$ are 1×1 matrices,

we can generalise to

$y = ax$, where y, a and x are $n \times n$ matrices.

Here the first case is isomorphic to and is a sub-class of the second case.

The psychological aspects of the process of generalisation may suggest a partial explanation, or give us a small clue to the build up of intrinsic motivation. Perhaps we can take a tiny step forward to stating why some children "like" mathematics. Dienes argues that a mathematician experiences the feeling of power and is exhilarated when he completes the generalisation. The extension of the rule to an infinite number of cases is followed by a sense of wonder and the release of energy. There is a renewal of "mathematical libido". The successful attempt at generalisation enables the child to write a simple statement containing an infinite amount of information. And this success is the power-house.

Without invoking a "ghost in the machine", the achievement of a generalisation may free storage space in the central nervous system, as generalisation may be looked upon as one aspect of a more efficient coding. A large amount of information occupies a correspondingly large space. But at the moment of success storage space is freed for the retention of other information and the act of creating space may give rise to fresh impetus. The central nervous system, intolerant of unused capacity moves in a direction to re-use the unfilled space, because of the release of pressure on the available space. Equally a Gestalt explanation would appear promising. The moment closure and structure appears from an apparently unconnected series of items of information, power is generated.

To illustrate the dictionary definition we first generalise the modulus inequality (3) to the following:

$$|z_1 + z_2 + \cdots + z_n| \leqq |z_1| + |z_2| + \cdots + |z_n|. \tag{7}$$

Inequality (7) is certainly true when $n = 2$ [cf. eqn. (3)]. Suppose it is true for $n = k$. Then by eqn. (3)

$$\begin{aligned} |(z_1 + z_2 + \cdots + z_k) + z_{k+1}| &\leqq |z_1 + z_2 + \cdots + z_k| + |z_{k+1}| \\ &\leqq (|z_1| + |z_2| + \cdots + |z_k|) + |z_{k+1}| \\ &= |z_1| + |z_2| + \cdots + |z_{k+1}|. \end{aligned}$$

Hence if the theorem is true for k it is true for $k + 1$. But it is true for $k = 2$ and so is true for $k = 2 + 1 = 3$, etc., and so is true universally.

The generalisation of the mean value theorem, p. 96, to Taylor's theorem requires a somewhat different technique, perhaps one might say it needs the use of some degree of low cunning. (Examples of "low cunning" are to be found distributed more or less uniformly throughout this chapter!)

Suppose $f(x)$ is a function which has derivatives of orders up to and including the nth at all points within an interval $a < x < b$. First we introduce a function

$$F_n(x) = f(b) - f(x) - (b - x)f'(x) - \frac{(b-x)^2}{2!}f''(x) - \cdots - \frac{(b-x)^{n-1}}{(n-1)!}f^{(n-1)}(x),$$

so that

$$F_n(b) = 0, \quad F_n'(x) = -\frac{(b-x)^{n-1}}{(n-1)!}f^{(n)}(x).$$

Next we introduce a function $\varphi(x)$ to which we can apply Rolle's theorem, viz.

$$\varphi(x) = F_n(x) - \frac{(b-x)^n}{(b-a)^n}F_n(a).$$

Clearly $\varphi(a) = \varphi(b) = 0$ and differentiation gives

$$\varphi'(x) = F_n'(x) + \frac{n(b-x)^{n-1}}{(b-a)^n}F_n(a)$$

$$= \frac{n(b-x)^{n-1}}{(b-a)^n}\left[F_n(a) - \frac{(b-a)^n}{n!}f^{(n)}(x)\right].$$

By Rolle's theorem $\varphi'(x)$ must vanish for at least one value $x = \xi$, where $a < \xi < b$. The factor $(b - x)^{n-1}$ cannot vanish for $x = \xi$,

$$\therefore F_n(a) = \frac{(b-a)^n}{n!}f^{(n)}(\xi). \tag{8}$$

Substitution of the full value of $F_n(a)$ into eqn. (8) gives

$$f(b) = f(a) + (b-a)f'(a) + \frac{(b-a)^2}{2!}f''(a) + \cdots + \frac{(b-a)^{n-1}}{(n-1)!}f^{(n-1)}(a) + R_n, \tag{9}$$

where $R_n = \dfrac{(b-a)^n}{n!}f^{(n)}(\xi)$ and $a < \xi < b$. This is the generalisation of eqn. (5) and is one form of *Taylor's theorem*.

The binomial theorem for a positive integer n, i.e.

$$(1+x)^n = 1 + nx + \frac{n(n-1)\,x^2}{2!} + \frac{n(n-1)\,(n-2)\,x^3}{3!} + \cdots + \frac{n(n-1)\ldots(n-r+1)\,x^r}{r!} + \cdots$$

can be generalised (to exactly the same form) for other values of n, provided $|x| < 1$ but the proof for non-integral n is essentially distinct from the (mathematical induction) proof for the case when n is an integer.

However, the idea of generalisation is not merely restricted to the extension of theorems. Consider the following illustrations from differing definitions of functions used in various branches of mathematics.

It is well known if n is an integer,

$$n! = \int_0^\infty t^n e^{-t}\,dt.$$

We generalise this and define $x!$ for x non-integral and $x > -1$ by the integral

$$x! = \int_0^\infty t^x e^{-t}\,dt.$$

This is an excellent illustration of generalisation. *The $x!$ formulae includes the $n!$ as a special case.* Similarly, the functions of a complex variable are obtained from functions of a real variable merely by replacing a power series in the real variable x by the same power series in a complex variable z. For example,

$$\sin x = \sum_{r=0}^{\infty} (-1)^r \frac{x^{2r+1}}{(2r+1)!}$$

is generalised to

$$\sin z = \sum_{r=0}^{\infty} \frac{(-1)^r z^{2r+1}}{(2r+1)!}.$$

Similarly

$$e^x = \sum_{r=0}^{\infty} \frac{x^r}{r!}$$

is generalised to

$$e^z = \sum_{r=0}^{\infty} \frac{z^r}{r!}$$

It must be emphasised that *these are definitions, they are not proofs in any sense of the word.*

The concept of generalisation is extremely well illustrated by the definition of the word "function". For example, one may define a function to be $y = x^2$ for all x, or $y = x^2$ from $-3 \leqq x \leqq 3$. These are simple functions: in each one when x is known y is also known. (This is the definition of a function.) But there are more difficult cases. The equation $\sin y = x$ defines a perfectly clear function, i.e. the inverse function $\sin^{-1} x$ or arc sin x. This is a generalisation of the definition of a function (when x is known, y is known) but it needs more elaborate calculation.

It may be necessary to define a function by means of several different formulae over a certain range. For example, we may have

$$y = \begin{cases} x & \text{for } 0 \leqq x \leqq 1, \\ 1 & \text{for } 1 < x < 2, \\ 3 - x & \text{for } 2 \leqq x \leqq 3. \end{cases}$$

The graph of this function is defined by three different equations and this graph is shown in Fig. 5. Note that the function is only defined for $0 \leqq x \leqq 3$. This does not matter in physical problems; our beam (in an elastic problem) may only extend from $x = 0$ to $x = 3$.

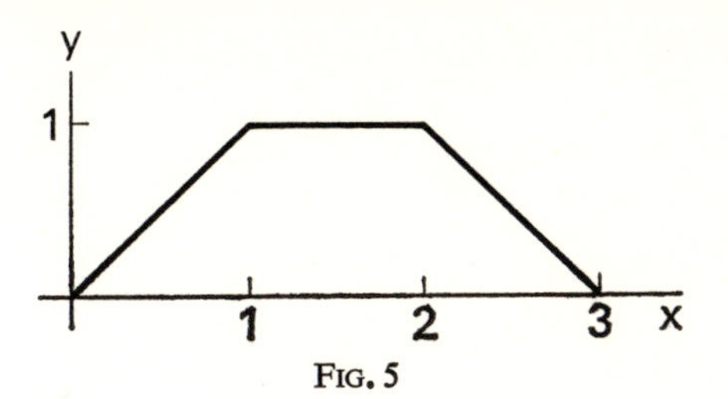

FIG. 5

More generally we may define a function by means of a series; for example, the series for e^x, or $\sinh x$ or $\cos x$. We may even define a function as the solution of a differential equation; for example, the function $\sin x$ could be defined as solution of the differential equation

$$d^2y/dx^2 = -y \quad \text{where} \quad y = 0, \quad dy/dx = 1 \quad \text{when} \quad x = 0.$$

If this were our definition, we could discover all the properties of the solution ($\sin x$) showing that it vanished when x was somewhere between 3·1 and 3·2, etc. It is possible to show this function is periodic and in fact possesses all the well-known properties of $\sin x$.

In more advanced mathematics it is not possible for us always to solve our differential equations in closed form (that is in terms of known functions), and so we may have to elaborate our ideas and define a new function as a solution of a differential equation and investigate its properties. The Bessel functions, Legendre's functions, the Hermite polynomials, all spring to mind in this way. Returning to the subject of integrals, pure mathematicians define $\ln x$ by the equation

$$\ln x = \int_1^x \frac{dt}{t}.$$

They find this more convenient than using the exponential function. Many other functions are defined by means of integrals. For example, the Beta function (of the two independent variables x, y) is defined by

$$B(x, y) = \int_0^1 t^{x-1}(1 - t)^{y-1}\,\mathrm{d}t.$$

Sometimes the variables arise in the integrand, sometimes a variable occurs as one of the limits of integration as, for example, in the error function

$$\operatorname{erf} x = \frac{2}{\sqrt{\pi}} \int_0^\infty e^{-t^2}\,\mathrm{d}t.$$

In some cases operations, such as definite integration, are used in order to define a class of functions. The integral transforms so obtained are of great importance in modern applied mathematics as, for example, the Laplace transformation $L\{f(t)\}$ defined by

$$L\{f(t)\} = F(p) = \int_0^\infty e^{-pt} f(t)\,\mathrm{d}t.$$

For example

$$L\{t\} = \frac{1}{p^2}, \quad L\{\sin \omega t\} = \frac{\omega}{\omega^2 + p^2}.$$

In fact we obtain a set of functions of p from a set of functions of t.

Remember that in mathematics we define things to mean what we want them to mean. Be careful when you see a definition to understand exactly what it means.

Other important illustrations are the generalisations of the vector, regarded as a triad of numbers to a tensor. We define a vector in the following way:

We suppose that we have two rectangular frames of reference $Ox_1x_2x_3$ and $O\xi_1\xi_2\xi_3$ which are both right-handed, and a vector $\mathbf{a}$ which has components (a_1, a_2, a_3) and $(\alpha_1, \alpha_2, \alpha_3)$ in the respective frames. We let the unit vectors $\mathbf{e}_1, \mathbf{e}_2, \mathbf{e}_3$ denote the directions of the axes of the frame $O\xi_1\xi_2\xi_3$ referred to the axes $Ox_1x_2x_3$. The components of these three unit vectors in the frame $Ox_1x_2x_3$ are

$$\mathbf{e}_1(l_{11}, l_{21}, l_{31}), \quad \mathbf{e}_2(l_{12}, l_{22}, l_{32}), \quad \mathbf{e}_3(l_{13}, l_{23}, l_{33}).$$

The first suffix of l_{ij} denotes the axis in the x-frame, Ox_i, and the second suffix denotes the axis in the ξ-frame, $O\xi_j$, which together enclose the angle whose cosine is l_{ij} $(i, j = 1, 2, 3)$. By definition of the components of a vector as orthogonal projections on to the coordinate axes it follows that

$$\alpha_1 = \mathbf{a} \cdot \mathbf{e}_1, \quad \alpha_2 = \mathbf{a} \cdot \mathbf{e}_2, \quad \alpha_3 = \mathbf{a} \cdot \mathbf{e}_3.$$

Therefore

$$\alpha_1 = l_{11}a_1 + l_{21}a_2 + l_{31}a_3 = \sum_{k=1}^{3} l_{k1}a_k,$$

$$\alpha_2 = l_{12}a_1 + l_{22}a_2 + l_{32}a_3 = \sum_{k=1}^{3} l_{k2}a_k,$$

$$\alpha_3 = l_{13}a_1 + l_{23}a_2 + l_{33}a_3 = \sum_{k=1}^{3} l_{k3}a_k.$$

These can be summed up in the single relation

$$\alpha_j = \sum_{k=1}^{3} l_{kj}a_k \qquad (j = 1, 2, 3).$$

A vector is defined as an entity having three components in any rectangular frame of reference, the components in any two frames being related by the above transformation.

A tensor is defined by a generalisation of the equation which expresses the transformation definition of a vector.

A set of quantities $t_{ikm\ldots}$ (for n suffixes there are 3^n quantities in the set) are the components of a tensor in the x-frame when the relations

$$\tau_{jln\ldots} = l_{ij}l_{kl}l_{mn} \ldots t_{ikm\ldots}$$

give the components $\tau_{jln\ldots}$ in the ξ-frame. (We usually say that $t_{ikm\ldots}$ 'is a tensor'.)

The generalisation from a function of one to a function of many variables comes rather insidiously when students first meet the idea of partial differentiation. In fact, in physics, pupils in sixth forms meet this concept in the equation of state of a gas. The ideas of partial differentiation and multiple integration, simple though they may be, nevertheless, cause considerable difficulties among those who have little concept of visualising things in three-dimensional space. One can draw two-dimensional figures on two-dimensional paper but when three-dimensional figures are drawn on two-dimensional paper there is a considerable demand on the intellectual ability of any student. This, together with the extension into many dimensions through the arts of analysis and coordinate geometry forms a major stumbling-block to all but the very best students.

In the following I put forward some ideas of generalisation as seen from the point of view of an applied mathematician. What does the applied mathematican do? He formulates physical problems mathematically, makes the appropriate manipulative calculations and then interprets the answer physically.

The gradual extension of his problem by the applied mathematican way not be generalisation according to the strict definition of the term but in practice operates in the same way (see section 3).

For example, consider the simple problem of a uniform rigid rod leaning against a wall. He draws the figure (see Fig. 6) and puts in the various forces; say, for simplicity, that the wall is smooth and the floor is rough. He takes two resolutes and moments as illustrated in the following equations, obtains the forces F and R, and from this he derives an inequality thus:

$$\text{Resolve} \rightarrow,\ F = S,$$

$$\text{Resolve} \uparrow,\ R = mg,$$

Moments about contact with wall, $S = \frac{1}{2}mg \tan \theta$.

The law of friction is $F \leqq \mu R$ and so we have finally

$$\mu \geqq \tfrac{1}{2} \tan \theta.$$

He interprets this result by stating either

(i) that the coefficient of friction must not be less than a certain value, or

(ii) that the inclination of the ladder to the wall must not exceed a certain value.

This is a mathematical simplification of a very difficult problem. What happens if both the wall and the floor are rough and the rod is not in limiting equilibrium? There are four unknowns, the two components of normal reaction and the two forces of friction (see Fig. 7). There are three independent equations of equilibrium and four unknown forces cannot be obtained from these; the problem is therefore indeterminate. In fact, equilibrium is possible provided the line of action of the weight goes through the shaded area as shown in Fig. 8;

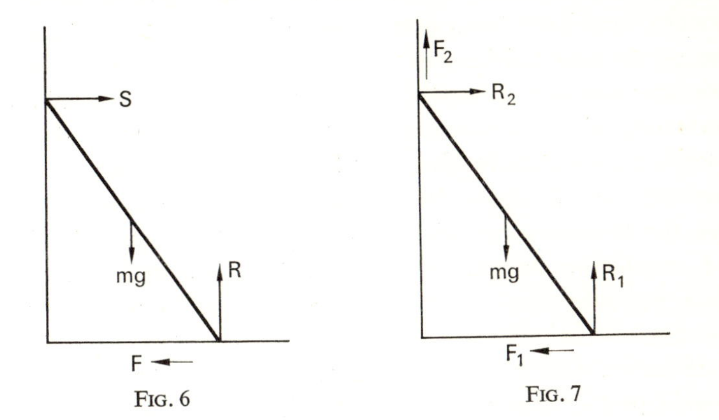

FIG. 6

FIG. 7

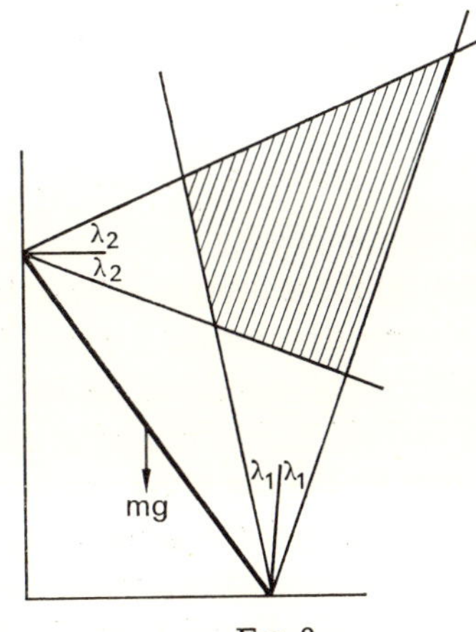

FIG. 8

the shaded area is obtained from the angles of friction on either side of the normal. How can this problem, which should be physically determinate, be solved? It has to be generalised. We have to take account of the elastic properties of the beam thereby obtaining further equations which suffice for the problem to be solved. Naturally, this is much more difficult.

Note, once again, *that the generalisation of some particular result should include this result itself as a special case.* In the example quoted above when the rigidity of the beam tends to infinity or becomes very large the solution should become somewhat akin to that of the rigid rod leaning against a wall. However, considerable difficulties can arise in applied mathematics as, for example, in fluid mechanics. If we use the equations for a perfect fluid, then the boundary conditions involve no relative normal velocity at a rigid boundary. But, if we use the equation of motion for a viscous fluid, then the boundary conditions involve no relative normal velocity and, in addition, no relative tangential slip at a boundary. However small we make the kinematic viscosity, if we keep the non-slip boundary conditions, the viscous solutions do not, in general, tend to the perfect fluid solutions. When applied mathematicians generalise physical problems they frequently introduce higher order differential equations with completely different solutions, producing complications which do not usually appear in elementary pure mathematics.

3. The Evolution of a Subject

In applied mathematics or physics, mathematics comes to the aid of the scientist by means of the process usually called model building. The mathematician attempts to construct a physical situation through his equations and differential equations which give an idealised approximation of the real thing. For example, models of the earth must have the same mass as the earth, the same moments of inertia as the earth and the same travel times for earthquake waves, etc. (These properties of the earth are known from experimental results.) On a smaller scale, mathematical models are built to represent the flow of compressible air past aircraft wings. The solution of such problems is the basis of classical applied mathematics. Historically, subjects develop in a manner which is very similar to the one applied in the classroom. For example, in any field of continuous mechanics, theoretical order may be obtained by placing problems into the following convenient classes:

1. static and steady state problems;
2. small oscillations about these states, including results concerning stability;
3. general motions and interactions.

For example, the study of hydrodynamics has progressed through hydrostatics, a study of steady flows of incompressible fluids, discussions of sound waves, viscosity and stability of laminar flows, to the more modern problems of high-speed compressible flow and turbulent motion.

All subjects pass through this stage. Naturally, when a new field is opened by a research worker he does the easy examples first, i.e. the easy steady state or static problems. He then investigates stability, and finally the very difficult generalised problems. This is the situation today in plasma dynamics.

In all problems we have to balance reality against the complexities of analysis and at the same time to realise that with modern computing tools we can be tempted to apply numerical methods without really attacking the problem analytically. Unfortunately, many of the apparently very powerful analytic tools do not give solutions unobtainable by elementary methods. For example, the Laplace transformation—a very cunning way to solve differential equations—may lead to certain solutions but it rarely leads to results undiscovered by other methods. There is a tendency for many people to regard applied mathematics as an art. I regard it as a science.

The above ideas describe *the essence and structure of applied mathematics*. It consists essentially of *generalisation from simple problems to the realistic ones*. In our models of the earth itself we should allow for deformations due to the tides, for all the complexities observed in the terrestrial magnetic field, the dynamo maintenance of this field, and its reversing polarity; all these phenomena have to be involved gradually and it is the essence of the structure of the problem that they are introduced one by one into new models, which gradually supersede and overlay the old. No theory remains intact for long, but every contribution, provided it is sensible and realistic, does produce worth-while advances, if only to stimulate other workers in the field.

4. Approximations

An important aspect of modern mathematics is the art of approximating. Few problems in modern applied mathematics can be solved exactly. Accordingly we must have resort to numerical work or approximations. The first thing we do is to look at the equations of a problem, compare the orders of magnitudes of the relevant terms and discard the terms which (we hope!) are negligible. Then we solve the (remaining) problem and verify that the discarded terms are in fact negligible. Thus in the modern subject of magnetohydrodynamics we usually neglect displacement currents, often assume infinite electrical conductivity and neglect compressibility and viscosity when it is convenient to do so.

However, most of our approximations are much more simple but still terrifying to students. An allied fear is mistrust of the symbols $>$ or $<$ or, even worse, of $\approx$. Students regard the signs $>$, $<$ with reserve and the sign $\approx$ with disgust. It was once remarked that in the Cambridge Mathematical Tripos the lowest marks were obtained for questions involving approximations. With the advent of modern computers this should be no longer true. In fact, computers work on the basis of approximations and, of course, by trial. After all, how does the computer divide one number by another? As far as I am aware, by taking away successive multiples of the divisor until the remainder is no longer positive.

The average student first meets the idea of approximation when he learns how to express a number "correct to so many decimal places or so many significant figures". For example:

The number 26·460734 can be written as

26·46073	correct to five decimal places, since it is nearer to 26·46073 than to 26·46074,
26·4607	correct to 4 decimal places,
26·461	correct to 3 decimal places,
26·46	correct to 2 decimal places,
26·5	correct to 1 decimal place.

Thus 26·460734 can be written

26·46	correct to 4 significant figures,
26·5	correct to 3 significant figures,
26	correct to 2 significant figures, since the original number is nearer to 26 than to 27,
30	correct to 1 significant figure.

Seldom does a student undertake the *calculation* of π. The following series are useful examples when considering the rapidity of convergence of series and accuracy of approximation:

$$\frac{\pi}{4} = 1 - \frac{1}{3} + \frac{1}{5} - \frac{1}{7} + \cdots,$$

$$\frac{\pi}{4} - \frac{1}{2} = \frac{1}{1\cdot 3} - \frac{1}{3\cdot 5} + \frac{1}{5\cdot 7} - \frac{1}{7\cdot 9} + \cdots,$$

$$\frac{\pi^2}{6} = \frac{1}{1^2} + \frac{1}{3^2} + \frac{1}{5^2} + \cdots,$$

$$\frac{\pi^3}{32} = 1 - \frac{1}{3^3} + \frac{1}{5^3} - \frac{1}{7^3} + \cdots.$$

Rarely are these or similar problems suggested to schoolchildren, but as exercises in the art of approximation they could be most rewarding!

Again, all experiments involving measurements of continuous variables must be approximations. We must estimate lengths to the nearest (smallest) unit of our measuring scale and the accuracy of the parameters deduced from our observations is not increased by a very large number of observations. You cannot increase the accuracy of your results one-hundredfold by taking 100 times as many measurements!

In mathematics a most important part of approximation work lies in the field of iterative methods which are capable of generalisation in many ways and have several additional advantages. The aim of the method of *iteration* is to obtain a sequence of numbers $x_0, x_1, x_2, \ldots, x_n, \ldots,$ each member of the sequence being closer than its predecessors to the quantity being evaluated. A method is an iterative method when it relies upon the repetition of a single formula to obtain one member of the sequence from its predecessors, i.e. the sequence is generated from a formula which gives x_{n+1} in terms of x_n (and perhaps of $x_{n-1}, \ldots$). If the method is not strictly iterative in this sense but still provides a sequence of approximations, it is better described as a method of successive approximation. A satisfactory iterative process is self-correcting; a small error in the calculation of one member of the sequence will not prevent subsequent members calculated from the wrong one from converging to the exact quantity. An iterative method is usually very suitable for use on a computer because it consists of the repetition of the same sequence of calculations.

Newton's method is a typical example of an iterative method of approximating to the roots of the equations. May I remind you that Newton's method implies that if a is an approximate root of the equation, $f(x) = 0$, then a better approximation is

$$a - f(a)/f'(a).$$

It follows that the sequence of numbers $x_1, x_2, \ldots, x_n, \ldots,$ defined by

$$x_{n+1} = x_n - f(x_n)/f'(x_n),$$

with $x_1 = a$, should converge rapidly to a root of $f(x)$. This method is particularly useful for a computer because the instructions are simple and repetitive. Iterative methods can almost be invented at will. For example, suppose one wishes to find the small positive root of the equation $x^3 - 3x + 1 = 0$. If we write the equation in the form

$$x = \tfrac{1}{3} + \tfrac{1}{3}x^3,$$

and then define a sequence of numbers $x_1, x_2, \ldots, x_n, \ldots$ by the equation

$$x_{n+1} = \tfrac{1}{3} + \tfrac{1}{3}x_n^3,$$

it is clear that if the sequence converges it will converge to a root of the original equation. Try it and see, starting off with some appropriate value of x_1; $x = 0$ will do.

Iterative methods are very suitable for the solution of linear equations especially when the number of equations is large and a computer is available.Although this is a simple case, consider the following equations:

$$6x + y = 9,$$

$$x + 3y = 4.$$

First write them in the form

$$x = \tfrac{1}{6}(9 - y),$$

$$y = \tfrac{1}{3}(4 - x).$$

Then we define a sequence of ordered pairs (x_1, y_1), (x_2, y_2), ..., (x_n, y_n), ... defined by

$$x_{n+1} = \tfrac{1}{6}(9 - y_n),$$

$$y_{n+1} = \tfrac{1}{3}(4 - x_n).$$

If this sequence converges it will converge to the solution of the original equation. This is Jacobi's method for the iterative solution of simultaneous equations. Can you see how to generalise it? This technique may be further defined by using the latest available data by writing, instead of the second of the above equations,

$$y_{n+1} = \tfrac{1}{3}(4 - x_{n+1})$$

and so use the following equations to derive the solution:

$$x_{n+1} = \tfrac{1}{6}(9 - y_n),$$

$$y_{n+1} = \tfrac{1}{3}(4 - x_{n+1}).$$

This is the Gauss–Seidel method and the convergence is rather more rapid than in Jacobi's method.

Just as instructive is Picard's method for the solution of an ordinary differential equation for small values of x. This is very simple but seems very confusing. Students look for a mystique which is not present; the application of a little common sense shows that the solution is simple in structure and operation. Consider the differential equation

$$dy/dx = 1 + y^2 \quad \text{where} \quad y = 0 \quad \text{when} \quad x = 0.$$

You can solve this equation by separation of the variables to show that the solution is $y = \tan x$.

But let us look at it in another way. Integrate the given equation between 0 and x, and you find

$$y = x + \int_0^x y^2 \, dx.$$

This is called an integral equation because the unknown y appears under an integral sign. This suggests that if we define a sequence of functions $y_1(x), y_2(x), \ldots, y_n(x), \ldots$ and relate these functions by the relation

$$y_{n+1}(x) = x + \int_0^x \{y_n(x)\}^2 \, dx$$

and this sequence of functions converges, it will converge to a solution of the original equation. In fact we find, taking $y_1(x) = 0$, a trial solution to fit the condition prescribed at $x = 0$,

$$y_1(x) = 0,$$

$$y_2(x) = x,$$

$$y_3(x) = x + \frac{x^3}{3},$$

$$y_4(x) = x + \int_0^x \left(x + \frac{x^3}{3}\right)^2 dx,$$

$$= x + \int_0^x \left\{x^2 + \frac{2x^4}{3} + 0(x^6)\right\} dx$$

$$= x + \frac{x^3}{3} + \frac{2x^5}{15} + 0(x^7).$$

(Note that we add one term only to our solution at each iteration.) The next iteration gives

$$y_5(x) = x + \frac{x^3}{3} + \frac{2x^5}{15} + \frac{17x^7}{315} + 0(x^9).$$

You may recognise the right-hand side of this last equation as the start of the power series for tan x.

The art of approximation is a major part of the structure of applied mathematics. Iterative techniques can usually be applied to the problems of perturbation theory, stability investigations and so on. This, of course, necessitates considerations of convergence, but, in general, if problems of this kind are put through a computer, then, provided the usual restrictions on the number of operations are placed upon the computer, the convergence is usually established without recourse to high-powered theorems.

5. Problems of Applied Mathematics and Technology

In my opinion two of the dominant aspects of applied mathematics are boundary or initial value problems on the one hand and relativistic problems on the other. In most applied-mathematics problems concerning continuous media the physical properties and relations such as conservation of matter, equations of motion, mathematical representation of physical phenomena, involve partial differential equations. But since the media can only be influenced either by pressing in some way, physically or electromagnetically, on its boundaries, and/or by starting it off in an initially prescribed manner, the values of the variables at the boundaries and/or at the start of the problem usually influence the nature of the solution. There are two consequences. If the differential equations are of elliptic type the solution on the boundary determines the values of the dependent variables everywhere within the domain under consideration. If the differential equations are of hyperbolic type then the disturbances are propagated along characteristics into the medium, whatever it may be.

This sounds very complicated indeed, but it is really very simple. For example, if a radio aerial is broadcasting, then what happens in the space outside the aerial is essentially dependent upon the oscillations fed into the aerial by the transmitting system. This is a problem of wave propagation.

On the other hand, if the floor of a building is loaded in some way, then the deflection of the floor is essentially dependent on the way in which it is supported on its edges and in which it is loaded. This is natural common sense, but when done mathematically the analytical complexities make it seem rather remote from physical reality. The mathematical problem is simple. We have to try and fit known solutions to known boundary conditions. Exact solutions are rare but variational methods give excellent results.

On the other hand, relativistic ideas are becoming ever more important. The Lorentz transformation, of course, shows the invariance of the Maxwell's equations and leads to quite complicated mathematical manœuvrings. But what of the much more simple Galilean transformation? The basic ideas of the Galilean transformation are as follows:

Consider two rectangular Cartesian frames of references $Oxyz$, $Ox'y'z'$ such that the frame $Ox'y'z'$ is moving with constant velocity $\boldsymbol{V}$ relative to $Oxyz$ which an observer at O regards as "fixed" in space. Let $\boldsymbol{r}$, $\boldsymbol{r}'$ be the respective position vectors of a particle P, of mass m, referred to O, O' as origins. Then the triangle of vectors gives

$$\boldsymbol{r} = \boldsymbol{r}' + \boldsymbol{V}t + \boldsymbol{a},$$

where t is the time and $\boldsymbol{a}$ is a constant vector (equal to $(\boldsymbol{r} - \boldsymbol{r}')_{t=0}$). This is the Galilean transformation.

Differentiation gives $$\frac{d^2\boldsymbol{r}}{dt^2} = \frac{d^2\boldsymbol{r}'}{dt^2}$$

and so the forces $\boldsymbol{F} = m\dfrac{d^2\boldsymbol{r}}{dt^2}$, $\boldsymbol{F}' = m\dfrac{d^2\boldsymbol{r}'}{dt^2}$ measured by observers at rest at O, O' relative to the two frames of reference are equal. Generalising, *in other words force systems measured by observers in uniform relative motion are identical.*

For example, the Galilean transformation tells us that if we keep an aircraft wing stationary in a wind tunnel and send air past it, then the forces acting on the wing are exactly the same as if we sent the aircraft at the same relative speed through stationary air. This is the way in which designers can carry out successful experiments. Careful considerations of relativity simplify many of the peculiarities of life around us.

But what is the relevance of the material discussed in this section to the title of this chapter? As far as applied mathematics is concerned, ideas such as those discussed above fundamentally lie at the core of the subject. More recent applications and developments in such fields as variational methods (control theory), non-linear differential equations (shock waves, hydromagnetic waves and finite strains) and stability investigations (plasmas, etc.) deal with extensions and generalisations of the simpler problems studied by undergraduates. However, it is essential that we should not overwhelm the young student with too early a statement of a problem in all its generality. He needs time and experience of formulating simple problems before he acquires the background of physical intuition which is so necessary before the important features of modern "real" problems can be disentangled. For example, although it may be fashionable to teach electromagnetism to undergraduates by starting with Maxwell's equations, experience has shown that it is best to follow the more traditional evolutionary (historical) structure, and then, in a later course, recast the whole subject starting anew from Maxwell's equations.

As a practical philosophy for teaching the structure of applied mathematics the above ideas work!

6. Some Miscellaneous Topics

In the following I do perhaps move away from the main title of my contribution, but I feel very strongly about a number of points. They are pertinent to the whole field of learning. Rarely have I heard them stated clearly and convincingly.

(i) *The technique of lecturing needs very careful thought*, especially in these days of large classes. Although disciplinary control of a large class is sometimes rather difficult to achieve and retain, nevertheless, there are occasions when the lecturer must stop and look out of the window or relax in some way thereby giving the students time to collect their thoughts. The tendency for lectures in England to consist of students writing down rapidly, without thought, anything a lecturer says, including the rather poor jokes, is to be deplored. Either a good book used as a text, or well-prepared hand-outs should be readily available; the lecturer can then concentrate on emphasising the major facets of the subject. The students should be able to listen without writing down any substantial amount of material and, let us hope, emerge from the lecture knowing something about the subject and having had at least an explanation of the more important ideas of some topic. It is wrong for a lecture, or even a sixth-form lesson, to consist of vast numbers of trivial examples done on the board; a few perhaps may be done but not many. The idea of Mr. A. at the front talking or writing on the board (and facing it), whilst the students spend all the time writing without appreciation or understanding of what they are writing, is ridiculous. Further, many scientists, and technologists in particular, have very high-pressure timetables and do not really have time to look at notes at all!

(ii) *The object of a university course, and even of a sixth-form course, is to teach people to teach themselves*. The way to do this is to learn from books. This is what one has to do in after life and without this understanding of how to organise one's own studies, a course at university, polytechnic or College of Education is wasted.

(iii) *At professional level it is very difficult to distinguish between mathematicians, physicists and engineers masquerading as one another*. The formal status of many dons gives little indication of their early academic life or, in some cases, their research interests.

(iv) *It is most important in mathematics to really distinguish between definition and proof*. Much difficulty arises in the early stages when immature students fail to observe that certain things are defined to have certain properties. Consider, for example, the properties of complex numbers. Students are deceived and they deceive themselves when they look for proofs of certain axiomatic properties such as "if $a + ib = 0$, then $a = 0$ and $b = 0$".

(v) *The technique of examinations.* As with most facets of our lives examinations are changing in shape and outlook. With the advent of courses in so-called modern mathematics, schools syllabuses are changing. So are schools examinations! The ideas of introducing groups, sets, matrices, vectors to the 12–16 years age-groups have become fashionable. Supplementing these topics are all the interesting ideas of rotation, reflection, enlargement, shearing, etc., in geometry together with elementary ideas of topology, statistics and probability. It seems certain that these subjects will take their place in our school syllabuses but nevertheless this must not be done at the expense of manipulative ability. It would be an absurd situation when students are unable to solve a quadratic equation but can talk about the eigenvalues, etc., of a quadratic form. Clearly mathematics, and I include computer science and statistics under this title, will have to be even more intensively taught in the next 20 years than in the past 20. Can we give up applied mathematics? For the benefit of engineers and technologists I think not. Many engineers spend much of their time investigating statical problems. How can we expect them to consider statics in three dimensions or elastic systems if they are incapable of thinking about the simple problems of two-dimensional statics? I am aware that elementary statics is unfashionable and is being discarded from some examination syllabuses, but this is because these syllabuses have been evolved for mathematicians as such and not for other scientists and engineers. It seems very likely that, although syllabuses will change in content and the examinations will change in format and motivation, nevertheless a considerable amount of manipulative ability will still be required. Naturally, the mysterious properties of triangles such as nine-point circles, etc., can be discarded and left where they belong in the nineteenth century, but the solution of triangles will remain of importance to the surveyor and the civil engineer, even perhaps to people who design any mechanical structure involving moving parts. This is not a plea for more and more extensive examinations but for a realisation that mathematics must be given even greater prominence in this time and age than ever before.

Although many people would be rid of examinations altogether, they remain a useful check on whether students have worked, give some estimate of ability and, by and large, sort out people into groups of common and relatively measurable ability. I once heard Professor H. Bondi say that in an examination a good question was one in which if you saw how to do it, there was very little to write down. On the other hand, a bad question is one in which, if you saw how to do it, you still had several pages of closely written analysis before you finished.

Conclusion

To summarise, it does seem to me that mathematics, when properly taught, can be stimulating, interesting, experimental and even, occasionally, slightly naughty. There is no reason why students at quite a high level, just as schoolchildren below the age of 10, should not be taught to discover results for themselves.

The classic illustration which springs to mind is discovering by experiment and then a logical proof that the angle sum of a triangle is 180°.

In a way one is reminded of Oliver Heaviside who produced some really remarkable results but was rather disregarded by the professional mathematicians of his time because he did not really consider the problems of convergence correctly. Nevertheless, many of his results stand today and are at the core of our understanding of the physical world. Surely Newton, and other outstanding early mathematicians, did much intuitive thinking and it is quite clear that such an eminent mathematician as Poincaré, with his enormous output, must have had great intuition. Harold Jeffries in his books continually refers to the patterns of intuitive thinking and it is a pity that much mathematics of great interest and intuitive simple background should be kept for the few specialists because it is shrouded in elegance, rigour and deductive thinking.

Throughout this chapter I have tried to emphasise continually the experimental nature of mathematics through the importance of trial processes in attempts to solve problems and in generalisations of simple ideas and results. This is immediately tied with the necessity for the educated man to teach himself from books which is surely at the heart of education; teaching people to educate themselves is essential and fundamental. Perhaps one could say that the most important aspects of mathematics education are trials and reading, and by reading mathematics one does not mean "read a book" but "look at a book and work out the problems concurrently with the author". Perhaps in time you will work it out ahead of the author, even better and more correctly. Books are rarely 100 per cent correct. "Deliberate mistakes" are there for all to find. How refreshing it is to find that you are as competent and understanding of the material as the author and even better at doing your arithmetic. If you use your native ability assisted by a little low cunning from time to time, you may surprise yourself and, even more, your teacher.

Attempts to Investigate the Learning of Mathematics

L. R. CHAPMAN

DURING the last few weeks you have listened to several lecturers talking about mathematics. Professor Kilmister attempted the difficult tasks of definition; he went to some lengths to explain how he viewed his own particular task and stressed the generality of his materials. He said that he looked at his assignment "What is mathematics?", and decided that the important questions fall broadly into two categories; those we could put to the mathematician/philosopher, and those, usually the more difficult ones, put to the psychologist.

The task is demanding. The elements of mathematics and the roles of intuition, logic, generalisation and structure are all difficult areas to delimit or to explore. No doubt you noticed the shadowy areas fringing the algorithmic processes slotted into the proof structures used during the lecture on the role of logic. These are left most cunningly for the psychologist to explore. Further, the mental processes of intuition were not mentioned; in this field we can add little as we do not possess the techniques to investigate the unobservable. The mathematician/psychologist can talk with some authority about a number of relatively simple learning situations, but the development of theories of learning in mathematics is comparatively unexplored.

C. A. Mace, for many years Professor of Psychology at Birkbeck College, used to say that the difficulties of structuring a lecture can be partially minimised if you first tell the students what you are going to tell them, then tell them, and finally tell them what you have told them. What am I going to tell to you? My aim is clear, yet difficult to attain. I want to try to help you to listen critically to the lectures which follow. For example, when Dr. Skemp talks about his work on reflective thinking, you will be able to compare his theoretical position with that of Piaget, or Dienes. I wish to present an overall picture which will help you to listen to, and read about, the various approaches to research and theorising in this area of mathematical learning.

It is important to know about "How?" as well as "What?" and "Why?" so I shall consider some methodological as well as some theoretical aspects. As the field is large, I hope you will agree when I say that another equally important objective is to give a "Gestalt" or "structured picture" of this rapidly changing, and to some extent, piecemeal literature. Wallace (1965) in his book *Concept Growth and the Education of the Child*[1] assembles a bibliography of some 530 papers, of which a high proportion are directly relevant to the teaching of mathematics. Clearly I cannot deal with more than a small part of the large literature.

If you ask the question, "Why do we need a theory of learning or a theory of mathematical education?", a story taken from Peterboroughs' column in the *Daily Telegraph* may help you to answer your own question.

> "Estimate Confirmed".
>
> After one customer had asked for four loaves a Sussex baker delivering by van asked his boy assistant how many were left.
>
> "Lots", was the reply.
>
> "That's no good", said the driver, "count them".
>
> After a while the boy reported back: "I warn't far off, we got best part of a tidy few".

This story highlights the problem. Assuming that the boy has been exposed to the normal processes of education, he has received approximately 1300 hours of mathematical education and yet apparently experiences difficulty with the counting numbers. If as a German mathematician, Kroneker, is reputed to have said, "God gave us the counting numbers, all else is the work of man", we must ask why man has done so little. Why does the boy possess so small a residue of mathematical knowledge? Which suggests the further question: how shall we find out why his knowledge is so scanty?

In order to find out we plan an experiment using a design that is acceptable to other workers in the field. The methods can be stated:

1. We have a "hunch", formulate a null hypothesis, design an experiment to test the hypothesis and go on from there.
2. Observe children or students in learning situations, ask questions and build from there.
3. We can introspect the process during our own mathematical activities.
4. We can borrow and adapt techniques from other disciplines.

Unfortunately progress in building a theory of mathematical education has been hampered by the practice of, until recently, experimental psychology to exclude work not based on the underlying design:

$$D = f(\text{I.V.s})$$

where D is the dependent variable and I.V.s are the independent variables. In other words, if the independent variables are known and controlled under experimental conditions, then it is possible to predict the value of the dependent variable. However, Rozeboom (*Psych. Rev.*, 1956) points out the fallacy of attempting to write down equalities to represent learning situations. The argument is that we cannot use the form

$$y = f(x)$$

but are restricted to a probability statement. Or the value of y can be predicted with greater certainty when x is know than when it is not known.

It is only in the last 20 years that we find influential experimental psychologists such as Osgood[2] including the work of Piaget in his standard textbook *Method and Theory in Experimental Psychology*. In the 1958 edition, Piaget is given not quite a full page in a large book of 727 pages and the material appears under the title "A non-experimental approach". Piaget did not follow the accepted design of American child-psychologists' technique.

Unfortunately this influential American preoccupation with experimental respectability limited the kind of work undertaken; rats are easier to control in an experimental situation. Hence Tolman's cry, "Rats not men". However, today psychologists are more electic; introspection and anecdotal evidence are again fashionable and permissible. We can choose among the rival claims of the Behaviourists or Gestaltists or the numerous theories set out in such standard works as Hilgard's *Theories of Learning*(3), and are free to employ those techniques most suitable for our investigation. A good case can be made for building a theory or theories from the collective anecdotal wisdom of teachers of mathematics; perhaps it is true that the learning process may be best observed in the mathematics classroom.

We are unlikely to gain understanding and insight into the learning process if the child is merely asked to repeat a previously learned response; recall of learned material is interesting, but we are not considering theories of remembering or short- or long-term memories, we are concerned with learning. It is exciting to observe a good infant schoolteacher working with a small group; the discussion exposes the steps in the mathematics. You can listen to children "homing" towards insight, you can observe and often feel the emotional changes; and note the thrill as a child takes a large step forward, often way beyond our expectation. Afterwards if you talk to the teacher, it becomes obvious that she is aware of every nuance of change in the child. But how are we to assemble all this expertise and filter out a theory?

Methods of Investigation

A closer examination of the large literature on mathematics learning reveals that many studies are based on variations of a small number of methods.

Methods with younger children

1. *The interview–questionnaire design.* Piaget is the best known exponent of this technique, but he is followed by most researchers attempting to validate his findings. The earlier cruder methods of questioning about responses to mathematical stimuli have been refined in an effort to meet criticism. Interviews are carefully controlled; each child is asked the same questions in *exactly* the same way. The apparatus is used to demonstrate mathematical "action" and the child is invited to comment, or the experimenter may control the language variable through use of a questionnaire. There is striving towards greater objectivity and standardisation; research is of small value if it cannot be replicated.

2. *The performance method.* In a situation involving discrimination learning the child is helped to form a conditioned response, to indicate discrimination between shapes such as the square and the circle; one shape being rewarded so as to form an association between the shape and the reward. In most cases the teacher has to supply the initial motivation by teaching the association. The posting box used in infant schools is an example of discrimination learning. The learning process can advance a stage further and children can learn how to learn (Harlow, 1959)[(4)]. This is very much a condition of practice but "what kind" and "how much" have to be considered.

Methods with older children and adults

1. *The learning method.* Here the subject has been taught a perceptual clue or cue which leads to the solution of a class problem. This cue facilitates the recognition of further members of the class. An example will make the point clearer.

Say the subject has learned, either by rote or with understanding, that an integral of the form

$$\int \frac{f'(x)}{f(x)}\,\mathrm{d}x \dashrightarrow \log_e kx$$

then he uses the cue, he literally sees that $\frac{d}{dx}$ (denominator) ----→ some form of the numerator and hence can solve

$$\int \frac{2x\,dx}{x^2 + 1}$$

2. *The problem-solving method.* The adult or child is asked to attempt a mathematical problem and at the same time records his thought processes. You can read brilliant accounts of the method by Wertheimer[5] and Polya[6].

It may be helpful to ask of the various theories:

1. Is the theory behaviouristic or neo-behaviouristic in origin? Is some variation of the Greek hedonistic pleasure/pain principle used? For instance, is it a reinforcement-type theory? Is motivation supplied through a mechanistic law of effect? Does the theory ignore the development aspect of concept formation in the life history of the individual? Does the theory use words like stimulation, motivation, reinforcement and how does it use these words? In simple terms, is the motivating force success or pain?

2. Have we a stated homeostatic mechanism? Cannon (1939) described body systems under the control of either the central or autonomic nervous system whereby the body, in a state of disequilibrium, tends to act or change in a direction to restore equilibrium. Bugelski (1936)[7] in a deceptively simple theory states a necessary condition for learning is to make the learner anxious then allow him to reduce the anxiety, whereupon learning takes place. However, the whole of motivated behaviour cannot be explained by drive reduction. But it is useful to ask ourselves, is the theory one which postulates a movement away from disequilibrium towards balance.

3. A young science such as psychology borrows and adapts from older more established disciplines. E. L. Thorndike's (1874–1949) connectionism implies the use of learning bonds paralleling the idea

of chemical bonds fashionable in the early part of this century. Explanations of learning are sometimes a product of the Zeitgeist; theories from the physical sciences being taken over and adapted by learning theorists. Information theory and cybernetics are current examples.

4. Is the theory orientated around a combination of the above? Many workers use some variation of a "black-box" theory.

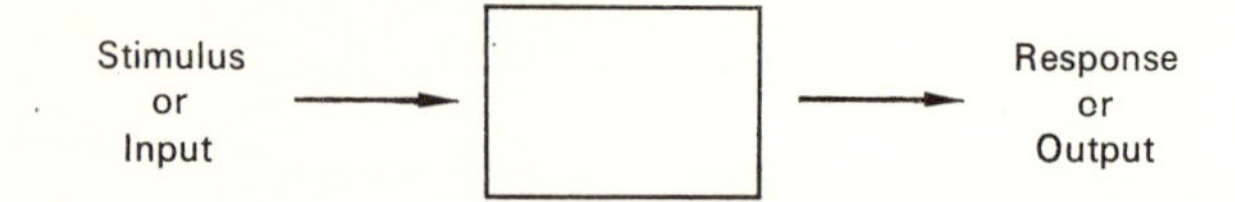

They employ either intervening variables or hypothetical constructs. Examples of intervening variables are "*g*", the gravitational constant and the organism itself. But there is no need for the variable to exist independently provided that it has properties implied in its units of measurement. The variable is an integrating intermediary. The hypothetical construct implies the possession of further properties than those implied in the theory. Examples of this type of intermediary are:

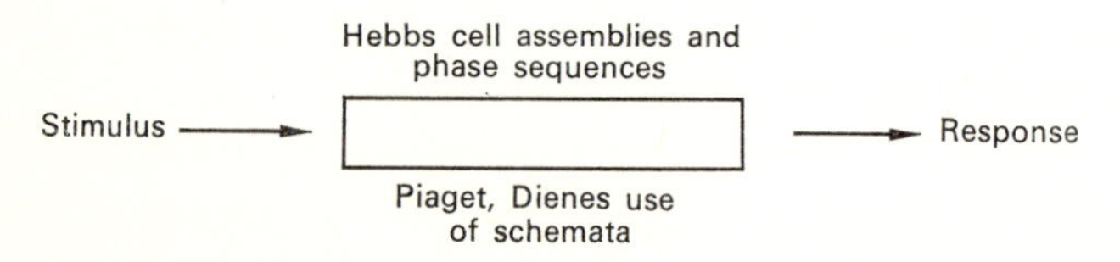

It is often worth while examining a theory to separate out the intermediaries. Constructs should not be allowed to assume a mantle of reality.

Enough theoretical orientation has been given; it is time to sketch some of the practical aspects of learning theory. E. L. Thorndike dominated for half a century the work of other researchers. As Tolman (1938)[8] says: "The psychology of animal learning—not to mention that of child learning—has been and still is primarily a matter of agreeing or disagreeing with Thorndike." Thorndike's work, a fusion of the new biology of Darwin and the older observationalism and association of ideas of Wundt, was initially with animals, but in line with the Darwinian thesis, was tranferred to children. In *The Psychology of Arithmetic* (1922)[9] he uses his stimulus/response model as the central mechanism in learning arithmetic. There is association between sense impressions or stimuli and the resultant action. A bond or connection or association is formed between S and R.

We have Stimulus—Response where $R \in \{R_1, R_2, R_3, \ldots, R_n\}$. If we want to learn some mathematics then we must connect the correct response to the appropriate stimulus.

A necessary, but not sufficient, condition to promote the connection between S and R was contained in his "Law of Effect". Thorndike[10] stated his law: "any act which in a given situation produces satisfaction becomes associated with that situation, so that when the situation recurs the action is more likely than before to recur also." To this law, which is similar to the Pavlovian reinforcement, he added the "Laws of Readiness and Exercise". These laws, or more correctly principles, therefore dealt with initial conditions for learning to take place, the sustaining of effort and the role of practice. His "Law of Readiness" was not the modern "number readiness", but referred to the physiological state of the child, as a necessary condition for the operation of the "Law of Effect".

The essence of Thorndike's work is an almost automatic strengthening of bonds through practice which is why it is called connectionism. Reward is central for Thorndike. Children need to practice in order that reward can act upon ill-formed connections. After Watson coined the term behaviourism in 1913, the work of Thorndike can be said to have become part of the S–R tradition.

Say the child meets:

3×7 and the connection with 21 is made.

1. Given situation ——————————— act.
2. Satisfaction is the reward, internal or external.
3. $3 \times 7 = 21$ becomes known and secure when $3 \times 7 \blacksquare 21$ or the bond is strengthened through repetition under reward.

He added other motivational features in his "Interest Series" and these soften the stark mechanical aspects of the laws. Thorndike[10] listed:

1. Interest in the work.
2. Interest in improvement.
3. Significance.
4. Problem-attitude.
5. Attentiveness.
6. Absence of irrelevant emotion.
7. Absence of worry.

And now his emphasis upon an active child is clear and modern! His theory has become old fashioned for another reason. John Biggs[11] puts it "the individual profits more from learning meaning and structure rather than responses or actions". Or again, "educators require a simpler less analytic theory, we need more of the technology to produce learning". Today extrinsic rewards are not emphasised; we are more interested in intrinsic motivation. But whatever the comment, if we reflect for a moment, we acknowledge the importance of rewards. The child does work for reward in the form of praise or stars or house points. Perhaps Thorndike[12] should have the last word: "... when he is confronted with a new situation his habits do not retire to some convenient distance while some new and mysterious entities direct his behaviour. On the contrary, nowhere are the bonds aquired with old situations more surely revealed in action than when a new situation appears."

At the moment large all-embracing theories are not in fashion. Dienes[13], Bruner and his co-workers at Harvard are content to observe mathematical activity, as it occurs in the classroom and then ask relevant questions. The implications is that the answer given may lead to a small hypothesis which can then be tested through controlled experiment. One school of thought holds that a new starting-point for a psychology of mathematics is the vast collective body of knowledge held by teachers of mathematics. An Aristotle sitting in an Institute of Education ought to collect this information, sift, classify and assemble into a small theory: this is a complete reversal of Tolman's cry "Rats not men".

The change to investigating the mental processes involved in mathematical activity has brought many difficulties. The study of the optimum conditions for rote learning the number bonds could be considered a simple aspect of cognitive process. The nature of thought itself is more complex. Dewey said, "thought can be regarded as problem solving behaviour", thus an insight might be achieved through the study of mathematical problem solving. Gilbert Ryle said, "thinking is partly a complex of skills and drills". So it would appear worth while taking a look at the view that thinking is at least in part a complex of skills and drills.

Bartlett[14] in his book *Thinking* sets out the development of thought, on a continuum of increasing complexity as we move from *S–R* units on to intelligent problem solving behaviour.

- → stimulus-response unit
- → chaining of S-R units to form habits
- → strengthening of habits through conditioning
- → habits chained into skills
- → complexes of skills or the whole approach

At this level of organisation thinking or intelligent insight appears.

Perhaps we can observe the theory in action:

Sometime ago I was watching television when this series of numbers was given and viewers were invited to find the next term.

$$5 \quad 10 \quad 70 \quad 4690 \quad ?$$

Before you read my solution try to find one yourself. This is important because we all attempt to short circuit the learning process. Bartlett would say that intelligent behaviour is the seeking of the direction from the chaotic stage towards the whole approach. Finding the direction is the learning.

Here is a solution:

5 × 2 → 10	5 × (5–3) → 10 *****
10 × 7 → 70	10 × (10–3) → 70
70 × 67 → 4690	70 × (70–3) → 4690
4690 × ? → Solution	4690 × (4690–3) → Solution

Were you too lazy to try? If you were then Andree[15] says you have a problem on your hands and your progress is inhibited. Now go back to my solution. Do you agree that the step marked with an asterisk is the moment of solution? The solution is built on:

1. Receptors and effectors were linked in order of succession to give direction. There was a rapid scanning of answers to questions such as what kind of series?, an A.P.?, A.G.P.? No, the terms increase rapidly, therefore we may have a squared function and so on.
2. The use of earlier conditioned habits.
3. The early fluid searching towards a goal ceased at the point marked with an asterisk. Insight was achieved and the solution apparent.
4. This search for directionality lead through the chaotic stage to the analytic and finally to insight.

But I did not achieve the perfect whole approach; I failed to reach the standard set out by Leibniz when he wrote: "A method of solution is perfect if we can foresee from the start, and even prove, that following that method we shall attain our aim." (*Opuscules*, p. 161.)

I failed yet again when I attempted to generalise my solution. There have been numerous efforts to isolate the cause factors resulting in mathematical failure. Froebel and Montessori, Renwick[16] and Holt[18], have reported their observations; John Biggs[17] and Dienes[13] report more classically orientated studies. Holt is concerned with "avoidance strategies", that is, the strategies used by children in a stress situation. Frequently this avoidance behaviour blocks satisfactory attainment. This is an important area of research and it would be a good step forward if we could match these avoidance strategies with the stimulus properties of mathematics and the teacher of mathematics. Depth and social psychology, the social sciences in general can contribute towards an eclectric attack.

Most teachers asked to write down causes of failure in mathematics would come up with rather similar lists. Many would agree to:

1. Fear.
2. Avoiding ridicule.
3. Attempting to avoid punishment.
4. Hiding ignorance.
5. Professing to be stupid.
6. Pleasing teacher.
7. Guessing.
8. Opting out.

The depth psychologists must surely have something to say about the ultimate sanction of staying away from school to avoid mathematics! Holt lists these and other causes. His collection of empirical data from direct observation, as it happens, and the conclusions he draws, sometimes appear to carry greater weight than the more classical approach of the null hypothesis followed by a well-designed experiment and the use of an appropriate statistical technique to accept or reject the hypothesis.

The penetrating observations of Renwick, culled from a long experience of teaching mathematics, are informative and sheer delight to read. In her book she reports twenty-nine stories of children learning mathematics over a wide range of processes. Many of the conclusions she draws have implications for teachers of mathematics. One example will suffice. The little girl in the nursery who gives the wrong answer when asked how many pieces of bread and butter remain out of the original three pieces, shows the confusion between the various uses of, say, two. There is linguistic and mathematical confusion in the use of two as a label or a natural number. Again is the conclusion to assemble the vast amount of expertise in the hands of teachers?

Polya[19] introspects into his own mathematical activity during problem solving. He says: "solving problems is a practical art, like swimming or ski-ing or playing the piano; you can learn it only by imitation and practice. If you wish to become a problem solver you have to solve problems." He appears to be fascinated by the

Cartesian method of problem solving and tries to extend this method, which is the basis of his imitation and practice.

René Descartes (1596–1650) tried to formulate a universal method for the solution of problems. His scheme was:

1. First, reduce any kind of problem to a mathematical problem.
2. Second, reduce any kind of mathematical problem to an algebraic problem.
3. Third, reduce any algebraic problem to the solution of a single equation.

It is this scheme that interests Polya and to it he adds an experimental technique. This technique or procedure is applied to his classification of two kinds of problems:

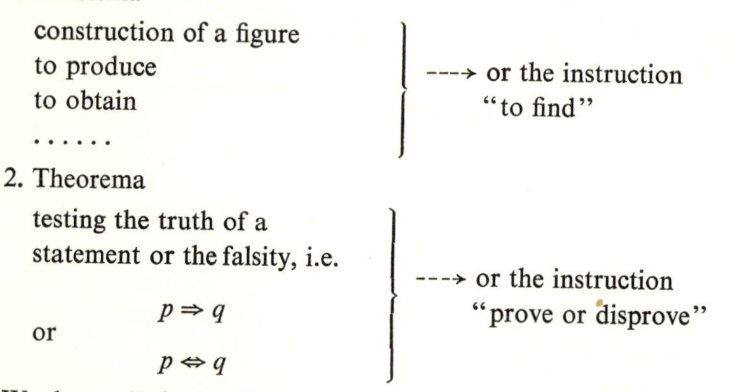

We have: Polya + Descartes + An experimental method, or in diagrammatic form:

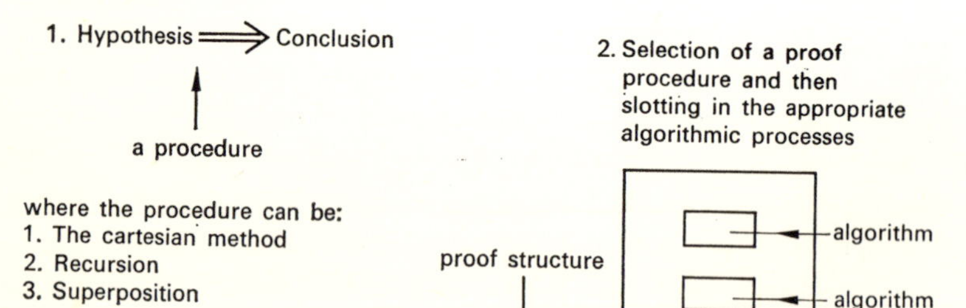

Some further explanation is necessary. The word hypothesis refers to the "if" part of the proposition to be tested, and the conclusion to the "then" part. The models refer to problems where we are asked to prove or disprove some proposition, but the three procedures can be applied to problems which instruct us "to find", i.e. the demand for a proof. It is not necessary to elaborate the Cartesian method as we are familiar with the technique. An example of an electric switching circuit is sufficient. If we are asked to simplify a circuit, we reduce the problem to an algebraic one by writing the appropriate Boolean expression. If it is possible this expression is simplified and then translated back into circuitry.

The usual method employed to find the sum of the first n terms of the series:

$$S_1 = 1 + 2 + 3 + 4 + \cdots + n$$

is to write

$$S_1 = n + (n - 1) + (n - 2) + (n - 3) + \cdots + 1$$

adding gives

$$2S_1 = n(n + 1)$$

or

$$S_1 = \frac{n(n + 1)}{2}$$

Since $S_0 = n$ and $S_1 = \dfrac{n(n + 1)}{2}$, can we find S_2?

or

$$S_2 = 1^2 + 2^2 + 3^2 + \cdots + n^2.$$

The road to a recursive solution, as often happens in other methods of attack, follows a perceptual reorganisation, that is, we "see" the way to success through a rearrangement of the perceptual cues. When, however, we attempt to use the same technique on S_2 we fail. We now have to use what many children say is a trick. We use the formula:

$$(n + 1)^3 = n^3 + 3n^2 + 3n + 1$$

or

$$(n + 1)^3 - n^3 = 3n^2 + 3n + 1$$

or

$$(n + 1)^3 - 1 = 3S_2 + 3S_1 + S_0$$

by giving n the values 1, 2, 3, 4, ..., n and adding all the n equations, since we know the sum of S_1 and S_0 we find S_2.

Similarly through the use of the formula

$$(n + 1)^4 = n^4 + 4n^3 + 6n^2 + 4n + 1,$$

we obtain S_3.

The method can be generalised: If

$$S_k = 1^k + 2^k + 3^k + \cdots + n^k$$

by a similar procedure we can obtain

$$(n + 1)^{k+1} - 1 = (k + 1)\, S_k + \binom{k + 1}{2} S_{k-1} + \cdots + S_0$$

thus linking $S_0, S_1, S_2, \ldots, S_k$.

Then as Polya[20] says, "we can find the terms, one after the other, successively, *recursively*, by going back or recurring each time to the foregoing terms". Or (say) we can find S_4 by going back to S_3, S_2, S_1, S_0.

Possibly the young Carl Gauss used a pattern of perceptual reorganisation when his teacher asked the class to add:

$$1 + 2 + 3 + 4 + \cdots + 19 + 20.$$

Polya[21] and Wertheimer[5] give full accounts of this well-known story. Polya argues for a discovery approach incorporating the perceptual reorganisation of the original data with the work of Wallas (1921) who suggested that creative solutions pass through four stages:

(i) preparation,

(ii) incubation,

(iii) illumination,

(iv) verification.

If the solution was perceived then the perceptual reorganisation could be explained by successive scanning of either, then both ends of the series, followed by closure. Possible stages of reorganisation are:

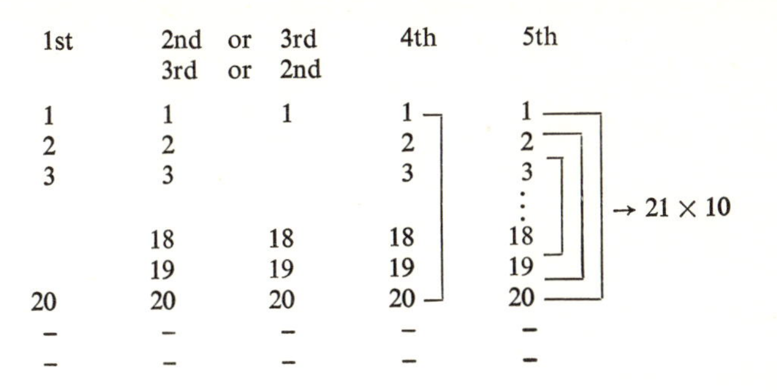

Or he may have flashed the answer. Perhaps he just saw how to do it in the following way. He used our technique for summing our simple series thus:

$$S = 1 + 2 + 3 + 4 + \cdots + 17 + 18 + 19 + 20$$

or

$$S = 20 + 19 + 18 + 17 + \cdots + 4 + 3 + 2 + 1$$

and

$$S = \frac{20}{2} \times 21$$

and

$$\underline{S = 210}$$

but we shall never know!

The procedure called superposition can also be best explained through a simple example. The theorem which asks us to establish that parallelograms and rectangles on the same base and between the same parallels have the same area can be shown to hold for the special case, Fig. 1. This case is the most easily shown. If we now superimpose the particular cases of Fig. 2 and Fig. 3, we obtain the general solution, through an algebraic operation.

Only in the special case shown in Fig. 1 do the points B and F coincide and the proof is established through the congruence of triangles ABD and BEC. In the other two cases it is crucial to show that $AF = BE$ and this is done algebraically.

In Fig. 2:	In Fig. 3:
$AF = AB - FB$	$AF = AB + FB$
$BE = FE - FB$	$BE = FE + FB$
$AF = BE$	$AF = BE$

or combining both cases:

$AF = BE = AB \pm FB$ and in Fig. 1 we have a special case $FB = 0$.

The algebraic operation leads to generality.

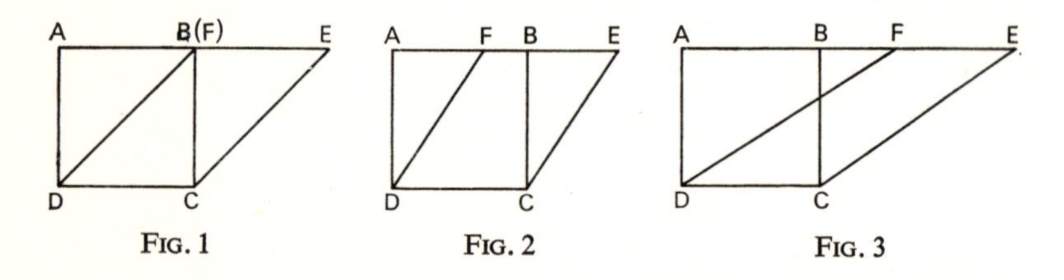

FIG. 1 FIG. 2 FIG. 3

There are a number of questions to ask of Polya about this heuristic approach. It is a personal approach and although he may be able to structure his own introspections into a general technique, how do these theories fit into the classroom? Is it a technique for any age? Do teachers agree with his findings or are his solutions applicable only to Polya? What energy, drive state, curiosity—call it what you will—initiates the problem-solving behaviour and what are the intrinsic processes that sustain the action? Is the large body of experience or expertise in the hands of teachers of mathematics, the acute observation yielding empirical data, in agreement with Polyian theorising? Perhaps it is time to remind ourselves of an earlier point which John Biggs made, namely that teachers need educationally orientated theory, or as Wheeler[(22)] puts it, "we require a theory of mathematical education not divorced from general theories of education and epistemology".

C. Fleming[23], writing in the *University of London Bulletin* (1968), makes the point that educationalists have looked at "ends" and have forgotten to ask themselves:

(a) How was the investigation planned?
(b) Is the researcher justified in drawing the conclusions he has stated?
(c) What statistics, if any, were used in the analysis of the raw data?
(d) Intervening variables, classified as sensations, percepts, concepts are passing out of fashion.
(e) Is it important to remember that we have process of learning not a biological development?

This plea for a more rigorous approach is timely but is not denying the value of the work of the Geneva School. Bruner[24] argues that the quality of thought of Piaget in his theorising is of a different order than say the logical thought of a 15-year-old boy. According to Bruner, Piaget is a genius who "makes science, myth and magic". But the advice of Fleming will help us to achieve perspective, when we read the relevant literature. However, a return to the Kantian notion that science is always characterised by mathematical as well as empirical observations and descriptions might endanger the point made by Peters[25] when he writes: "The advance of science depends upon the development of imaginative assumptions as well as upon exact techniques for testing them. To invest in the latter without the development of the former is like buying a combine harvester for use at the North Pole".

Piaget trained as both zoologist and mathematician; he and his co-workers in Geneva have investigated the whole development of organised mental logical structures; his cross-sectional, developmental studies stem from a lifelong interest in genetic epistemology, that is, how the child builds up his knowledge of the outside world. The studies and data gathered by his "clinical" method reflect the interest in the activities of the child as being central to intellectual development. Piaget has reserved the more rigorous experimental methods for the study of perceptual development, concluding from his work that conceptualisation and perceptualisation follow separate and independent courses.

His stage theory is applicable only to conceptual development; he does not deny that concept and percept are related, but he suggests that perception subserves intellectual development. Piaget sees his subjects as being in interaction with the outside world; knowledge derives from this activity. His subjects are not passive agents being acted on by external stimuli. To paraphrase Nathan Isaacs[26], "For Piaget the child makes himself—he is the architect of his own intellectual development".

Piaget is not primarily concerned with teaching but his elucidation of stages of mental growth is of direct educational concern and certainly relevant to the issues being considered here. Piaget's theory of stages implies that the child has a different kind of "mind" for each stage.

The five main stages of intellectual development are:

1. The sensori-motor stage. From birth to 18 months. During this period the child relates the various sensory "inputs" through his actions on the world; initially, what is himself and what is the object are not discriminated. Towards the end of the period the child comes to see himself as one object among many and also that he has his own space-filling properties and his own movement in the common spatial field. Objects can now be seen as things apart, each having its own laws of displacement; later he will come to learn more about objects, especially how they can vary in visual appearance, but their substantiality is no longer in doubt.

2. The stage of intuitive thought, from 18 months to 4/5 years. The child develops language and the ability to move through his world. Language enables him to label objects and thus frees him from the limiting use of imagery; his experience becomes more articulated. Through imaginative play, exploration and experimentation, questioning, listening and talking, the various kinds of objects and events he can adjust to, anticipate, imagine and remember,

free him increasingly from the "here" "and now". Space and time relationships become more elaborated. Yet there is much vagueness and instability in his ideational and imaginal world; he is very limited in his capacity to move away from the immediate situation.

3. Towards the stage of concrete operations, from 4/5 years to 7/8 years. This stage corresponds to the first years of schooling and so is of great importance to teaching mathematics. Here the concern is with the development of such basic concepts as space, time, movement and speed, number and measure, whole-part and serial order. It is at this stage that children will confuse the number of counters or coins according to their spatial distribution; the heaped coins will be considered as less than the same number spread out across the table; there is considerable confusion between other conceptual categories. There is, however, a consistent progress towards the grasp of simple concepts so that by the end of the stage the child can deal with the concrete experimental situation very much in the manner of the adult. Now the child attains the level of "conservation". He will dismiss problems almost contemptuously which he once treated seriously; the pint of liquid, however, contained will be stated to be the same quantity. The child hardly attends to the problem; he knows what conservation implies. More importantly he can relate the various concepts to form new schemata of operational thinking.

4. Concrete operations stage. From 7/8 years to 11 years. Each basic structural concept is clear and stable; in various sets taken together these concepts come to form distinctive schemes of operational thought. The stage is so called because the child seems to need the reinforcement of practical activities to deal with his conceptual world, although the later stages of abstract levels of operation are preshadowed and becoming increasingly functional.

5. The formal operations stage. From 11 to 14 years. The child has now attained the full stage of thinking abstractly; his basic conceptual world is one of functioning ideas. Time, space, number, logical and mechanical relations are now serviceable to him as internalised processes.

Piaget's work has led to the realisation that attempts to teach the child at a level beyond his stage of development end in rote-learning of ill-understood ideas and material; clearly we must teach with regard to the child's age and stage, that is, by bearing in mind functional concept of "readiness". It is necessary to delimit the material, so we will examine only one of the many concepts which Piaget has concerned himself with, that of number, and see how it fits into the pattern of intellectual growth. Piaget's five stages can be aligned alongside this model of how the number system is built up.

Number and a Number System

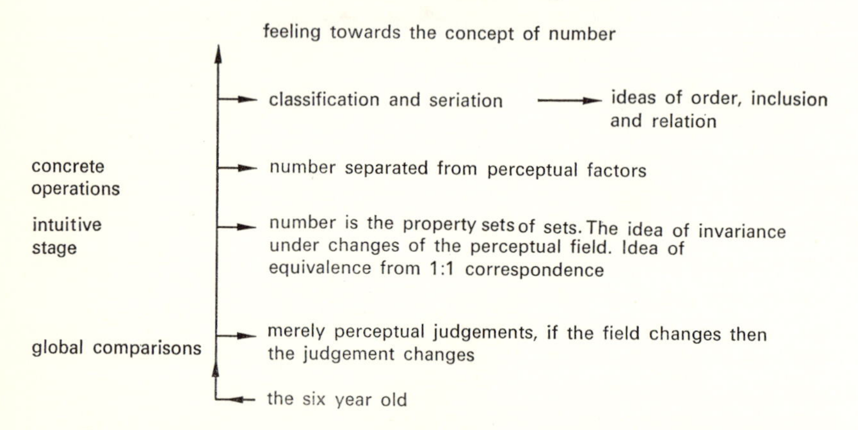

In an article in the *Scientific American* (Nov. 1953) Piaget[27] writes:

> . . . a child of six and a half or seven often shows that he has spontaneously formed the concept of number even though he may not yet have been taught to count. Given eight red chips and eight blue chips, he will discover by one-to-one matching that the number of red is the same as the number of blue, and he will realise that the two groups remain equal in number regardless of the shape they take.

The implication is clear; children develop mathematical concepts independently and spontaneously. If we as teachers impose mathematical concepts on a child at too early an age or stage, then there is little real understanding. Note the parallel growth of conservation which is a logical notion and the mathematical concept of number. The logical notion precedes the concept.

Other statements that catch the eye are:

A. "The ability to co-ordinate different perspectives does not come until the age of 9 or 10". This insistence that such and such an ability appears within a fairly closely defined age range has lead to criticism. It is suggested that the stages are inherent in the mathematics itself and the remarks of Sigel[28] appear pertinent. Do you remember the points made by Fleming? Sigel criticises stage development theories on the following grounds:

(a) In most areas of performance overlap can be found among chronological age groups.
(b) The apparent continuity of development may result from "gaps" existing between the stages. If we fill in these observation gaps we may find discontinuity.
(c) He finds the responses of his subjects unstable and argues that this instability makes it difficult to determine which stages have been reached.

Berlyne[29] says that sampling techniques are suspect. For example, the data on the sensori-motor stage is obtained from observations of Piaget's three children. Few statistics are available and possible variance is ignored while tests of significance give way to categorical assertions that *at such and such a stage a particular ability appears*. Since the original work was translated many of his experiments have been replicated. If we allow for ethnic, cultural and class differences, there seems little doubt that Piaget has given us a brilliant working model, which will continue to provide hypotheses for testing by psychologists and mathematicians.

B. Piaget states that the psychological order for development of geometries reverses the historical order. The child's discovery of spatial relationships follows the pattern:

nineteenth-century Topology .... seventeenth-century
Projective .... Euclidean.

He concludes from his experimentation that the child at the age of 3 can distinguish between "open" and "closed" figures long before he draws a rectangle and can talk about its Euclidean properties. The first geometrical discoveries are topological ones; a long interval must elapse before the child develops his notions of projective and Euclidean geometry. The evolution of the geometries matches the growth of the child's logical power; we find this correspondence of psychological order and logical growth repeatedly in the work of Piaget and his co-workers. The results of this experimentation point to the need for reappraisal of our teaching of geometry and suggest that the teacher should construct learning situations involving ideas of openness, closed, interior and exterior, before introducing sets of shapes and their properties.

Piaget (1950) followed Sir Henry Head, the neurologist, and Sir Frederick Bartlett, the psychologist, in his use of the terms "schema" and "schemata" as providing a "physiological" model of cognitive structure. Skemp and Dienes have extended the use of this hypothetical construct. Cognitive structures are built through the child interacting with his environment and are enlarged through further experience. The necessary conditions leading to the acquisition of new responses are handling concrete things, discussing what has happened, and discovery through investigation, leading to a logical notion which in turn allows the existing schemata to enlarge through the assimilation of the new material. The schemata has accommodated the fresh material. In diagrammatic form:

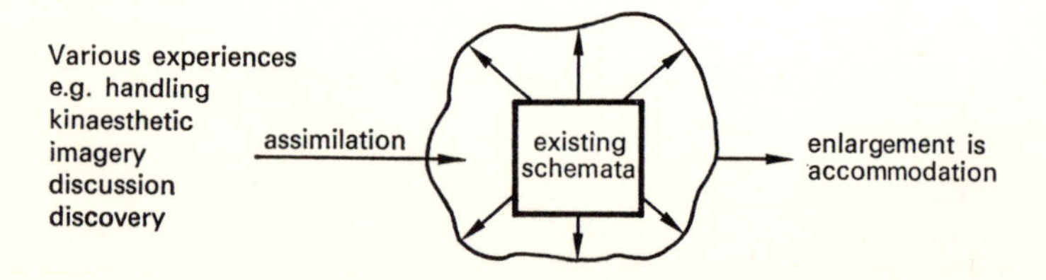

An example will make the point; the teacher and a small group of children have played a game of partitioning a set and have taken out various subsets, afterwards reversing the action to re-form the set. They have looked at, handled and discussed sets of nesting trays, they have examined growth rings of a tree section and they have built concentric rings of clay or plasticine. The notion of inclusion appears spontaneously or after directed discussion. This logical notion of inclusion, with further suitable experience, will be assimilated into the existing number schemata and the concept of number resulting from the consistent cognitive growth.

Dienes also used the term; beginning with the collection of empirical data from observations of children learning mathematics in the classroom. He erects quickly a theory, compounded from elements of Piaget, Eysenck and Lewin, which soon becomes a model: That is:

Description of data ---→ Explanation ----→ Theory ---→ Model.

Many questions occur as we read Dienes, but one is predominant. Alex Comfort talks about "hard"- and "soft"-centred explanations. "The hard-centred explanation is predictive and creates its own questions whereas soft-centred theorising forbids the asking of further questions." How shall we classify Dienes' work? A good model will help us to discover if the theory works. It may shift attention to more appropriate parts of the theory or learning process, and thus greater coverage than expected may result. But the danger exists that the model may lead to a false assumption of reality between the model and the event. A good model should have:

1. Deployability. This aspect is measured by the success with which terms used in the model can be transferred to a new setting.

2. A range of situations to which the model is applicable.

3. Precision or the ability of the model to predict from within its own structure.

These criteria may help you to assess the stimulating ideas put forward by Dienes.

His work is worthy of attention for many reasons, but I would argue that his theory of the functioning of play is possibly most important. Guided by knowledge gained from the Leicestershire Mathematics Project he was able to predict roughly what children could learn in open-ended situations. Observations, *in situ*, of children playing "mathematical games" or handling mathematical material led to the definition of three kinds of play, and the description of ways in which the energy released through play leads, or is transformed into modes of behaviour establishing higher cognitive processes. The energy released through the excitement of play is the motivating factor. Dienes has designed games and apparatus to facilitate the appearance and interaction of the three types of play. If his theory is valid then the means of controlling learning is available. No laws are formulated or indeed are known, but he argues that the transitions from a state of exploration or curiosity to cognitive activity is seen to happen. His theory can be expressed diagrammatically.

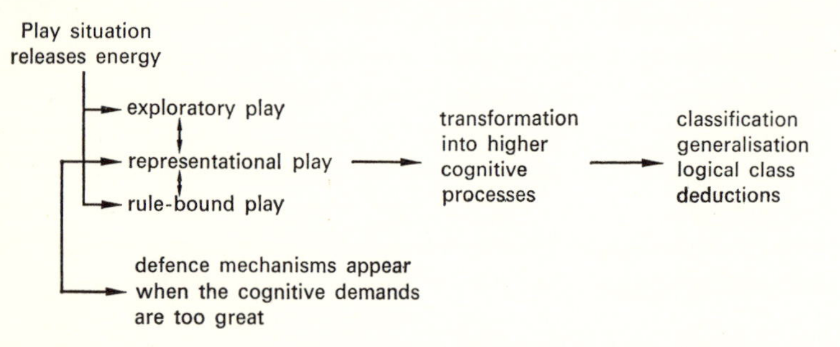

Representational play occurs if objects are given attributes different from those they normally possess; rule-bound play is play bounded in some agreed manner. The defence mechanisms or projections that appear when the cognitive loading is too great tie in with the avoidance strategies of Holt. The function of play in mathematical thinking is described by Dienes in Chapter 1 of his book reporting the findings of the Harvard Study. The title of the book is *An Experimental Study* of *Mathematics-Learning*[13]. But some mention of the Piagetian Cycle and his construct of a conceptual-modality-space must be made.

His well-known Multi-base Arithmetic Blocks lead children towards notational concepts, but in addition Dienes has designed apparatus and games which match his notions of the necessary experiences that promote both "pure" and "applied" concepts. He argues that there is a common pattern of experience, called Piagetian Cycles, leading to mathematical concepts. His apparatus and games aim to provide the structured experience matching his theorising. Some of the experiences are said to be artificial, but he argues that it is almost impossible to provide real situations that lead to higher-order mathematical concepts. But in the hands of Dienes the material is effective.

Four principles govern his learning theory. The first is the dynamic principle, the second the constructivity principle, followed by the perceptual variability and mathematical variability principles. Dienes has constant shifts in his theoretical position and he has replaced the first two principles with a single extended dynamic principle. A much simplified cycle might be:

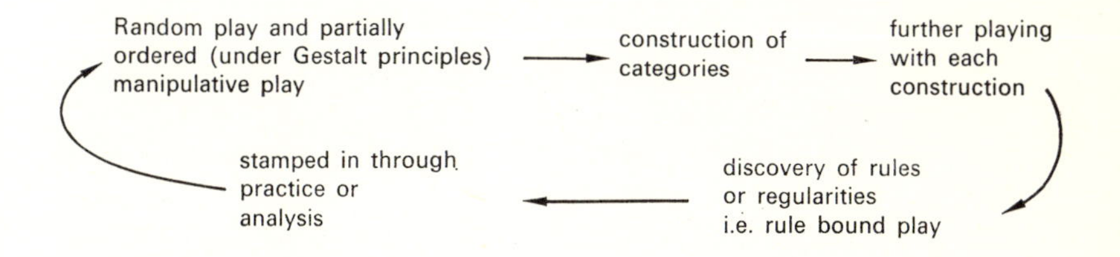

As the cycles progress more and more complex categories are constructed. An example may make the complex theorising clearer. In his report of the study of vector spaces through games, Dienes gives the following cycles:

1. Creating vectors from the objects.
2. Obtaining another space through the transformation of all vectors.
3. Start of the process of inverting the transformations.

The games apparatus is usually simple. Dienes uses open or closed boxes, cups right way up or inverted, and gloves inside out or right side out, to represent vectors and from the vectors children create three-dimensional vector space. It is this ability of Dienes to translate mathematical structures and ideas into embodiments within the cognitive grasp of the child that makes his work promising for the future.

The perceptual variability and mathematical variable principles are clearer. They are imbedded in the folk-lore of mathematics teaching. Mathematical structures must be perceived in different situations to ensure efficient abstraction. To acquire full generality, all the variables must be varied. The educational implication is, to quote Dienes: "... whether generalisation should take place simultaneously on a broad front, or on several narrow fronts followed by abstraction into a broader front later. Younger children seem to find generalisations on a narrow front, i.e. within certain well-defined fields, easier. This may be part of the development pattern." He points out that behind the language used by the child there lies an enormous number and variety of experiences and he asks that teachers should attempt to augment and enrich the range of mathematical experiences.

The influence of Lewin and Eysenck is seen in his model of what he terms a conceptual modality space. Most of us have memories of a good teacher and sometimes embody our feelings in a phrase such as "he taught me all the mathematics that I know". Why have we appreciated the efforts of one teacher rather than another? Dienes says that one reason is we possess the same conceptual modality space. The work of Lewin is reflected in the topological construction of the space. A given concept occupies space mapped out by cut-off points on a number of linear axes. The conceptual modality space is defined by points or weightings on these axes. The nature of the axes or dimensions, though operationally defined, has not been experimentally established. And the description of the axes is vague. One dimension is labelled open–closed, or the tendency for a concept to assist in or to prevent the formation of further concepts. Another, the explicit–implicit dimension, is scaled to explain the usability or not of a partially formed concept.

All this is most vague, but if the weightings assigned to the various dimensions could be experimentally determined, we could compare the child and teacher. Intersection of modality spaces leads to insightful learning while, if there is no overlap, rote learning occurs. This is clearly an extremely difficult area of research, but if the dimensions were experimentally determined, the existence of a common pattern of organisation demonstrated and correlated with personality variables, then improvement in mathematics teaching would follow. Dienes suggests that a lack of sympathy between the modalities of the teacher and the child can be overcome through the use of appropriate material and apparatus. Further he suggests an answer to the difficulty of teaching a large class of children. If each child possess a differently organised cognitive structure, then children ought to be provided with games, apparatus and structured mathematical situations in order that they may extract in their own individual manner.

The way to an understanding of the work of Dienes is to read his published work, and try to assess for oneself the potential of this original approach. Better still, is to work through some of the games with a partner and trace the Piagetian cycles.

A comparative approach may hold some value and expose a little of the dynamics of learning. If it were possible to map out or define the boundaries of existing schema or cognitive maps, rather in the way the ophthalmic surgeon maps out the field of vision, then a new starting-point would be available. Attempts should be made to plot schema of, say, (i) $\int \sqrt{(x^2 - 4)}\, dx$ and (ii) the definition of a circle in terms of a stationary and of a moving point. Given the necessary motivation, in terms of curiosity, exploration or some drive state, observations and introspective evidence might enable us to say how the arousal of one map "fires" another, much in the manner set out by Hebb in his theories of neurological functioning. In diagrammatic form the dynamics of the learning may operate:

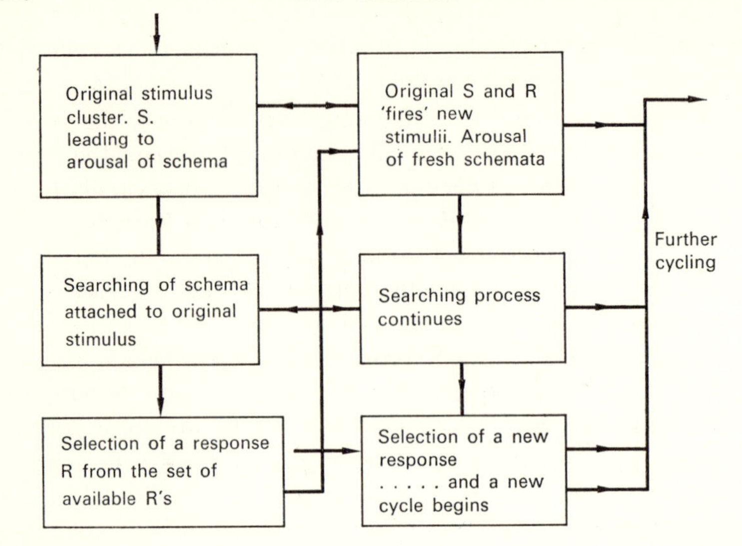

As indicated in the diagram, a stable system would involve cross links between the separate parts of the structure. In a new learning task the fresh element would be learned, that is there would be "closure" and only when the system was closed could we say that effective and permanent learning had occurred.

When a student is confronted with the above integral what action does he take? Probably he searches schemata attached to his ideas of integration. He asks himself questions. Is the integral in any way perceptually similar to one that he has previously solved? If he is forced, in trial-and-error fashion, to review his entire knowledge of methods of integration, then he will spend rather more time reaching a solution than the examiner allows, if indeed he does solve the question. The comparison of the weak and strong student appears to indicate that the weak begin at the beginning and roll the film on and on. He is not able to pick up cues by focusing his attention on a specific part or frame of a long film.

If we break the fundamental condition of the definition of a circle, interesting possibilities are open to us. We shall ask the question, "What happens if the moving point does not remain a fixed distance from the fixed point?" A number of systems can be aroused:

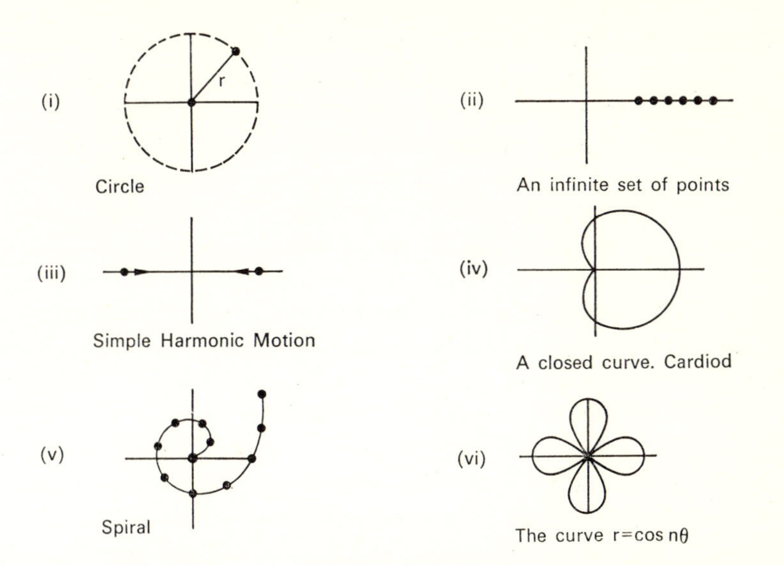

But if you are asking yourself some questions, the previous section is of some value. Where is the experimental data to support my theory?

Wrigley[(30)] writes in *New Approaches to Mathematics Teaching*, "... most of the important problems to be solved in connection with learning theory and mathematics are inherently difficult and in this respect the research worker may be in a dilemma. He can take an important and difficult one and spend a long time on it." As Wrigley uses the singular form of problem are we to infer that the erection of large-scale theories is premature? How many of you are thinking of Piaget and Dienes? Can we validate such theorising? The answer is the obvious one. We cannot, in the sense of providing an experimentally precise and controlled collection of data. As Wrigley argues, we must be willing to accept research of a less precise kind.

But are there any trivial problems? Wrigley himself is trying to answer a difficult problem, namely, the factorial nature of mathematics. He has attempted to answer the question "is there a unitary mathematical ability?" The evidence points to the existence of a group factor in addition to "*g*", the Spearman general factor of intelligence. Spatial ability, the ability to think and visualise in three dimensions, is another area of interest. Trivial or non-trivial work involves more than an experimental hunch, the subjects are usually children of varying intelligence and attainment. They belong to a certain culture; they are members of varying social groupings. Children are emotionally involved in the learning task. All this suggests that there are no trivial problems; workers may create seemingly simple situations by ignoring some of the variables involved.

The affective side of learning mathematics is important. It is not clear if mathematics is more anxiety provoking than other disciplines, but anyone engaged in teaching students is quickly brought fact to face with this problem. How can this lack of success of otherwise intelligent students be explained? Poffenberger and Norton (1959), in a U.S.A. study, asked 390 first-year university students about the factors which determine attitudes towards mathematics. As we might have guessed, the most important listed factors were the home influence, that is the expectation and attitudes of the parents, and secondly the influence of teachers. John Biggs (1962)[17] investigated the relationship between anxiety, different aspects of performance in arithmetic, and methods of teaching. He found that a child's performance depends upon anxiety levels, success, the way in which learning takes place, and the sex of the child.

Wallace[1] writes: "the chronic lack of funds to finance the large-scale inquiries essential to explore the basic problems in sufficient depth to yield authoritative findings has produced an endless series of small research projects often claiming to be dealing with the same topic and producing contradictory results." It follows that economy of effort and money would result from planned research programmes. The adoption of:

1. a clinical method,

2. some variation of the behaviouristic model,

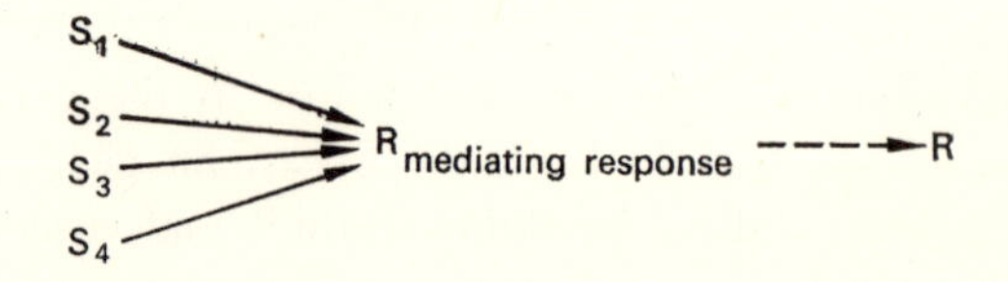

or some union of both, or a return to the observational tradition, is of lesser importance than reaching agreement on research priorities.

The following table showing programmes of research should be read with caution. Omission of an area of work from an individual column does not mean that the researcher is not interested in that area. No order of priority is implied. There is a degree of unanimity.

Peel	Wrigley	Dienes
Concept formation. Thinking. Reasoning.	Factorial nature of mathematical ability. Concept formation. The use of hardware. Problem solving.	Cognitive processes. Extending the curriculum research. Teaching methodology.

It is now more than 100 years since Wundt opened the first experimental psychology laboratory, but progress has been slow and our knowledge of mathematical learning processes meagre. Will time and an agreed research programme enable us to answer, with confidence, the question, "why has the baker boy such a small residue of mathematical knowledge?"

Selected References

1. WALLACE, J. G., *Concept Growth and the Education of the Child*, National Foundation for Educational Research, 1965. The book gives a critical appraisal of papers dealing with the work of Piaget and Dienes amongst others.
2. OSGOOD, C. E., *Method and Theory in Experimental Psychology*, p. 673, New York and Oxford University Press, 1959.
3. HILGARD, E. R., *Theories of Learning*, Methuen, 1958.
4. HARLOW, H. F., The formation of learning sets, *Psychol. Rev.* **56**, 51–65 (1959).
5. WERTHEIMER, M., *Productive Thinking*, New York, Harper, 1945.
6. POLYA, G., *Mathematical Discovery*, Vols. I and II, Wiley & Sons, 1962.
7. BUGELSKI, B. R., *The Psychology of Learning*, New York, Holt, 1956.
8. TOLMAN, E. C., The determiners of behaviour at a choice point, *Psychol. Rev.* **45**, 1–41 (1938).
9. THORNDIKE, E. L., *The Psychology of Arithmetic*, New York, Macmillan, 1922.
10. THORNDIKE, E. L., *The Psychology of Learning* (Educational Psychology, II), pp. 217–26, New York: Teachers College, 1913.

11. BIGGS, J. B., *Mathematics and the Conditions of Learning*, pp. 10–51, National Foundation for Educational Research, 1967.
12. THORNDIKE, E. L., *The Psychology of Learning* (Educational psychology, II), p. 29, New York: Teachers College, 1913.
13. DIENES, Z. P., *An Experimental Study of Mathematics-Learning*, Hutchinson, 1963.
14. BARTLETT, Sir FREDERICK, *Thinking*, Allen & Unwin, London, 1958.
15. ANDREE, R. V., *Selections from Modern Abstract Algebra*, p. 15, Holt, Rinehart & Winston, Inc., 1962.
16. RENWICK, E. M., *Children learning Mathematics, Classroom Studies*, Stockwell, 1963.
17. BIGGS, J. B., *Anxiety, Motivation and Primary School Mathematics*, Occasional Publication No. 7. London: National Foundation for Educational Research, 1962.
18. HOLT, J., *How Children Fail*, Pitman Publishing Corporation, 1964.
19. POLYA, G., *Mathematical Discovery*, Vol. I, preface v, Wiley & Sons, 1962.
20. As above, pp. 65–68.
21. As above, pp. 60–62.
22. WHEELER, D., *Mathematics Teaching*, Autumn 1968, No. 44, p. 40.
23. FLEMING, C. M., Retrospect, *University of London Institute of Education Bulletin*, Summer 1968, pp. 30–33.
24. BRUNER, J. S., *The Process of Education*, Harvard University Press, 1960.
25. PETERS, R. S. (Ed.), *Brett's History of Psychology*, p. 509, Unwin, 1953.
26. ISAACS, N., *New Approaches to Mathematics Teaching*, ed. F. W. LAND. Contribution—Mathematics: The Problem Subject, pp. 20–27.
27. PIAGET, J., *How Children Form Mathematical Concepts*, reprinted from *Scientific American*, November 1953. W. H. Freeman & Co.
28. SIGEL, I. W., *The Attainment of Concepts*. In M. S. HOFFMAN and L. W. HOFFMAN (Eds.), *Review of Child Development Research*, pp. 209–48, New York: Russell Sage Foundation, 1964.
29. BERLYNE, D. E., Recent developments in Piaget's work, *British Journal of Educational Psychology*, **27**, 1–12 (1957).
30. WRIGLEY, J., *New Approaches to Mathematics Teaching*, ed. F. W. LAND. Contribution—Some Programmes for Research, pp. 30–38.

The Growth of Logical Thinking

RUTH M. BEARD

CHANGES in the teaching of mathematics since the middle 1950s have arisen not only from a consideration of the relevance of subject matter to the modern world and to current growth points in the subject, but also from an increasing knowledge as to how children acquire skill in mathematical thinking. There can be no doubt that the psychologist who has had the greatest influence on our understanding of children's thinking is Piaget. Although he did not set out to improve teaching as some innovators have done, his impact on teaching methods in Britain's primary schools has been considerable. Perhaps this is partly because the earlier work of Susan Isaacs and others who promoted activity methods for younger children in the 1920s and 1930s had created a climate in which his views of development were particularly acceptable. Advocates of discovery methods were quick to see the relevance to their views of his findings that children's thinking was closely related to their activities, at least up to 12 years or so, whilst those who were beginning to use structural apparatus could point to the increasing importance of structuring activities to aid children's thinking, particularly during the years from 1 to 12. Interest among teachers of adolescents has been less widespread; but perhaps this was because Piaget's books are difficult to read and the earlier popular texts outlining his findings were restricted to accounts of concept formation among children between 5 and 11 years.

Piaget's main contribution has been to provide a model of qualitative changes in thinking during development from birth to adolescence. But in the course of his extensive investigations, he has also initiated new lines of inquiry into the growth of thinking in infancy, the role of language, formation of concepts, and modes of thinking which first appear in adolescence.

Processes in the Development of Thinking

Like a number of outstanding innovators, Piaget used insights gained in other fields of inquiry. At the age of 16, when studying the behaviour of molluscs, he defined processes by which they became adapted to their environment. In common with many earlier psychologists who experimented with animals, he assumed that the processes he observed would apply equally throughout the animal world but, unlike many of them, he devoted his subsequent investigations to the study of the complexities of human thinking.

In observing the development of his own three children in infancy, Piaget concluded that, at birth, children were endowed only with a number of reflexes and an innate tendency to organise their actions; thus they had no ready-made mental abilities but only a manner of responding to their environment. Very soon, however, habitual actions developed, such as a tendency to seek with the mouth anything which made contact with the lips or to grasp an object which touched the palm of the hand. Within a few weeks the infants turned towards the light or towards the origin of a sound and developed distinctive habits such as clasping the hands. Piaget termed these well-defined sequences of actions *schemas.* Their chief characteristic, whatever their nature or complexity, is that they are organised wholes which are frequently repeated and can be easily recognised among other diverse and varying behaviours.

When an organism uses already established schemas on new objects or in new situations Piaget speaks of a process of *assimilation*; if familiar schemas will not suffice to solve a problem so that new schemas have to be developed by combining or extending those already learned, Piaget calls the process *accommodation.* It is through the interplay of these two processes that an organism becomes *adapted* to its environment. Whereas in a simple organism such as the mollusc this is usually a once-for-all experience, in the case of human beings, maturation of new capacities, development of abilities and changes of environment—or within one environment—oblige each individual to make many adaptations throughout a lifetime.

The process of *assimilation* may have a distorting effect as *representative schemas* develop throgh the use of words and symbols. Children, and sometimes adults, attempt to fit new experiences into schemas which relate to a representation of their own personal and limited worlds. Thus children often attribute life and feeling to objects, or suppose that natural phenomena such as lakes and mountains are man-made. Much misunderstanding which occurs in learning is of this kind, as children try to force new experiences into their existing repertoire of schemas, or, from lack of understanding, develop faulty schemas. Thus a 7-year-old will attempt to take 82 from 17 giving the answer as 35, because his schema consists simply in application of a mechanical process. Or, at 10, on meeting a quadratic equation for the first time, he may attempt to apply the schema that he knows for linear equations, unless he appreciates that there is a problem and seeks advice.

The complementary process of *accommodation* is seen in active exploration, questioning, trial-and-error experiments and, increasingly among older children, in reflection. Combinations of schemas are tried out or new schemas are arrived at from a process of physical or mental experiment and, of course, through experiences and information provided by other children and adults.

Parallel with the process of *adaptation* which results from the interaction of *assimilation* and *accommodation* is the process of *internalisation.* Whereas an infant's world appears to be primarily one of actions and transient perceptions, an older child learns increasingly to represent the world mentally by means of memories, imagery, language and symbols until, in adolescence, thought may proceed entirely in symbolic terms without recourse to action.

Periods of Development

A brief outline of qualitative differences in behaviour observed by Piaget and his collaborators among children and adolescents provides a basis for discussing methods in teaching mathematics.

They define three main periods of development, of which the first, the *sensorimotor* period, corresponds roughly with the first 18 months of life; this is the time when schemas of action are developed and the first notions of space, time and physical causality are achieved.[26] In addition, Piaget has shown by a number of experiments that infants gradually come to realise that objects have permanence. At 3 months, if an object which a child is holding is removed from view he ceases to take any interest; at 6 months he may look in the direction in which it was moved or feel for it where he previously found it. If at this age he sees part of it hidden he may retrieve it, but this will depend on which part he sees. If he views the most familiar part, such as the teat of his bottle, he reaches for the bottle; but if the other end is presented he may fail to recognise it. By 1 year he will retrieve a toy which is hidden behind a cushion while he looks on; but he seeks the toy behind the same cushion when it is hidden behind a different one in his full view; for he is still influenced more by past successful actions than by visual cues.

A second period of development begins in the second year when a young child realises that one object or action can be used to represent another, and that words have meaning. This is the period of preparation for, and achievement of, *concrete operations*. It extends (again, very roughly) from 18 months to 12 years, the sub-period of preparation lasting approximately to 7 years. In the earlier part of this sub-period young children learn rapidly that they can pretend to eat or sleep, they recognise words and then use them appropriately, imitate the actions of adults, other children or animals and represent objects with their toys. Within a few more years they add drawing, games of pretence, reading and writing to these accomplishments. The schemas of action acquired during the first period are gradually incorporated into *representational schemas*. But this cannot happen immediately. Although young children are capable of an action and can interpret other people's words it may be several years before they can coordinate actions in response to commands or provide their own directions to guide their actions. During the years up to about 5 or 6, children often accompany their play by oral description as though needing a verbal commentary to aid in the production of a moderately complex construction or drawing. Much of this speech is monologue, not seriously addressed to a listener, despite exhortations to "Look what I'm doing", etc.; pauses for response, if there are any, barely interrupt the solo performance. Piaget has called this egocentric speech(20) and sees its function partly in directing the child's own actions before interiorised verbal directions lead to an ability to think silently.

Despite increasing verbal fluency, Piaget finds no evidence that children form true concepts before about 4½ years. They do not make comparisons except in absolute terms such as "He's small, I'm big", nor do they see relationships or make categories. If asked to sort bricks or toys so that the same kinds go together they are apt to construct something which interests them, such as a train or a rocket. For this reason Piaget speaks of a *preconceptual* stage until about 4½ years.

But at about this age, or soon afterwards, a further change is observable in that children go half-way to forming concepts, or to understanding relationships. From their capacity to observe similarities and differences, they develop an ability to make single comparisons. Thus they will put a series of sticks in order by comparing pairs of them. It tends to be a tediously long process and may well involve errors which would be obvious at once to an older child from scanning the whole sequence. Children in this transition stage also fail to understand conservation of substance when shape is distorted—for example, when one of two equal balls of plasticine is rolled into a "sausage"—for they make comparisons of the lengths or breadths only and are incapable of seeing how the dimensions compensate each other. When invited to put together "the ones which are the same" among various coloured shapes they make a sequence in which successive pairs have a common attribute, such as red square, yellow square, yellow circle, blue circle, and so on. It does not occur to them to classify consistently by colour or shape and, still less, by both. It is this stage which Piaget terms the *intuitive* stage. As we shall see, thinking of this kind tends to persist until much later with unfamiliar materials, where more complex relationships are involved or, if relevant experience is lacking or the individual is retarded, it may continue throughout life.

The two preceding stages together constitute the sub-period of *preparation for concrete operations* in which children develop a repertoire of schemas which enable them to organise their experiences in such a way that, by the end of it, they can perform actions mentally. When this occurs, Piaget speaks of an *anticipatory schema*. Children who have achieved it will rapidly put a sequence of rods in order, evidently seeing "in the mind's eye" how to do it. Moreover, it enables them to coordinate actions without the distorting influence of direct perception and thus enables them to perform increasingly complex mental operations such as classifying objects in a diversity of ways. Children who think operationally can also explain reciprocal relationships such as friend, brother, adjacent, etc., inclusion in classes of higher order, such as birds and fishes among animals, or complex relationships such as those of the family tree which are reciprocal between siblings but asymmetrical between parent and children. It is probably no accident that the concept of "fairness" becomes so important at this stage, for the common relationship of children with parent or teacher and their reciprocal relationship with each other can now be appreciated.

Despite a variety of increasingly complex conceptual developments, children still show certain areas of limitation in their thinking, at least until adolescence is attained. The complexity of relationships they can handle, for example, does not extend to relations between relations such as are required in dealing with proportionality, probability or correlation. They tend to take a one-sided view of a situation or character, seeing its good or bad characteristics but not its contradictions or its full diversity. They are incapable of making assumptions in order to see what will follow in hypothetical cases and, in consequence, they have no systematic method of solving problems; rather, they say, "Let me see", and proceed by trial and error. Teachers of mathematics will recognise this limitation if they have asked a class to early to prove geometrical properties which they already "know" from experience; for example, that two sides of a triangle are equal when given that its base angles are equal. For, they say, "We know this, you don't have to prove it. Why should we pretend we don't know it when we do!" A few years later they make similar objections to proof by *reductio ad absurdum*. Pupils protest "You can't argue first that the corner of the quadrilateral is outside the circle, and then that it's inside; it must be one or the other." Thus their thinking is tied to the concrete. They are concerned with what they can see, handle, or do, and a large part of the world of imagination is still inaccessible to them. For this reason Piaget speaks of a stage of *concrete operations* to distinguish it from the adolescent stage in which possibility rather than reality becomes the main preoccupation.

Piaget found a change of this kind at about 12 years. In Britain a mental age of about 13 seems to be critical[17,19]. Assumptions are more readily made, explanations are sought, general laws are obtained and quoted, whilst common meanings like common principles are understood and explained. In addition, adolescents learn to discuss the intangible, the indefinitely large or small and purely imaginary worlds. They also begin to appreciate that different viewpoints arise from different experiences and thus that behaviour which is right for them may be wrong for someone else. Moreover, their characteristic form of argument is in propositions. They may say "If *X* were true, then *Y* (some unobserved consequence) would follow"; and they employ such propositions in thinking out scientific or social problems.

A single example showing development of one kind of mathematical concept throughout the entire range may be instructive: we will take the ideas of a *set* and of *intersections of sets*.

An Illustrative Example

An infant of some 8 or 9 months already makes collections of objects. These are "sets" in their most general sense. He does not classify them, of course, but he may have some notions that certain objects go together such as parts of a toy which takes to pieces. However, he does not normally make permanent sets or recognise particular collections as such. Before he speaks, he may be able to point out the cars in a picture; yet, in doing so, he is simply recognising each object as a car and gives no indication that he conceives them as "cars" versus "houses" or "people". He may also collect together bricks of a similar kind to make a transitory construction, and in doing so he shows ability to recognise similarities and differences between shapes; but he does not normally act consistently in putting together sets of objects. At a mental age of about 4 years there is a noticeable change in his behaviour because he has learned to direct his actions verbally. He may now say "a square", "another square", etc.; but it is highly improbable that he will say "I'll take all the squares" as an older child would do. He still recognises and takes each object individually. He may also select the same bricks on different occasions to build some construction which appeals to him, showing a capacity which with a little prompting will enable him to build up consistent collections or categories. At 3 or 4, children readily identify boys and girls, babies and so on. At 5 or 6, average children will find all the bricks of a kind or may even learn to build simple graphs by adding sticky named coloured squares to make blocks corresponding, perhaps, with children whose birthdays are in the different months of the year. In this way they can visually represent sets which they are able to name, e.g. "children with birthdays in January".

Nevertheless, their thinking is still substantially "pre-operational". They make most classifications step by step, failing to see them as a whole and have difficulty in relating "sets" with their "subsets", or in understanding what are their intersections. Experiments made by Piaget illustrate this difficulty. When showing children pictures of ducks (which children agreed were birds), other birds and animals (the children agreed that all birds were animals) and asking "If all the birds are killed, will there be any ducks left?", many children under 9 believed that there would be. Similar questions relating to flowers, and experiments requiring children to classify objects, indicated that up to this age, many of them treated subsets as though they enjoyed a separate existence[27]. The difficulty of Genevan children in admitting that they were in Switzerland as well as in Geneva, mentioned in an earlier book[21], shows the same partial understanding. Only three-fifths of the 9-year-olds questioned by Piaget had achieved this. However, among Londoners aged 6 and 7 some 25 years later, half a sample of children I tested already understood this kind of inclusion. The same children had no more difficulty in putting a farthing inside both both of two overlapping circles[2]. Of those who failed, some protested that there would have to be two farthings, others said that it would be impossible, while others unhappily put the farthing on the perimeter of one circle although they knew that this was not "inside". When, prompted by their headmistress, boxes and a marble were substituted, their difficulties became greater. They realised that the marble could be put inside the smaller box which, in turn, could be enclosed in the larger one; but, to them, the marble was inside the smaller box only. The solution of a number of children who managed to insert the small box—with its lid open—inside the larger one and who sat back beaming with satisfaction, since the boxes now shared an inside, clearly illustrated the limitation in their comprehension.

It is of interest that where "sets" are taught, teachers commonly report that thinking seems to become operational in this respect earlier than formerly. However, we need to know to what extent this advance is specific to the situations in which the relationships were taught. In dealing with a line, the conception of a line as consisting of a set of points may itself present difficulties. Piaget found that initially children accepted lines as straight only if they were parallel with the edge of the paper; later if they could be drawn with a ruler at any angle across the page, and last of all if they consisted of a sequence of dots[28]. Perhaps this concept too may be achieved earlier as a result of the graphical work in junior schools. Nevertheless, it is not until about 11 years that most children are able to conceive of a line as an infinite set of points, although they may be able to think of the intersections of sets of points whether the notion of an infinite number is attained or not. And they need to be about 11 years old before they conceive of the equation of the line as embodying a general law which is satisfied by a particular set of points. At an even higher level of abstraction, and of mathematical sophistication, which depends largely on teaching, children of some 15 years or more may appreciate the construction of a class, or family, of curves $S + \lambda S' = 0$, where λ varies, which share the intersections of the curves $S = 0$ and $S' = 0$.

The higher levels of understanding appear to be dependent increasingly on capacity for verbal argument and symbolic thinking. Although Piaget's findings relating to activity are being increasingly well applied among junior school children, it is at least questionable whether the links between activity and its verbal and symbolic expression are adequately stressed in the upper stages of the junior school and in secondary schools. An outline of Piaget's findings and those of other investigators concerning the function of language, or of other representational modes in thinking, may serve to indicate areas of comparative neglect.

The Function of Language in Thinking

Although language is only one means of representing the world it is essentially an important one, for it is the chief means of communicating with other people and of clarifying thinking. Nevertheless, that the ability to represent one thing by another is primary, and that language is simply one expression of it, seems to be well established; but there is still some doubt whether, as development proceeds, each successive mental structure must develop before the relevant language is employed or whether language itself can initiate advances to later structures. Piaget's view is that mental structures are the source of further developments; he sees language as one means of gradually internalising experience to the point where actions can proceed in imagination without recourse to their physical repetition.

Experiments by Luria in Russia show, on the one hand, how slowly the directive function of language develops and, on the other, suggest that training in the use of language may facilitate a variety of modes of thinking including some which are normally considered to be "non-verbal". American authors, however, incline rather to sharing Piaget's view of the primacy of structure; but they find a very close relationship between the kind of vocabulary at a child's command and the concepts he achieves.

The slow development in infants of ability to direct their actions by verbal instructions was demonstrated by Luria in a recent sequence of experiments(15). In the first experiment infants over 1 year would obey a simple command unless conflicting stimuli misdirected their attention. By the end of the second year they followed visual stimuli, retrieving an object they had seen hidden under a cup, for instance, but failing to follow a verbal instruction to take the object from beneath the cup unless they had seen it concealed there. Evidently, it took longer to master speech which preceded an action and organised it than to employ a visual organisation. Experiments with children throughout the third and part of the fourth years showed that ability to act, or to inhibit action in different circumstances, existed only if distinct positive commands were given in each case, and that a child would expend so much energy in giving the commands himself that he had none available to perform the actions at all. By $3\frac{1}{2}$ years most children could themselves give and obey a positive command, but they still tended to repeat the same action when giving a contrary command. It was not until $4\frac{1}{2}$ years that children could both give commands and follow them consistently. At this point Luria observed that external speech became superfluous, for the directive role was taken over by inner connections.

In observing children at play, however, one finds that the monologue which Piaget refers to in *The Language and Thought of the Child*(20), often accompanies their actions until 5 or 6 years. A large number of factors may be operating; Piaget commented on two 6-year-olds whose speech was recorded over a period of time, that the one who used more egocentric speech was less emotionally stable but also had an exuberant imagination. In contrast, in an unpublished study at Birmingham University, among six nursery school children whose speech was recorded for 6 hours, an intelligent 3-year-old already addressed most of her speech to adults or other children with a view to obtaining answers to questions, giving directions, co-operating in play, and so on; whilst children from a poor area talked less but largely in a manner that was socially oriented although they were all between 2 and 4 years. It seems probable that personality, cultural background, the nature of an activity, age and familiarity of companions may all influence the ways in which speech is employed. Where monologue is used, one of its functions may be to give expression to internal fantasies and ideas, thus serving to create a better relationship between the internal and external worlds.

A second experiment undertaken by Luria and Yudovich[16] was made with identical twins who, on arrival at nursery school at 5 years, were severely retarded in language although fairly normal in other respects. After a few weeks, the twin who was judged slightly less able was subjected to a period of training in language. At the end of 10 months, when both twins were tested again, the trained twin was more successful not only in talking but in a variety of verbal and non-verbal skills. He could build more complex models, had a greater capacity to play a role or to employ objects in doing so, and was more successful in classifying objects; in describing a picture, he told a connected story whereas his twin simply named individual objects. Some caution must be exercised in drawing conclusions from this experiment for the training procedure is not described in the English version and, in any case, only one pair of abnormal children was involved. Nevertheless, one may say, with reservations, that it suggests that advances in language can facilitate development towards higher levels of thinking.

A contrary finding is suggested in an American study with children. Sinclair[33] found, in the first place, that vocabulary differed between those who were already able to conserve and those who could not and, in the second place, that training in vocabulary made little difference to children's ability in this respect. Of children who already conserved, 70 per cent used comparative terms such as "The boy has more than the girl", whereas of children who did not, 90 per cent used absolute terms, e.g. "The boy has a lot, the girl has a little". An interesting point was that 20 per cent already used comparatives for discrete units (marbles) whereas they did not do so for continuous quantities (plasticine). Of the children with conservation, all used different terms for different dimensions, using two distinct pairs of opposites, e.g. big–little and fat–thin, whereas three-quarters of those who did not conserve used a single word for different dimensions, such as "fat" for long or thick. Of those with conservation, 80 per cent were capable of describing objects differing in dimensions while coordinating the dimensions, e.g. "This pencil is long(er) but thin(ner), the other is short(er) and fat(ter)". Whereas 90 per cent of those without conservation described only one dimension or used four separate sentences, e.g. "This pencil is long, that one is short; this one is thin, that one is fat".

In a second series of experiments in which children were "taught" the appropriate vocabulary of conservation, it proved easy to teach them to use differentiated terms, more difficult (about a quarter did not succeed) to teach them to use the comparatives "more" or "less" in these situations, and still more difficult to teach a coordinated structure such as "long and thin", "short and fat". Despite their advance in vocabulary, only 10 per cent of the sample acquired conservation in comparing contents of water in dissimilar glasses, although about half of them now observed and commented on each dimension individually. But, as we shall see, this concept is attained later by English children than are many other conservation concepts.

The function of language in summarising experience and assisting in its recall is shown in another Russian experiment. Galparin and Talyzina[12] used the same method both with normal children of

6 and with backward adolescents of 15 and 16 in learning generalised concepts. The children were initially engaged in activities with concrete materials, then progressed to making audible descriptions and instructions—the concrete aids being withdrawn gradually—until, finally, the concepts were interiorised in verbal form. With this approach concepts in elementary geometry which had previously proved impossible to backward adolescents were formed almost faultlessly from the beginning. The physical responses needed to be fully developed initially and every step in forming the concept had to be worked through; but in learning subsequent related concepts some steps could be omitted and students could begin at the stage of thinking aloud or even at the purely mental level. Thus language is seen as a means of storing experience and, if this is sufficiently general in nature, of facilitating future problem solving.

Advancements in thinking due to recognition of categories expressed verbally has been discussed by Brunner *et al.*(8) They point out that one effect of categorising objects and events is to reduce the complexity of the environment; secondly, it enables us to say "this is an '*X*' again" or "there is another '*Y*'", thus reducing the necessity for constant learning—once we know what a square is, we look for its defining characteristics in identifying another quadrilateral as a square. But, ability to categorise provides a means of ordering and relating entire classes of events or of inventing categories with attributes previously considered incompatible or impossible—the neutrino, for example, was postulated before it was "found".

This takes us out of the concrete world of children into the adolescent and adult realm of possibilities. Mastery of activities alone will not achieve this. It is essential to develop language or other symbolic representation and it is, perhaps, in making this transition that teachers are less successful. In many primary schools activities are performed but concepts are not fully explored nor are general laws obtained. These children "complete the square" using a square wooden block of side x, strips of unit width and length x and unit squares. The expression $x^2 + 8x + ?$ is easily represented and the square completed by putting half the eight strips on adjacent sides of the square and filling in the "hole" with unit squares. Children will complete a dozen examples in very few more minutes. But they may never be asked to say what they do and are therefore perplexed by a request to complete the square in the case of $x^2 + 3x + ?$. They have acquired a method of acting but not a formula for action; and it is the latter they need to develop to attain the level of formal operations. In secondary schools they may find themselves already expected to deal with abstractions without concrete aids, and, having built no links with past experiences, they revert to rote learning or become lost because, as they say, they suddenly do not understand. But it is the teacher's task to find out whether the laws have been verbalised, or expressed and understood in formulae, and, if not, to see that appropriate links are forged between concrete experience and the corresponding abstractions.

The Formation of Concepts

In addition to drawing attention to the part played by language in development of thinking, Piaget has developed new approaches to the study of concept formation. Where other psychologists concentrated on an individual's ability to discriminate between different entities, he began by drawing attention to ability to conceive transformations of an object and thus to distinguish between real and apparent changes. In following up his work, Wallach(34) suggests that it is in comprehending that the same operation will result in the same changes in equal quantities, that the idea of conservation is born; she therefore dismisses social or verbal learning, absence of addition or subtraction, ability to deal with two dimensions simultaneously or reversibility of action as mainsprings of the ability to conserve in children. However, this is still a matter for discussion and experiment. Whatever is the main cause of the development of the notion of conservation, the ability to conceive transformations undoubtedly plays a part; but this seems to depend to a greater degree on a child's experiences than Piaget's views on development of mental structures might lead us to believe. For example, understanding conservation of substance proves easier when a biscuit is divided than when a ball of plasticine is deformed or broken up. In a study among English children, it was with more familiar substances that the majority of children first formed the concept, although experiments with liquids proved more difficult. No less than 40 per cent of 10-year-olds supposed that if one of two equal quantities of liquid was poured into a number of smaller glasses the quantity would change[(3)]. In this case, since water is familiar, the cause of difficulty must be sought elsewhere. Certainly many parents and teachers seem reluctant for children to play with water but, in addition, changes in quantity are particularly difficult to follow in a fluid—and we recall that Sinclair found more children who understood conservation of substance with discrete quantities (and used appropriate vocabulary) than with a continuous one (plasticine) where the difficulty of demonstrating the reversal of a transformation is somewhat greater. Nevertheless, in arriving at conservation of number, individual children do not necessarily follow the logical order of difficulty. If the correspondence is very obvious as in matching eggs to cups or flowers to vases, the majority of infants believe that the numbers of each remain the same despite rearrangement of the eggs into a heap or the flowers into a bunch. If the initial correspondence is not predetermined, e.g. in making a line of counters match one already laid out, this task itself presents problems and many of the younger children put too many or too few counters. If they succeed in the matching task, they may nevertheless believe, when one set of counters is rearranged into a shorter line, that the numbers become unequal. In the most difficult case, where the correspondence disappears at once, i.e. where a child and experimenter simultaneously put beads into a flat dish and a glass tube, matching the beads in pairs, few infants are convinced that the numbers are equal in the end. Most assert that there are more in the tube "because it is tall". But it is of interest that if these three experiments are performed with a large number of children, there are individuals who realise that there will be conservation in any one or any two of these cases[3]. Thus it cannot simply be a matter of understanding transformations of quantities. Other determining factors must be operating such as familiarity with the materials, perhaps through play, in these instances; and the possibility of making and maintaining a one-to-one correspondence.

Appreciation of conservation of area comes earlier on the average than that of liquids. When children are asked to cut one of two equal squares into halves by a line parallel with a side and to rearrange these halves into a long oblong, most children of 8 realise that the amount of paper in the unaltered square and the oblong is the same. Perhaps the ease with which this transformation can be reversed and children's experience, in an industrial society, with scissors and paper, contribute to its early achievement. Indeed, in experiments with 6- and 7-year-olds, I found some children who, in comparing areas, already claimed to "imagine little squares all over them"[2]. Probably they had achieved some idea of measurement from experiences in playing with bricks and squared paper.

Where an additional complexity is added, involving co-ordination of relationships, conservation may not be achieved until several years later. For example, if three equal squares are presented and two are halved, one diagonally and one by a line parallel with an edge, some 40 per cent of children fail to realise that two halves of dissimilar appearance combine to form an area equal to that of the unaltered square. They quote "two halves make one whole", and explain that either a triangular or an oblong shape is half a square, but insist that a house (an oblong with a triangle above it) contains more, or less, paper than the square[5].

Perhaps the difficulty in comprehending conservation of volume, using displacement of water, is due to a similar complexity. Piaget found that when a rectangular block was stood on end, or on one broad face, beneath the surface of a trough of water, children asserted that the block contained the same amount of substance, or the same "bricks", but supposed that it displaced more or less water according to its position. Most children failed to co-ordinate these conceptions until about 10 years. In an investigation undertaken by students in a large number of schools[3] only 37·5 per cent of children aged between about 9 and 10 years had attained conservation in a similar experiment despite its achievement by 26 per cent of 6-year-olds.

An appreciation of conservation is, of course, necessary to understanding of measurement. But, with different content, development in both respects overlaps. Many children solve problems involving measurement of length and area before they appreciate that volume is conserved and, in understanding of measurement as with that of conservation, there is a long period of growth. Thus, young children asked to build a tower of equal height to one already constructed look only to see whether their tops are level, ignoring the levels of their bases[28]. At a somewhat later stage, children make rough measurements of height by placing their hands at the two extremities of the tower to be copied. Only those who have already attained operational thinking employ a unit of measurement. And whereas children appreciate at about 6 years that a distance remains the same in whichever direction it is measured, the notion of measurement with respect to two axes, like most notions requiring co-ordination of two relationships, does not usually occur until about 10 years[5]. It will be interesting to see, therefore, to what extent graphical and other practical work in primary schools advances these kinds of development.

Ability to apply measurement intelligently implies, of course, that some elementary spatial concepts have been attained, for example, a concept of order. Most children can make an identical copy of beads on a wire by 5 years, but to reverse the order or to copy it from a circular wire to a straight one continues to present difficulties until about 6 or 7 years[31].

Predictably, co-ordination of relationships shows a similar sequence of development in acquisition of spatial conceptions as in that of measurement or conservation. Children begin to interpret right and left from another point of view at about 7 years, but co-ordinating right and left with before and behind in reversing a map is not achieved until 9 or 10 years and ability to draw someone else's view of a tilted rod or circle is rare even at 10 years[6,31]. The majority of children draw their own view or a mixture of views which they fail to co-ordinate.

There is some evidence that spatial concepts may be peculiarly subject to influences from everyday experience. For example, in comparing Ghanaian and English children I found, on the whole, growing deficiencies of the former between 8 and 10 years. But Ghanaian children usually lack constructional toys, travel very little, live in one-storey houses, have no thread or bricks with which to make patterns nor waste materials to use in model making, and they do not make good these deficiencies in school. Where their experience equalled that of English children, for example, in recognising the right and left hand of a boy in a picture, their performance was equal[6]. A further finding of interest in this study was an observed change in English children as compared with a sample tested 12 years before. In 1953 I found among a rather advanced sample of 6- and 7-year-olds that no child with a mental age of less than 9 succeeded in drawing a bird's eye view. By 1965, however, practically all children of 8 had attained this concept[6] and teachers reported that even nursery school children of 4 years drew such views spontaneously. Yet there has been no remarkable gain in other spatial concepts so far as I know. Teachers suggest that seeing aerial views on television or using simple constructional toys which are accompanied by instructions "in plan" possibly account for this development.

In contradiction to these findings in England, some Americans claim that current educational programmes designed to advance concept formation are proving ineffective. There is, of course, the possibility that school programmes are less effective in influencing a child than is his total environment. Young children spend at most 25 hours per week in school—hardly more than one day—and occupy very little of this in activities or discussion relating to any particular kinds of concepts. Some may spend many more hours in viewing television. Thus it does not seem unreasonable if they learn more from everyday experiences. Moreover, the suitability of school programmes to the interests of pupils can be critical. Churchill[10] found that introduction of number experience in stories and play was effective with 5-year-olds in developing number concepts. But an unpublished investigation at Birmingham shows that 6-year-olds are not influenced in this way although it confirms that 5-year-olds are. It is true that Piaget's investigations assist teachers in finding out at what level each pupil should begin to learn and in devising suitable teaching materials; but many of the answers remain to be found either through experiment or by trial and error in the class-room.

Adolescent Thinking

It is not until children have built an orderly representation of their own world that they are capable of realising the extent to which other views differ from theirs. In Piaget's opinion it is this realisation, which arises through co-operation and discussion with their contemporaries, that initiates the developments of adolescent thinking. Growing capacity to view a situation from different angles leads to an ability to make assumptions, and to a changed emphasis in thinking from the limited reality or extravagant fantasies of childhood to a diversity of possibilities or to consistent imaginary worlds. It leads also to an ability to deal with greater complexities in co-ordinating relationships and to increasing ability to seek and to formulate general laws.

When, in teaching mathematics, we ask such questions of children and adolescents as: "If I have more than 6*d*. I buy a cake or a peach. If I am hungry I buy two buns or a cake. Today I have a shilling and I am hungry. What shall I buy?" we find that children do not accept the speaker's premises. They make judgements from their own experience, saying, perhaps, "If I was hungry I'd buy two buns" or "I'd buy the peach; I like peaches". Thus they do not make assumptions for the sake of argument for they are too involved in their own world. Adolescents, on the other hand, normally solve this problem easily enough for they are more able to make judgements independent of personal biases. It will be appreciated that this is not the only complexity to the problem. A co-ordination between two different conditions is also required and we have already seen that concepts which require co-ordination of relations tend to be achieved later.

Piaget has identified a group of problems which involve co-ordination of relations and which, in logical terms, he identifies as belonging to "the group of four". Problems of this kind are not mastered until well into adolescence. An example which seems deceptively simple was devised by Mealings[17]. He asked pupils of a comprehensive school to use a balance to sort red and blue blocks into "light" and "heavy" ones, the light and heavy blocks being of roughly the same weights but distinctly different from each other. Some subjects had a clear and concise procedure from the beginning and immediately used one block as a standard; these had mental ages over 16 years. Others proceeding more by trial and error arrived at a satisfactory solution. These had mental ages over 13 years. Those with mental ages less than this made aimless comparisons, failing to use the balance at all or deciding that blocks of one colour were heavier. When the three red and two blue blocks which were "heavy" had been separated from the "light" ones the children were asked "Which are there more of—red blocks or light blocks?" Those with mental ages over 16 years 3 months answered correctly; twelve children with mental ages less than 13 years 8 months were unable to comprehend the question; while between these mental ages children restricted their replies to a single comparison, for example, "There are three reds in the heavy pile but only one in the light" or "There are more heavy blocks than there are light".

Correlation, as tested by Inhelder and Piaget, presents similar difficulties[14]. He used cards with pictures of children belonging to four categories: fair hair with blue eyes, fair hair with brown eyes, brown hair with blue eyes and brown hair with brown eyes. Lyn (12·4) was shown a set of cards with (6, 0, 0, 6) in each category respectively and was asked "Can you find a relationship between hair and eye colour in the cards?"—"Yes, these (the dark ones) have the same colour eyes as hair."—"No—here?"—"Here it's only brown."—"And here?"—"It's blue, they are all blue." "And here (6, 2, 2, 4), is there a relationship?"—"No. Yes, the four (sub-sets) separately, but not when they are together."—"Why?"—"Because some are yellow (blond) and blue and some are yellow and brown." Thus her comparisons always fall within one sub-category: the blonds. Vic (14·6) at a somewhat later stage, given sets (4, 2, 2, 4) and (3, 3, 1, 3), on each occasion compares the blue- and brown-eyed first among the blonds and then among the brunettes separately. Asked "But for the whole set?" he replies "There are 4/12 outside the law and 8/12 covered by it. It would be the same for the whole set." Cog, however, (15·2) can state a law: "Most of the people with brown eyes have brown hair and most of those with blue eyes have blond hair."

The operations involved in understanding proportionality require a similar balancing of relationships. In an experiment with shadows cast by a number of circular rings of different sizes by a point of light into a screen, Inhelder and Piaget[14] asked children to arrange a number of rings of different sizes so that their shadows coincided. Children under 7 failed because they did not realise how a shadow was formed; up to 11 they proceeded by trial and error; by 11 and 12 years some argument was used while, at 15 or more, children succeeded in working out a law. Typical arguments serve to show that "four operations" are involved. A child may say "If the shadow of the second ring is too big I can reduce it by moving the ring further from the light; if it is too small I can enlarge it by moving the ring nearer the light; alternatively, if the shadow is too big, I can replace the ring by a smaller one or, if it is too small, I can replace the ring by a larger one." This "group of four operations" occurs in all problems of proportionality and in a number of related problems. Characteristically, those who reach the stage of formal operations obtain a law, e.g. Huc (15·6) says "For the light to make twice the size (of shadow) it takes twice the distance".

Laws are likewise obtained in other situations. In classifying floating and sinking objects Jim (12·8) says "If you take metal, you need much more wood to make the same weight than metal". Or in attempting to solve probability problems when coloured balls are drawn from a bag Laur (12·3) says "It's always two green ones which have more chance". In every case, then, an assumption is made as to a possible action and a consequence is mentioned; thus each assumption serves as a hypothesis which can be tested by an experiment. This is an approach which is needed for systematic problem solving. An adolescent will determine and test every possibility. A child may hit on a solution by trial and error; the existence of a number of equivalent solutions does not occur to him. Although he solves problems, such as determining which of two glasses holds more, his procedure—e.g. filling one and pouring its contents into the other—depends on conceptual knowledge, i.e. that the quantity of liquid is not changed, and does not involve him in setting up hypothetical conditions.

Since adolescents see more possibilities, the definitions they give spontaneously, like their experimental procedures, are also richer; and, as they learn to understand relationships between various properties of matter or objects, their definitions gain greater conciseness since they can recognise and mention essentials. The definition "A square is a parallelogram with one right angle" which seems so misleading and even foolish to children becomes meaningful to him.

In addition, adolescents' growing capacity to understand continuity, "infinite" numbers and imaginary systems adds to their repertoire of mathematical skills. Thus Bet (11·7) who has already reached the level of formal operations in this respect says in answer to "How many points could I put on this line?" "One cannot say, they are innumerable. One could always make some smaller points" and, asked to make the shortest line it is possible to make, replies "No, one cannot, because it is always possible to make a still shorter one". Duc (12·9) sees at once that a line of points will be a continuous line if there is an indefinitely large number of them[28]. And, although adolescent pupils usually share the surprise of mathematicians who first postulated "imaginary" numbers, they are intrigued by the notion of "the square root of minus one" and enjoy using it.

Finally, the capacity to see other people's views makes each individual more conscious and critical of his own judgement. Perhaps this is a process which is never wholly completed. Abercrombie[1] found with first-year university students that although they were well grounded in the concepts of biology, physics and chemistry, they were often unable to use their information to solve slightly unfamiliar problems or to defend a view in argument, and they tended to observe what the textbook said should be there rather than what was actually on a slide or X-ray. But in an experiment following free discussion of their own observations, definitions, evaluation of evidence, or views on causation, they became more aware of unjustified assumptions they made, and of bias in handling evidence or describing an exhibit. As a result, students who took part in this experiment learned better than those who attended the normal course of lectures and demonstrations to discriminate between facts and conclusions, to avoid false inferences, and to consider more than one solution; and they were less adversely influenced in their approach to a problem by their experience of the preceding one.

Conclusions

This outline of Piaget's studies of the development of children's thinking and those of some other investigators which, in the main, support his findings, should suffice to persuade any subject-centred teacher to consider the level of thinking at which his pupils operate. A brilliantly well-organised lesson is useless to children who have not acquired the mental structures which will enable them to assimilate the material without seriously distorting it. Moreover, since children of the same age differ considerably in their level of understanding, there is a need for more individual work, or work in small groups, especially in the secondary schools (in English primary schools some work based on individual or group activities is now almost universal). Although thinking becomes operational "at about 7 years" it is important to realise that at a chronological age of 7 we may expect one-sixth of children to have mental ages less than

5 years 10 months, one-fortieth having mental ages lower even than 4 years 8 months, whilst a further sixth have mental ages above 8 years 2 months and one-fortieth exceed 9 years 4 months. And at about 13 years when formal operations are developing, the range in mental ages is almost twice as great. When to these considerable variations we add the well-known differences in special abilities—numerical speed and accuracy, spatial imagination or verbal fluency, for example—in addition to differences in interest between the sexes and effects of different social environments, the case for collective class teaching seems to be almost entirely destroyed. The main argument remaining in its favour for mathematics teaching is its convenience for secondary school teachers who have not been trained to deal simultaneously with pupils engaged in a variety of activities.

A difference which remains to be resolved by further experiment is that between teachers and psychologists who incline to a maturational view of development, believing that experience in the classroom plays a minimal part in advancing "stages" in thinking, and "environmentalists" who consider that such experience is a determining factor. Piaget's view that the development of mental structures is primary and cannot be hastened by verbal training suggests that he inclines to the former view; but his idea that a child organises his experience in adapting to his environment suggests that the sum total of experience affects this organisation.

In so far as school activities are seen to be relevant to a child's attempts to adapt to his environment they are therefore likely to reinforce his efforts. In either case, since we cannot yet claim that mathematical teaching is ideally suited to children's needs, there is room for improvement and the enrichment of mathematical experience at every level. Those who take a maturational view will usually agree that stages in thinking can be delayed, especially in a subject like mathematics where missed steps may be critical, or due to emotional inhibitions aroused by frequent inability to understand. They will therefore suit activities to a child's developmental stage. However, there seems to be a danger in holding this view, since teachers may wait for maturational changes to bring new abilities and fail to help children to build bridges to the next developmental stage. Classes of children who contentedly perform activities without comprehending the full range of mathematical ideas which these are intended to illustrate sometimes result from thinking of this kind.

Over-simplified activities and artificial "problems" which require that children should juggle with the language and symbols of modern mathematics have inspired one Oxford don to a felicitously expressed outburst entitled "On the enfeeblement of mathematical skills by 'Modern Mathematics' and by similar soft intellectual trash in schools and universities"[13]. He mentions an experiment in Oxford University in which students who attended problem classes where they were set very difficult problems in pure mathematics considerably excelled those who attended their usual tutorials. No doubt in attempting difficult problems students tried many unsuccessful paths but, in so doing, gained knowledge which they could turn to good use on later occasions. This underlines the stimulus to activity which difficulty can provide and, incidentally, shows the value of widely ranging activities in improving problem solving skills.

The "environmentalists" are more likely to provide such stimuli; and, although no one would wish to enter children in a race to achieve higher levels of thinking as rapidly as possible, it seems likely that the need to adapt to an increasingly complex environment will result in continued challenges and corresponding advances in children's mathematical thinking. The value of Piaget's work to teachers in meeting these demands is that it enables them to gauge the needs of their pupils more skilfully, and it provides an enormous range of ideas for activities which can be used to aid pupils' learning throughout their school days.

References

1. ABERCROMBIE, M. L. J., *The Anatomy of Judgment*, Hutchinson, London, 1960.
2. BEARD, R. M., "The nature and development of concepts, II", *Educational Review* **13**, 12–26 (1960).

3. BEARD, R. M., "The order of concept development: studies in two fields", *Educational Review* **15**, 105–17, 228–37 (1963).
4. BEARD, R. M., "Further studies in concept development", *Educational Review* **17**, 41–58 (1964).
5. BEARD, R. M., "Educational research and the learning of mathematics" in *Aspects of Education. II: A New Look in Mathematics Teaching*, edited by LAND, F. (*Journal of the Institute of Education*, University of Hull), 32–45 (1965).
6. BEARD, R. M., "An investigation into mathematical concepts among Ghanaian children", I, *Teacher Education*, May 1968; II, *ibid.*, Nov. 1968.
7. BEARD, R. M., *An Outline of Piaget's Developmental Psychology*, Students' Library of Education, Routledge & Kegan Paul, London, 1969.
8. BRUNER, J. S., GOODNOW, J. J. and AUSTIN, G. A., *A Study of Thinking*, Wiley, 1956.
9. CHARLESWORTH, W. R., "The role of surprise in cognitive development" in *Studies in Cognitive Development*, edited by ELKIND, D. and FLAVELL, J. H., Oxford, 1969.
10. CHURCHILL, E. M., "The number concepts of the young child", *Leeds University Researches and Studies* **17**, 34–49 and **18**, 28–46 (1958).
11. DIENES, Z. P., *Building up Mathematics*, Hutchinson, London, 1964.
12. GALPARIN, P. Y. and TALYZINA, N. F., "Formation of elementary pupils" in *Recent Soviet Psychology*, edited by O'CONNOR, N., Pergamon, New York, 1961.
13. HAMMERSLEY, J. M., "On the enfeeblement of mathematical skills by 'Modern Mathematics' and by similar soft intellectual trash in schools and universities", *Bulletin of the Institute of Mathematics and its Applications* **4** (4), 1–22 (Oct. 1968).
14. INHELDER, B. and PIAGET, J., *The Growth of Logical Thinking*, Routledge & Kegan Paul, London, 1959.
15. LURIA, A. R., "Development of the directive function of speech", *Word* **15**, 341–52 (1959).
16. LURIA, A. R. and YUDOVICH, F. I., *Speech and Development of Mental Processes in the Child*, 2nd impression, Staples Press, London, 1966.
17. MEALINGS, R. J., "Problem-solving in science teaching", *Educational Review* **15**, 194–207 (1963).
18. PARSONS, C., "Inhelder and Piaget's 'The Growth of Logical Thinking'. II: a Logician's Viewpoint", review article in *British Journal of Psychology* **51**, 75–84 (1960).
19. PEEL, E. A., *The Pupil's Thinking*, Oldbourne, London, 1960.
20. PIAGET, J., *The Language and Thought of the Child*, Routledge & Kegan Paul, London, 1926.
21. PIAGET, J., *Judgment and Reasoning in the Child*, Routledge & Kegan Paul, London, 1928.
22. PIAGET, J., *The Psychology of Intelligence*, Routledge & Kegan Paul, London, 1950.
23. PIAGET, J., *Play, Dreams and Imitation in Childhood*, Heinemann, London, 1951.
24. PIAGET, J., *The Origins of Intelligence in the Child*, Routledge & Kegan Paul, London, 1953.

25. PIAGET, J., *Logic in Psychology*, Manchester University Press, 1953.
26. PIAGET, J., *The Child's Construction of Reality*, Routledge & Kegan Paul, London, 1955.
27. PIAGET, J., *La Genèse des structures logiques elémentaires*, Delachaux & Niestlé, 1959.
28. PIAGET, J. *et al.*, *The Child's Conception of Geometry*, Routledge & Kegan Paul, London, 1960.
29. PIAGET, J. and INHELDER, B., *Le Développement des quantités chez l'enfant*, Delachaux & Niestlé, Paris, 1941.
30. PIAGET, J. and INHELDER, B., *La Genèse de l'idée de Hasard chez l'enfant*, Presses Universitaires de France, Paris, 1951.
31. PIAGET, J. and INHELDER, B., *The Child's Conception of Space*, Routledge & Kegan Paul, London, 1956.
32. PIAGET, J. and SZEMINSKA, A., *The Child's Conception of Number*, Routledge & Kegan Paul, London, 1952.
33. SINCLAIR-DE-ZWART, H., "Developmental psycholinguistics", in *Studies in Cognitive Development*, edited by ELKIND, D. and FLAVELL, J. H., Oxford, 1969.
34. WALLACH, L., "On the bases of Conservation", in *Studies in Cognitive Development*, edited by ELKIND, D. and FLAVELL, J. H., Oxford, 1969.

Schematic Learning

RICHARD R. SKEMP

I FIRST became interested in this problem as a general one in human learning, quite apart from mathematics. Afterwards I found it specially applicable to mathematics, which I have come to regard as a paradigm of that kind of learning in which man most differs from other species, and which least resembles those kinds of learning which can be studied in the laboratory rat. So I shall outline the general problem first, which is how one's existing knowledge affects new learning.

We know that in many fields there are topics which we cannot learn effectively unless we know something else already. For example, in medical studies, one cannot progress far in Physiology without a knowledge of biochemistry; and this depends in turn on remembering one's school chemistry. The problem is related to that of concept formation; but we are now concerned not just with single concepts, but with structures of interrelated concepts: which is to say, with schemas.

The medical example just given is of a high-order conceptual schema. But any structure of knowledge, single or complex, is an example of a schema. Games like football and cricket are mainly sensori-motor schemas ("mainly", since one also has to know the rules); and these depend on abilities to kick, throw, run, follow a moving object—on movement and postural schemas. Speech, reading, and writing are basic mental schemas which are essential for the learning of school subjects. At the other end of the scale we might take as example the conceptual structures which an aircraft designer must have available. Among these is fluid dynamics, which will depend on a knowledge of calculus, which you cannot learn unless you know some algebra, and this in turn requires arithmetic.

By schematic learning I mean learning which uses existing schemas as tools for the acquisition of new knowledge. To develop this idea further, it will be helpful to look at an experiment done some years ago with some 12-year-old grammar school boys, for two reasons. First, it illustrates the principle of schematic learning in a simple form; and second, the results were clear and striking, and their implications are fundamental for the learning of mathematics.

The experiment was done in ordinary school periods, but it was explained to the boys that the results would not affect their school marks or promotion; they were taking part in a scientific investigation. On the first day they were asked to learn some symbols and their meanings, sixteen in all. Some examples are shown in Fig. 1. For the second day's learning task, these symbols were combined into pairs and threes; with new meanings which were derived from the meanings of the single symbols (Fig. 2). On the third day, the groups were of larger numbers of symbols. As before, the task was to learn the group of symbols representing each meaning. They had pencil and paper, and were advised that the best way to learn was by self-testing and correction (Fig. 3).

The subjects were now ready for the most important part of the experiment, which was to compare their success at learning material which did, and which did not, fit in with the schema they had learnt. To this end, different schemas had been given to two parallel first forms. (The examples given here are from the schema for form IA.) The symbols themselves were the same, but the meanings of the single symbols and consequently of the groups were different. On the fourth day each form was given a page of material to learn which was fairly complex, but which fitted the schema they had learnt. A few examples are shown in Fig. 4. They were also given another page of similar material which fitted the schema which the boys in the other form had learnt, but not their own. So what was schematic learning for one form was rote memorising for the other, and vice versa. In this way it was possible to compensate for any difference between the average intelligence of the two forms by averaging all the scores for schematic learning, and all the scores for rote learning. The results are given in the table in Fig. 5.

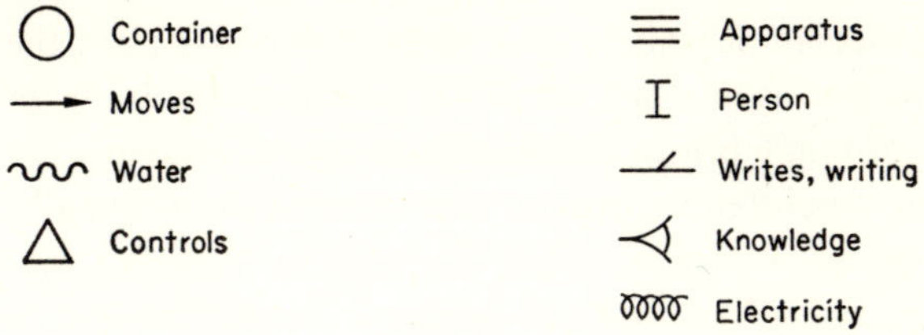

Container
Moves
Water
Controls
Apparatus
Person
Writes, writing
Knowledge
Electricity

FIG. 1

Vehicle
Ship
Message
Letter

FIG. 2

Driver
Crew member
Telegram

FIG. 3

Traffic lights
Teleprinter
Ship's captain
Telegraphist
Note: means 'several or many of'

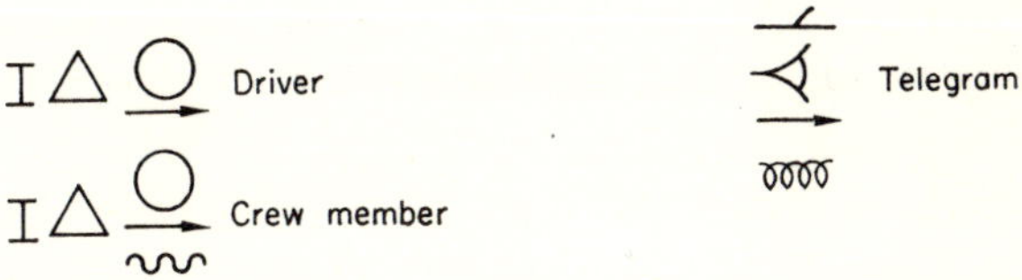

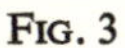

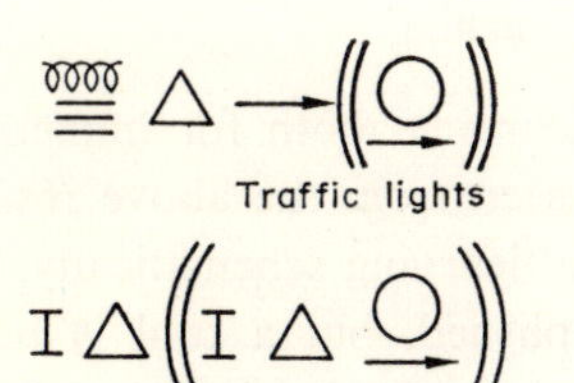

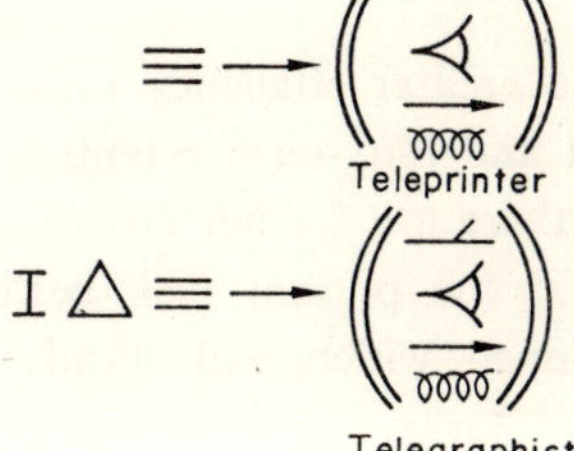

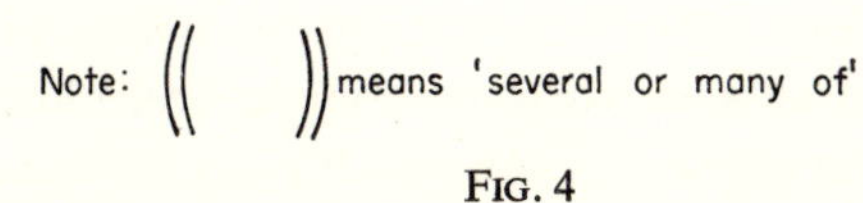

FIG. 4

The difference between the results of schematic and of rote learning is striking. The former resulted in twice the number of symbols being correctly recalled immediately after learning, while after 4 weeks it was seven times as many. The superiority of schematic learning for the immediate learning task had been expected; but the different rates of forgetting in the two conditions had not been anticipated. The table of Fig. 6 brings this difference out more clearly, by showing the amount forgotten as a percentage of what was originally learnt.

Percentage recall by all subjects (N=47)

	Immediate	1 day	29 days
Schematic learning	69	69	58
Rote learning	35	26	7

FIG. 5

Amount forgotten (as percentage of amount already learnt)

	First day	Next 28 days
Schematic learning	0	16
Rote learning	28	47

FIG. 6

Consequences for Teaching

The greater efficiency of schematic learning, both for immediate recall and long-term retention, can be seen from the above results. And there are further advantages. When learning schematically, not only is the present task better accomplished, but a tool is being developed which will stand the learner in good stead for future related tasks. By learning algebra schematically, one is not only doing a better job for algebra: one is preparing a schema which will provide the same advantages when one comes to calculus. And when using a schema for a given task, one is also practising and consolidating the earlier elements from which the schema has been built up. While learning to solve quadratic equations, one is practising one's basic algebraic manipulations; and also, to some extent, one's arithmetic. This gives schematic learning a threefold advantage over rote memorising, when considered in the context of a long-term learning situation.

For most people schematic learning is also more enjoyable. During the experiment described above, the boys appeared to enjoy every part of the experiment except the rote learning. Though they co-operated also in the latter, they did so more because they were specially asked to do their best at it, "in the cause of science". In the experiment here described, the difference in enjoyment by the subjects was a subjective observation by the experimenter. But this hypothesis has been supported by a later experiment, in which children were given rote learning and schematic learning tasks, and afterwards asked for volunteers to help with a further experiment. The number of volunteers for another schematic learning task was substantially greater than for another rote learning task.

Assimilation and Accommodation

Related to these very substantial advantages of schematic learning, there is a disadvantage which seems unavoidable. Since whatever fits into an existing schema is so much better taken in and retained, that which does not thus fit, however important it may be, suffers greatly by comparison. In his famous chapter on Habit, William James says that we must "... guard against growing into ways that are likely to be disadvantageous to us, as we should guard against the plague". Likewise, we ought to guard against the formation of schemas that are likely to be disadvantageous to us as we would guard against the plague. The parallel may be continued. In the case of children, it is not they who guard against the plague, but those directly or indirectly in charge of their health—parents, family doctors, public health authorities. Nor are children in a position to defend themselves against the formation of unsuitable schemas. This is the responsibility of those who are directly or indirectly in charge of their learning—parents, teachers, writers of textbooks, those who teach teachers.

The responsibility is a serious one, for as has been shown, an inappropriate early schema will make the assimilation of later material much more difficult—even harder, it is likely, than if no relevant schema were available. The learning task described on p. 184 as "rote learning" may have been made more difficult by the presence of an inappropriate schema—a hypothesis which has been confirmed by a subsequent experiment. By *assimilation* is meant the fitting-in of new material to an existing schema, with no great change in the schema itself. (Some change there must be, since after absorbing new material a schema cannot be quite the same as before.) An everyday example will serve to illustrate this idea, and also to introduce a new one. Imagine a child growing up in London—or Manchester, or Liverpool, or Birmingham. He encounters people who dress somewhat differently from his own family, who speak English less well and with a strange accent, perhaps whose skins are a different colour from his own. Collectively he learns that these are "foreigners". If he encounters a totally new person in this category—say, an Arab, or an Eskimo—he can without difficulty assimilate this experience to his existing schema for foreigners.

But it will be a different matter if, on a school trip to France, he hears himself described as *un étranger*. This does not fit in at all with his existing schema, according to which the French are still the foreigners, not himself. Before he can take in the meaning of this new experience, the schema itself has to change its basic structure. His idea of a foreigner must now become that of a person in a country different from his own, and this new schema will enable him to assimilate the new experience.

Notice also that the new schema is not so much an overthrow of the earlier schema as a major reorganisation of it. It contains the earlier schema, with its various particular examples, while giving it a wider meaning. This process we call *accommodation*; and while Piaget has pointed out that any assimilation implies an accommodation, for the reason indicated earlier,† it is perhaps useful to distinguish between these two processes by calling the change in a schema which accompanies straightforward assimilation "growth", and a reorganisation of its basic structure "accommodation". The importance of the distinction is in the fact that assimilation is usually easy and enjoyable, whereas accommodation may be difficult, and accompanied by feelings of insecurity.

Here are two mathematical examples. Pupils who have learnt the meaning of sine, cosine, tangent using a schema based on ratio and proportion, and the similarity of certain right-angled triangles, will assimilate problems of height and distance with little difficulty. If they are competent at heights and distances in two dimensions, progression to three-dimensional problems involves mainly assimilation, though with the added difficulty of imagining figures in three dimensions. But to understand a statement such as $(\cos + i \sin)\,\theta = e^{i\theta}$, accommodation of the schema is required. This accommodation will be easier if sine, cosine are already thought of as functions; and easier still if pupils are familiar with the idea that the sum of two functions is another function.

A more elementary example is that of numbers. Initially, to a child, "numbers" means counting numbers. He starts by learning to count to 5, then to 10, after which he rapidly progresses to 20. Of his own accord he is likely to extend his ability in counting to larger numbers; and when he encounters much bigger numbers, still no basic change of idea is involved. A number such as 1971 is still countable in principle, given enough time and patience. But fractional numbers are quite another matter. These are for the child an entirely new number system, not an enlargement of a system already known and understood. To learn these with understanding requires first a major accommodation of his schema—one of the most difficult which is likely to be required of him. A teacher who is to help him

† Page 188, lines 14 and 15.

effectively at this stage must both be aware of the psychological processes involved, and himself understand fractional numbers much better than is usually the case. It is small wonder that so many people go through life without ever understanding fractional numbers, or the difference between these and fractions. The difficulty of this particular accommodation is likely to be insuperable for most children without skilled assistance; and few are compelled by circumstances in adult life to subject these ideas to the prolonged activity of reflective intelligence.†

To the extent that a child is unable to make the necessary accommodation, whatever follows will not be understood. All that is possible is a reversion to rote learning, with all its attendant disadvantages. And this change for the worse in a child's learning method may not at once be apparent to the teacher, since we do not easily separate our own perception of objects from the objects themselves. For example, if we see a book, we tend to think of the quality of being a book as something residing in the book rather than in our own classification of this object. Similarly if a child writes or utters symbols which are meaningful to us, we tend to think of the meaning as something residing in the symbols themselves rather than in our own minds. So we may never question whether in fact the child understands them as we do. But, unless he has available similar schemas to those which we are using, he cannot have understood, although he may give the appearance of having done so.

Since it is not always obvious whether children are learning schematically or by rote, teachers need to take special care to ensure the former (i) at times when basic schemas are being formed, (ii) at stages when accommodation is being required. In both of these cases, the pace of learning needs to be adjusted, and understanding verified. Less is at stake when only assimilation is required, since if this fails, future progress is not being imperilled to anything like the same extent.

† See next chapter.

Basic Mathematical Schemas

It is not only a matter of a slower pace and more careful work at these critical stages of learning. We also have to look ahead to future stages, at which necessary accommodations may be much easier with some basic schemas than with others. Our choice of approach should be based not only on what best serves the needs of the moment, but on long-term planning.

This is one of the reasons why we as teachers have a much harder job than our pupils: we have to know far beyond what we are teaching. Another reason is that they have (if we plan properly) nothing to unlearn, whereas most of us have. So we are faced with just those difficulties of accommodation which have been under discussion, possibly with the additional problem that we are trying to teach the subject at the same time as we are rethinking our own ideas.

Part of our task is to enable our pupils to learn better and faster than we did; so that more of them can learn as much as we have, while some of them will go further than we have done. I have been arguing that to do this, we need greater understanding of the learning process; and this has been seen to imply particular attention to the basic schemas at all stages. We need to understand the foundations of our subject better than we have ever done in the past. When I was at university, I learnt higher and higher mathematics. Now I am exploring deeper and deeper.

When one starts doing this, one often finds that reflecting about methods which one learnt—and very likely taught—without understanding sometimes leads to mathematical ideas of great generality. Consider, for example, long multiplication.

$$\begin{array}{r} 328 \\ 47 \\ \hline 13120 \\ 2296 \\ \hline 15416 \end{array}$$

The validity of this method depends partly on the fact that $328 \times 47 = 328 \times 40 + 328 \times 7$ in this particular case; and in general that in the natural number system, multiplication is distributive over addition. To work out 328×40 without knowing our 40 times table, we multiplied 328×4 and multiplied the result by 10. This depends on the fact that (in this case) $328 \times 40 = 328 \times (4 \times 10) = (328 \times 4) \times 10$; and in general, that multiplication of natural numbers is associative—by which is meant that we get the same result whichever two we multiply first. Our method for adding 13120 and 2296 depends on the associativity of addition, since we add these two numbers by working out separately $(0 + 6)$, $(20 + 90)$, $(100 + 200)$, $(3000 + 2000)$ and $(10000 + 0)$. This is by no means the end. The reader may care to consider what mathematical principles are involved in evaluating 328×7, a process involving "carrying"; and the help we get in all this by place-value notation, which is by no means the only system which has been invented for writing numbers. Further analysis leads back to the ideas of number and counting, and to the discovery that even these are not the most basic ideas of mathematics. This further analysis is not given in detail here, since our present aim is to illustrate the dependence of quite elementary mathematics on still more basic schemas which are often not available, rather than to give a detailed exposition of the mathematics itself. But before leaving the subject, it is worth mentioning that all the manipulations of elementray algebra depend on the five properties of a number system, combined with the idea of a numerical variable. How many of us knew these ideas when we began algebra? And indeed, how many of our pupils do now?

The diagram below (Fig. 7) represents, in a very incomplete and simplified way, a schema for some of the more elementary parts of arithmetic, leading to algebra. The general direction of development is from below upwards.

It is not implied that all of these ideas must necessarily be made explicit to the pupils; but teachers should be aware of them, and how they relate to each other. Another point which the diagram brings out is that there is no one best order of teaching these topics.

Rather the opposite—we have to cover a network in some linear time-sequence, so whichever sequence we may choose, it cannot by its nature have the same arrangement as the network. The pupils have to reconstruct the network from ideas encountered sequentially, and we should help them to do this by calling their attention to the various interconnections.

This chapter began as a discussion of certain aspects of the psychology of human learning. The application of these principles has led directly to a study of the structure of mathematical ideas. And herein lies the paedagogic importance of "modern mathematics", a term which can be misleading. For what is important about it is the help it gives in understanding the basic, structural, ideas in mathematics: not its so-called modernity. (Little of it originated in the present century.) There is indeed a danger in over-emphasising the latter. The present rate of change of scientific and mathematical thinking is such that we simply do not know what mathematics will be like in 10 or 20 years, nor the uses to which our pupils will have to put the mathematics which we are now teaching them. Nothing

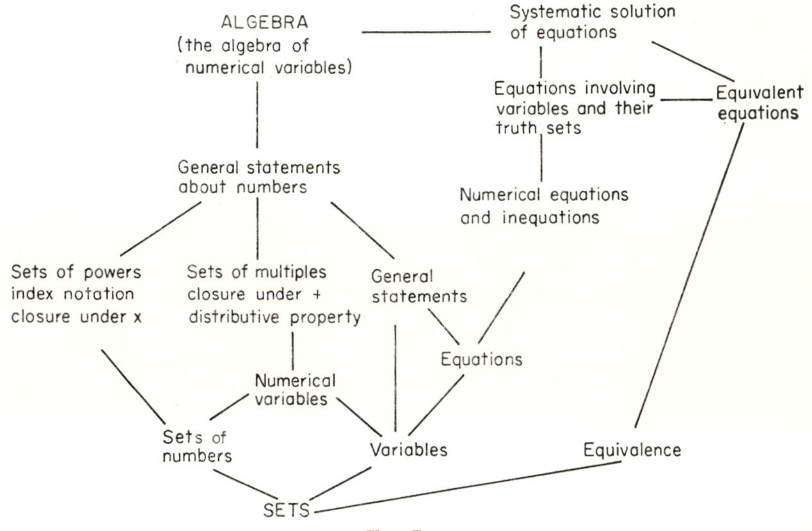

FIG. 7

dates more rapidly than that which is chosen simply because it is fashionable, be it dress, décor, or a topic in mathematics. And anything which we teach for this reason alone will be of little or no long-term value.

A modern approach to the *teaching* of mathematics is quite another matter, and I have tried to suggest a few of the ideas on which this may usefully be based. We have just seen that a further requirement is imposed by the rate of growth of knowledge, and rate of change of our scientific and technological background to our culture. This means that we have to teach not only a certain field of knowledge, but also the ability to adapt this knowledge to new tasks, and the ability to assimilate new knowledge in the future. In the light of the idea of schematic learning, I suggest that we may try to do this by (amongst other ways)

(i) finding ourselves, and helping pupils to find, basic patterns in mathematics;
(ii) teaching pupils always to be looking for these;
(iii) teaching them always to expect to have to expand, adapt, accommodate their schemas.

The first of these is teaching mathematics. The second and third are teaching pupils to learn mathematics; and it is only by doing this that we truly prepare them for the future.

Reflective Intelligence, and the Use of Symbols

RICHARD R. SKEMP

Sensori-Motor and Reflective Intelligence

So far as I know, the term "reflective intelligence" was first introduced by Piaget, in his difficult but important book *The Psychology of Intelligence*[1]. The most explicit formulation of this idea which he gives there is the following:†

> There are thus three essential conditions for the transition from the sensori-motor level to the reflective level. Firstly, an increase in speed allowing the knowledge of the successive phases of an action to be moulded into one simultaneous whole. Next, an awareness not simply of the desired results of an action, but its actual mechanism, thus enabling the search for the solution to be continued with a consciousness of its nature. Finally, an increase in distances, enabling actions affecting real entities to be extended by symbolic actions affecting symbolic representations and thus going beyond the limits of near space and time.

Using this as a starting-point, I have over the years been developing the idea that one may think of intelligence as functioning in two different modes. In the first, the sensori-motor mode, the objects of its attention and the activities which it directs are objects in the outside world, accessible to the senses of sight, hearing, etc., and manipulable via the activity of our muscles. In the second, the reflective mode, the objects which are perceived and acted on are mental objects, and although we are able to manipulate them in the

† Page 121 of the English translation.

activity which we call thinking, we have no knowledge of the physiological mechanism by which we achieve this. Since all mathematical activities are of the second kind, we would expect the ability to do mathematics to be closely related to the reflective functioning of intelligence. If this is so, then an important line of investigation is indicated for the teaching of mathematics; so fairly early in my researches I designed an experiment[2] to test this hypothesis. This was in two stages. First there was some preparatory work which involved the formation of certain class-concepts and operational concepts. Then came the main part of the experiment, in which the subjects were asked to do tasks involving some kind of reflection on these concepts. For the class-concepts, they had to distinguish examples having one, or the other, or both of two given properties: a reflective activity involving manipulation of these class-concepts. For the operational concepts, they had to combine two given operations, reverse an operation, or reverse and combine two operations. Their scores on these tasks were correlated with a mathematical criterion (scores on a G.C.E. trial "O"-level examination); and the results well supported the experimental hypothesis, that mathematical ability and the ability to reflect on one's ideas are closely related. The results also suggest that of the two, reflection on operational concepts is the more important.

I have found the following diagram helpful in thinking about these two modes of functioning of intelligence.

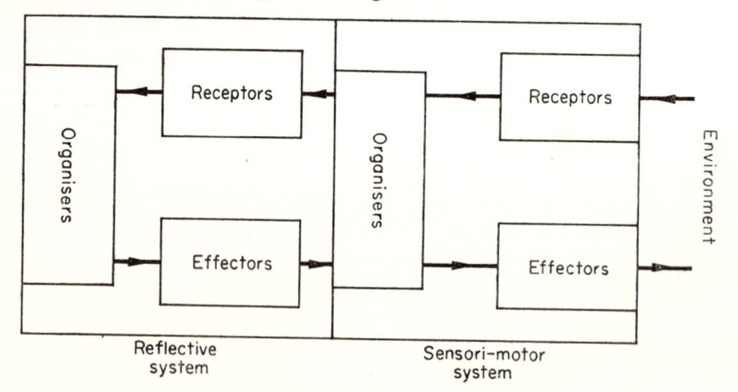

The right-hand system represents that which we use in our direct interactions with the outside world. Sense-data of various kinds—particularly visual and auditory—are received via receptor organs from the environment; and we act on the environment in various ways, by the use of our voluntary muscles. The nature of the activities is determined by a complex of intervening systems which are classed together in the diagram as "organisers". (Readers who are acquainted with cybernetic ideas will see that the diagram implies a feed-back system, but this aspect will not be pursued further here.) Let us bring this abstract diagram down to earth by a simple example. We are eating strawberries and cream. Each separate strawberry is classified as a member of the class "strawberries", and we act on each in the same way. Among them we now see a green object with a number of little points, which we classify differently, as a stalk. We behave differently towards this, depositing it on the side of our plate. Or it may be that, obscured by the cream, we have already taken it into our mouth. A different receptor organ is now involved, our tongue; but the function of classifying, leading to behaviour appropriate to the class, is the same.

The above is an example of sensori-motor activity; and since it involves classifying and behaving appropriately, it is an action involving intelligence, though not of a very high order. In contrast let us consider an elementary mathematical example: say, a pupil who persistently gets wrong a simple algebraic manipulation is told his mistake, corrects himself, and can then do manipulations of this kind correctly. In order to do this he has had to reflect on what he did—that is, information about the organisation of his responses is received by the receptor function of his reflective system. He then changes this organisation: an effector activity by the reflective system. If this change is to involve understanding, it must be based on an awareness of the relationships between this particular method and other relevant mathematical ideas: so the organisation of this reflective activity is categorised as intelligent. (There are other kinds of reflective activity such as guilt, narcissism, wish-fantasies, which do not concern us here.)

We use simple kinds of reflective activity continuously in our everyday life. Suppose that we have to post a parcel, collect a child from school, and call at the garage for petrol and air. This could be done in six different orders, and automatically we begin planning which would be the best, taking into account the positions of school, post office, and garage; whether parking would be on the left or the right-hand side of the road; the dislike of the child for waiting outside shops, and his liking for using the garage air-line. After arriving at a suitable mental organisation of these activities, we switch from the reflective to the sensori-motor modality and set off on our journey. We reflect in this kind of way so habitually that we fail to realise what a remarkable activity it is. We can see and hear because we have eyes and ears, manipulate objects because we have an articulated skeleton controlled by voluntary muscle. But what anatomist has yet revealed the means whereby we can see or hear our own mental imagery, and manipulate our own ideas? What is more, Piaget has shown[(3)] that a child aged 5–7 may be able to give the correct answer to a simple arithmetical problem, but be unable afterwards to describe the method by which he arrived at this answer.

So we must not take for granted that children can reflect in the same way as we can.

Reflective Activity in Mathematics

What are the differences between the simple everyday kinds of reflective activity like that described above, and those involved in mathematics?

An important one seems to be in the degree of abstraction of the concepts involved. In the everyday example, only primary concepts are involved—that is, concepts whose examples are objects or events in the outside world. In mathematics, even the simpler concepts are secondary—that is, they are concepts derived from other concepts, which may be derived in turn from other concepts.... Thus "pig" is a primary concept, representing what the animals in a certain set have in common. "Three" is a secondary concept, representing what a set of three pigs, a set of three pennies, a set of three pianos, and a set of three postage stamps, have in common.† This is quite an abstract concept, moreover, since the way in which these sets are alike is much less obvious than their differences. Yet most 5-year-olds have the concept of three.

"Three", like "pig", is a class concept. The "+" in $3 + 2 = 5$ represents a mathematical operation, abstracted from a variety of combining operations such as putting a set of three pigs together with a set of two pigs and counting the combined set. When we say that $a + b = b + a$, we are classing together all statements like $3 + 2 = 2 + 3$, $5 + 7 = 7 + 5$, etc. When we say "addition of natural numbers is commutative", we are classing together operations like $+$, $\times$, $\cup$, $\cap$. As we progress to more and more abstract ideas, so we are more and more dependent for their formation on being able mentally to scan a number of different mathematical ideas, in order to be able to perceive what they have in common: this common property being the new concept. In other words, I am suggesting that while low-order abstractions can be formed intuitively—that is to say, without reflection—we are dependent on reflective intelligence for the formation of the higher-order abstractions which constitute most of mathematics.

From being able to reflect on particular ideas, the next step is to reflect on our schemas. As teachers, we have to become aware of these in order to be able to communicate them. Often we try to make children conscious of a method which they use correctly in one context, in order that they can apply it deliberately in another. A particularly important case of the latter is the process of mathematical generalisation, where ideas developed first in a limited context are extended to a wider one. We do this when we extend the ideas of addition and multiplication, encountered first in the context of natural numbers, to integers, fractional and rational numbers, real numbers, complex numbers, matrices, vectors, functions, and other mathematical entities. We also do it when we extend index notation

† This is not intended as a definition of "three".

to fractional and negative integers; or when we extend the idea of a space to n dimensions. These generalisations constitute *accommodations* of the relevant schemas, as distinct from straightforward assimilations. Simple accommodations are possible intuitively; but a much more powerful way of making these possible is by a reflective process, in which we become aware of the essential ideas whose field of application we wish to extend. When generalising the idea of a number system, starting with the natural numbers and continuing as described earlier, these essential ideas are: that addition of numbers is commutative and associative; that multiplication of numbers is commutative and associative; and that multiplication is distributive over addition. These properties are by no means those which would first of all strike one as important or obvious, any more than threeness is the most obvious property of the collections of objects (pigs, pennies, pianos, etc.) listed earlier. It is, however, only by virtue of these properties that we can mix numbers from the different systems; and, still more important, that algebraic statements such as $(a + b)^2 = a^2 + 2ab + b^2$ are true for each and every one of the number systems listed. If any reader has thought these properties trivial, let him now reflect that their preservation through the successive generalisations of the idea of a number system has saved him from having to learn not one but six systems of algebra!

The Contribution of Symbols

Though it was said earlier that we did not know by what means reflective activity became possible, it does seem fairly clear that symbols make an important contribution. Years ago, and in quite a different context, Freud pointed out that the process of making an idea conscious was closely associated with the use of symbols. Concepts are elusive and inaccessible mental objects—no one can see or hear someone else's verbal thoughts or mental images. Symbols can be made visible or audible, and even when we do not write or speak them, but use them in our thinking, they are still much more concrete objects of thought than the ideas which they represent and to which they are attached. A symbol seems to act as a combined label and handle for its associated concepts. By the use of a symbol, we are able to evoke a concept from our memory store into consciousness—to make it available for reflection. Having done this with a variety of symbols, we are able to manipulate the associated concepts by manipulating their symbols. It is very largely by the use of symbols that we can arrange and rearrange our ideas, be these mathematical, or other.

Formulae offer a simple example of this. We wish, say, to find the simple interest on £150 for 3 years at 4 per cent. Immediately the formula $I = PTR/100$ comes to mind. Each of the symbols I, P, T, R has been chosen to evoke the appropriate concept (interest, principal, etc.), and the way they are arranged evokes the appropriate arithmetical operations by which we can obtain the answer to our problem.

By reflecting on one of our own schemas, and then making public the attached symbols, we can enable someone else to organise his own thoughts according to the same schema. "Telling someone how to do it" is, again, something we do so habitually that we take it for granted. But we should not, for it has taken millions of years of evolution to produce an animal that can do this, and we are the only animal that can. And a mathematical formula, be it a simple interest formula or that by which the trajectory of a moon rocket is calculated, is a particularly powerful and effective way of doing this.

Communicating and Reflecting

These two functions of symbols, namely communicating our mental processes to other people and becoming aware of them ourselves, are closely connected, and it is not always easy to know which comes first. With children, audible speech develops before the internalised speech which is one kind of thinking, so it is tempting to hypothesise that becoming conscious of one's own thoughts is an imaginary telling of them to some other person. This is supported by the help which we have all experienced at some time, when working on a problem, from speaking our thoughts aloud to a willing listener, even if the latter says nothing. The hypothesis applies only to verbal symbols, however. Many people think more in terms of visual imagery, and this is a more individual matter.

There can also be little doubt that the process can work the other way about; that is, that having to explain our ideas to someone else forces us to become more consciously aware of them ourselves, and sometimes to realise that we understand them less well than we thought we did. This may, with luck, lead eventually to greater understanding. Those of us who teach are constantly becoming aware of this; and to verify the hypothesis experimentally I arranged for two parallel classes of secondary school boys to be taught a mathematical topic—different for each class—by their regular teachers. Afterwards each class was split into two parallel halves, on the basis of a test on this topic. One half then had to teach it to their opposite numbers in the other class, while the other half spent the same time in practising it themselves. At the end, a further test was given, in which it was found that those who had taught the topic to others did better than those who had been practising it. This may have resulted both from the process of making conscious by communicating, already discussed; and also from the further stimulus to reflective activity arising out of questions from those to whom they were explaining. In either case, the result suggests that getting pupils to explain and discuss their ideas is one way of encouraging the development of reflective activity.

Some other Functions of Symbols

So far we have emphasised the importance of symbols for communicating with others, and for becoming conscious of, and manipulating, our own concepts. Closely related to both of these is their function for recording knowledge, both for the benefit of others and that of ourselves.

It may take years of work by a gifted individual to develop a certain area of knowledge. The number of those to whom he can communicate this directly is relatively small; but by recording his ideas they can be made available to as many as can read and understand his work. Also, as we have seen from the previous chapter, knowledge builds on knowledge. Each generation can learn the schemas of earlier generations, and among each generation will be those who develop these schemas further. This handing on of knowledge to contemporaries and successors is made possible very largely by the use of written or drawn symbols; and it is to me surprising that so little effort has yet been given to finding out what makes a good symbol.

We also use written symbols to aid our own memory, both short term and long term. The number of items which we can keep accessible to consciousness at one time is quite small. By "thinking on paper", we can make easily accessible the earlier stages of some piece of thinking, and free our attention to concentrate on one part of it at a time. The reduction of cognitive strain is enormous. Again, we can commit to paper the result of a period of concentrated reflection, and months or years later use these written or drawn symbols to recover the ideas from our own memory stores. This is but an extension of the use of symbols described earlier, as combined labels and handles to bring their related concepts into the focus of consciousness. Writing them is a precaution against losing the handles! Looked at another way, it is a device for ensuring that the activity of reflective intelligence is not made impossible by the unavailability of the objects (of thought) on which it acts.

I said earlier that little attention had been paid to finding out what makes good symbols. One useful quality is that they can be used not only to evoke individual ideas, but also to suggest the ways in which they are related—to remind one of a structure. Some symbols which do this well are

$$\sum_{a}^{b} \qquad f: x \to f(x),$$

and Venn diagrams. Some symbols which (in my view) do this badly are $\angle PQR$; $\mathrm{d}y/\mathrm{d}x$; and $\equiv$, which suggests a closer relation than that of equality but represents a weaker one. (On the whole it is easier to find good symbols than bad ones; it seems that an intuitive choice is more often successful than otherwise.) Another rather simple requirement is that they should be easy to write: some which give me trouble are $\mathscr{E}$ (for universe of discourse), and $\aleph$ (aleph). Younger students also have trouble with ξ and ζ. Also that they should be easily distinguishable: cf. v and υ (Roman vee and Greek upsilon).

This last requirement is less trivial than it may seem. For if symbols play a key part in bringing ideas into consciousness, and manipulating them, it is reasonable to expect that the clearer the symbols, the more effectively they will do their job. This is confirmed by experience. A typescript conveys ideas more clearly than a manuscript, and a well-designed page better still. One result is that when an author sees his own ideas re-presented to himself at the proof stage, he often sees ways in which they could be better expressed—though this is not always appreciated by the editor.

These are not the only functions of symbols, but they are the ones most directly connected with the main topic of the chapter. Others may be found in the book listed under "For Further Reading".

In this chapter, I have been inviting the reader to reflect intelligently upon the reflective activity of intelligence; and, by the use of symbols, have invited him to consider the function of symbols in making our concepts and schemas more available to our own consciousness and that of others, which the reader can only do by the further use of symbols. Having written this, I find it surprising that the task is a possible one. But I believe that it is; and I believe, further, that unless we think not only about mathematics itself but also about the psychological processes involved in learning and communicating it, we shall make slow progress in our efforts to improve the teaching of it.

References

1. PIAGET, J., *The Psychology of Intelligence* (English trans.), Routledge & Kegan Paul, London, 1950.
2. SKEMP, R. R., "Reflective intelligence and mathematics", *Brit. J. Educ. Psychol.* **31**, 45–55 (1961).
3. PIAGET, J., *The Child's Conception of Number* (English trans.), Routledge & Kegan Paul, London, 1952.

For Further Reading

SKEMP, R. R., *The Psychology of Learning Mathematics*, Penguin Books, Harmondsworth, 1971.

Motivation, Emotional and Interpersonal Factors

A. P. K. CALDWELL

"Most of us tend to replace understanding by explaining (the former requiring an open mind, the latter a theory)". CALEB GATTEGNO[6].

When G. H. Hardy, that doyen of English pure mathematicians, was writing his *Apology*[7], he said: "I do not remember having felt, as a boy, any passion for mathematics, and such notions as I may have had of the career of a mathematician were far from noble. I thought of mathematics in terms of examination and scholarships: I wanted to beat the other boys, and this seemed to be the way in which I could do so most decisively"; that, in his own words, is one man's motivation for learning mathematics, a man who became an outstanding mathematician. In Hardy's case he puts the matter firmly in a social context: he remembers no passion, it was a matter of a career, of competitiveness amongst his peers; the motivation was socially determined, the emotions involved were apparently hidden, and the interpersonal factors crude. Elsewhere in his *Apology* Hardy says "I hate 'teaching' ... I love lecturing", where again the emphasis is on denaturing the interpersonal aspect of the business. Perhaps we might have learned a great deal more on this matter if we had some of Ramanujan's thoughts on it; for Ramanujan was self-taught, an Indian railway clerk, with whom, later, Hardy was proud to have worked on an equal footing. But, even then, we know that a man is not always the best judge of his own motives.

Certainly mathematics is one of the most emotional of subjects, though its appearance belies this. Not only does it provide a beautiful safe haven from the fears and anxieties of life, but there is a quasi-divine afflatus in stepping out of the flux of time, concentrating within oneself and solving some problem, getting it right, unassailable by others, subject to no test of time or circumstance. It is small wonder that the uninitiate recoil at the mention of mathematics as from something icy-cold and a little menacing. No wonder that this power, power to "beat the other boys", gives a feeling of superiority; as a recent review of a mathematics-for-the-layman book had it, "Those of us who are non-mathematicians are constantly reminded that we are intelligent and constantly talked to as if we were not."[3]

The idea that mathematics is for the elect possibly goes back to the Pythagorean brotherhood; and maybe they have much to answer for. As Bertrand Russell put it:

> The doctrines of Pythagoras, which began with arithmetical mysticism, influenced all subsequent philosophy and mathematics more profoundly than is generally realized. Numbers were immutable and eternal, like the heavenly bodies; numbers were intelligible: the science of numbers was the key to the universe. The last of these beliefs has misled mathematicians and the Board of Education down to the present day.[10]

Hardy maybe was rather a special case. Here is a description of a bit of everyday life in a school classroom by Jules Henry, from his book *Culture Against Man*[8]:

> Boris had trouble reducing "12/16" to the lowest terms, and could only get as far as "6/8". The teacher asked him quietly if that was as far as he could reduce it. She suggested he "think". Much heaving up and down and waving of hands by the other children, all frantic to correct him. Boris pretty unhappy, probably mentally paralyzed. The teacher, quiet, patient, ignores the others and concentrates with look and voice on Boris. She says, "Is there a bigger number than two you can divide into the two parts of the fraction?" After a minute or two, she becomes more urgent, but there is no response from Boris. She then turns to the class and says, "Well, who can tell Boris what the number is?" A forest of hands appears, and the teacher calls Peggy. Peggy says that four may be divided into the numerator and the denominator. Thus Boris' failure has made it possible for Peggy to succeed.... To a Zuñi, Hopi or Dakota Indian, Peggy's performance would seem cruel beyond belief.... Yet Peggy's action seems natural to us; and so it is....
>
> Looked at from Boris' point of view, the nightmare at the blackboard was, perhaps, a lesson in controlling himself so that he would not fly shrieking from the room under the enormous public pressure.... It was not so much that Boris was learning arithmetic, but that he was learning the essential nightmare.

How often, alas, are the two things closely associated! Here again we see the same sort of factors at work as those to which Hardy drew attention. Huge social pressures are operating through the teacher, who may have to commit aggression, even when it is done in the nicest way, against his pupils, and to arouse their anxiety.

The aggression we speak of on the part of the teacher is something implicit in his role. Too often this may take the form of the teacher knowing something and conceiving it his job to see that his children acquire that same knowledge—something it is only too easy for the teacher of mathematics to fall into, when it seems to him that he has knowledge of absolutes and, moreover, that there is some necessary order in the way knowledge of these absolutes is to be acquired; but however it may be, a teacher seeks to promote changes in his pupils, and this very design to change them involves in principle a role of subordinator for the teacher. Not subordination of the children as such—we are not talking of the difficulties of large unruly classes—but a subordination of learning to teaching. Children are learning all the time, many things at many levels, and a teacher has to subordinate all this to what it seems to him a child needs to learn. A story from E. M. Renwick's *Children Learning Mathematics*[9] shows this happening so naturally even in the best of circumstances.

JOHN (age 6)

It was Friday afternoon. The teacher was ready with suggestions for any children who could not decide on some "occupation". She said to John, "You make a shop with the desks in the corner and I'll come and buy from you".

John set to work with zest and spent about half an hour collecting and arranging materials for his shop. Then he went to the teacher and intimated that he was ready for customers to bring some cardboard money and buy his goods. He showed her the "till" for money.

The teacher's first faux pas was to point to something and say "How much is that?"

John protested vigorously. "You've put your hand through the window! Come inside if you want to point to things".

Then came the second faux pas, a terrible one. The teacher chose two articles and asked "How much?"

John: "This is 5d, and that's 9d".

Teacher: "I want to pay for them both together".

John, after a pause for furious thinking: "It's one and tuppence". Then, shouting and almost hysterical: "And you're not coming into my shop again. I'm going home to my dinner. It's half-past five and everybody has to go home to their dinner at half-past five. I'm locking the door. Click!"

And the moral is: Don't shatter the child's make-believe world by thrusting the workaday world into it. It is an unforgivable intrusion. After all, work is work and play is play.

Of course it is one thing to arouse anxiety about failing, about not being able (or willing perhaps) to come up to what society, or teacher, expects; and another to arouse anxiety by dismantling or destroying someone's preconceived notion so that they are impelled to build afresh, to learn something new. To do this we still have to commit our aggression, but we may, if our relationship is good, provide the reassurance that there is something left to build on. This use of anxiety arousal can as we know provide a powerful local stimulus, motivation, for learning mathematics. Here is another unsophisticated and amusing example from *Children Learning Mathematics*[9]:

The class were considering a teasing question: A teacher collected some pennies from her small class of girls. Each girl gave as many pennies as there were girls in the class. If the collection came to 4s. 1d., how many girls in the class?

Philippa, usually a sedate, well-behaved little girl, grew angrier and angrier. At last she stood up and said, "It isn't adding, it isn't subtracting, it isn't multiplying, it isn't dividing"; then, almost screaming, "It's NO SORT of a sum".

Joan too was roused to fury. She said, "Tell us HOW to do it and we'll DO it".

Of course when we are thinking of mathematics learning in a classroom setting we know that the paramount factors are the social ones; for the most part the subject itself is incidental. These social factors which mingle in the classroom come from without and within. The confluence of historical and economic process in the classroom is overwhelming; and the weight a teacher bears, consciously or not, consequent upon the fact that we have "compulsory education" is nigh intolerable. Then within the classroom we have people, in a very highly structured situation, a social situation which, relative to the common ideal aims of education, is overstructured, but *people*, thrown together socially: this is the basic social situation.

Though a teacher has a very well-defined role with its implicit norms, easily recognised by all, still, within that rather rigid framework, interpersonal factors are powerfully at play; and for the great majority of children, and teachers, it is the quality of their relationships that makes the day bearable or unbearable in school. Looked at from this point of view, which subject we are talking about may be irrelevant; but it seems to be a fact that teachers whose main subject is mathematics are, generally speaking, less competent in the classroom than others; and the preliminary findings of a research I am currently doing, corroborate this. To try to see why this may be we must delve a little into the nature of the subject itself, considered not as a system of pure thought but as a human activity.

There are three separable, though not separate, things we may consider: doing mathematics, learning mathematics and teaching mathematics. We have looked briefly at the learner and the teacher in their social setting; now we consider what it means to do mathematics, what distinguishes that, so far as it can be distinguished, from doing anything else and what sort of emotions attend this activity or give rise to it.

In 1968 at Dillington House, Ilminster, G. Spencer Brown read a paper to the September conference of the Association of Teachers of Mathematics on a new approach to Boolean algebras. It formed the mathematical core of his book *Laws of Form*[2] published in the following year. He threw out many fascinating asides at that conference; some of which, along with others, appear in his book as a context for his mathematical ideas. These give us some of the most clear-sighted and penetrating comments we have on mathematical activity. One of his most pregnant passages is the following, from the Introduction.

> In arriving at proofs, I have often been struck by the apparent alignment of mathematics with psychoanalytic theory. In each discipline we attempt to find out, by a mixture of contemplation, symbolic representation, communion and communication, what it is we already know. In mathematics, as in other forms of self-analysis, we do not have to go exploring the physical world to find what we are looking for.

This characterisation of mathematics as a form of self-analysis is very revealing indeed. And again at the Association of Teachers of Mathematics Conference at Ilminster in 1968, Geoffrey Beaumont, in his paper pointed out some of the profound implications of this and drew our attention to the *locus-classicus* of reference here, Sandor Ferenczi's notes dating from about 1920 entitled "Mathematics", included in his *Final Contributions to the Problems and Methods of Psycho-Analysis*[(5)].

Before going any further, it is as well to remind ourselves that words cannot lay bare the mechanisms of reality, though we often talk as if we believed they could. What we provide for ourselves are metaphors or conceptual models whose justification is empirical, or sometimes perhaps aesthetic; they may provide good suggestions as to what to look out for or provide a more or less satisfactory rationale for action. Maybe it is nowhere so obvious, when we think of it, that we are using metaphor all the time, as it is in speaking of things of the mind. While on the subject it is worth pointing out that we can expect to learn very little indeed from "scientific" investigation here: in the attempt to be objective, to minimise direct observer effect, the structuring of the experimental situations is so great as to restrict the human subjects, out of all recognition one might say.

To get back to Ferenczi, let us distinguish three kinds of mathematical activity, which we may call "implicit mathematics", "mathematics with a small m", and "Mathematics with a big M". Ferenczi does not use these labels but we may find them useful. (For the big-M–little-m idea I am indebted to Jack Oliver.)

Implicit mathematics is what Ferenczi is referring to when he says:

> Skill demands an immensely precise calculation. Even a dog can do that.

And again further on

> the dive of the eagle on his prey, the spring of the tiger—demand calculation. Differential and integral calculus, geometrical functions, even though no *knowledge* of geometry.

There is a fundamental mathematical, with a small m, principle involved here which Ferenczi calls "condensation".

> The work of an acting man,

he says,

> is a magnificent *condensation performance*; the condensed result of a vast quantity of separate calculations and considerations;

and then he says:

> The mathematician is a man who has a fine capacity for self-observation of this condensation process.

This is the essence of Ferenczi's theory: that what is characteristic of the mathematically inclined is a talent for introspection which senses the "*formal* in the process of intra-psychic excitation", for "*self-observation* for the metapsychological process of thought and action".

> The mathematician [he says again] appears to have a fine self-observation for the metapsychic (also probably physical) processes and finds formulas for the operation in the mind of the condensation and separation functions, *projects them*, however *into the external* world, and believes that he has learnt through *external experience*.

Here we may say that we have arrived at mathematics, with a big M.

Two things we may ask about this are, what produces such a talent, and what is it that urges people to develop it. "What then is the real motive for learning mathematics"? we may say. What are emotions which go with it?

We are aiming to lay bare some of the sub-strata here; they may be near the surface for some but far below for others.

It seems clear that there are deep fears and anxieties about the outside world, occasioning the development of a symbolic representation of an inner world where a narcissistic omnipotence holds sway.

There are plenty of reasons for being fearful and anxious about the world in which we find ourselves and we have little hope of grasping a thread that may lead somewhere unless we look back to experiences common to us all. Wilfred Bion in his *Learning from Experience*[1] says:

> 1. H. Poincaré describes this process of a mathematical formulation thus: "If a new result is to have any value, it must unite elements long since known, but till then scattered and seemingly foreign to each other, and suddenly introduce order where the appearance of disorder reigned. Then it enables us to see at a glance each of these elements in the place it occupies in the whole. Not only is the new fact valuable on its own account, but it alone gives a value to the old facts it unites. Our mind is as frail as our senses are; it would lose itself in the complexity of the world, if that complexity were not harmonious; like the short-sighted it would see only the details, and would be obliged to forget each of these details before examining the next, because it would be incapable of taking in the whole. The only facts worthy of our attention are those which introduce order into this complexity and so make it accessible to us."
> 2. This description closely resembles the psychoanalytical theory of paranoid-schizoid and depressive positions adumbrated by Mrs. Klein.

Perhaps the significant thing here is that what Poincaré is describing is again condensation, to use Ferenczi's term. If we can elucidate Bion's insight we may have something more to work on.

Melanie Klein has produced some very influential work on the psychic life of early childhood based on her psychoanalytic work. Ferenczi himself was among those who encouraged her to take up this work. We may outline the part of her theory relevant to Bion's remark as follows: There exist for the newborn already pleasant and unpleasant things: if nothing before, then birth itself creates this distinction. Pleasure is soon associated with something outside itself, an object as opposed to itself as subject, the warmth and comfort and food, associated with the mother. The infant cannot avoid unpleasure too, discomfort sometimes and hunger, if nothing worse, these feelings are projected onto some object outside itself, another "mother", a bad mother. Thus we have two elements "seemingly foreign to each other", good and bad mothers. This in brief is the "paranoid–schizoid" position which every child experiences, and re-experiences. In this position both "mothers" are equally real to the infant. When development necessitates that he "realise" that it is really the same person who feeds and comforts him, the same person who is not feeding him when he is getting hungry, or comforting him when he is getting distressed, he has mentally to bring the two mothers together. There is ensuing tension, conflict, and anxiety that the good mother may be defeated; and guilt feelings are aroused because it is he who has brought the two mothers together. This state is what Melanie Klein calls the "depressive position".

Before the age of 6 months, then, every child has experienced the paranoid–schizoid position and has had to face the depressive position. At this time of life, we know, he is developing faster than he ever will again: as his faculties mature he is, almost literally, making up his own mind, ready to deal with thoughts and soon to use his native tongue. An infant is not playing a passive role in his own life.

We all have to live through the depressive position somehow if we are to survive as agents in the world. Our reactions to it at the time will depend on what we are making of our experience; and our mental strategems to defend ourselves against our guilt and anxiety are just as various. One kind of defence may involve idealising the situation, making it into a formal structure where the two elements, seemingly foreign to each other, are seen as symmetrically related, equal in their positive and negative qualities respectively; and their being brought together results in a neutralising of each or a mutual annihilation, leaving us alone, unscathed. This might be the way to mathematics, a retreat from the dialectic into symmetrisation. If only we could win through to accept the dialectic of good and bad in one person, of love and hate in ourselves!

> These are only hints and guesses,
> Hints followed by guesses; ...
> Here the impossible union
> Of spheres of existence is actual,
> Here the past and future
> Are conquered, and reconciled,
> Where action were otherwise movement
> Of that which is only moved
> And has in it no source of movement—
> Driven by daemonic, chthonic
> Powers. And right action is freedom
> From past and future also.
> For most of us, this is the aim
> Never here to be realised.[4]

References

1. Bion, W. R., *Learning from Experience*, Heinemann, London, 1962.
2. Brown, G. Spencer, *Laws of Form*, Allen & Unwin, London, 1969.
3. Caldwell, W. E., *Mathematics Teaching* **44** (1968).
4. Eliot, T. S., *The Dry Salvages*, Faber & Faber, London, 1941.
5. Ferenczi, S., *Final Contributions to the Problems and Methods of Psychoanalysis*, The Hogarth Press, London, 1955.
6. Gattegno, C., *Towards a Visual Culture*, Outerbridge and Dienstfrey, New York, 1969.
7. Hardy, G. H., *A Mathematician's Apology*, Cambridge University Press, London, 1940.
8. Henry, J., *Culture Against Man*, Random House, New York, 1963.
9. Renwick, E. M., *Children Learning Mathematics*, Arthur H. Stockwell, Ilfracombe, 1963.
10. Russell, B., *The Principles of Mathematics*, 2nd ed., Allen & Unwin, London, 1937.

Suggestions for Further Reading

Books

Caldwell, A. P. K., "I thought you were going to tell us about automorphisms", in *Mathematical Reflections*, Edited by members of the Association of Teachers of Mathematics, Cambridge University Press, London, 1970.

Gattegno, C., "The human element in mathematics", in *Mathematical Reflections, idem.*

Tahta, D. G., "Idoneities", in *Mathematical Reflections, idem.*

Journals

Caldwell, P., *A.T.M. Supplement* **11** (1969).

Tahta, D., *A.T.M. Supplement* **11** (1969).

Investigational Methods

EDITH E. BIGGS

I HAD expected a small group called the B.Ed. Group. I had been asked if I would talk on discovery methods. I wrote back to say, "By all means, but you cannot talk about discovery methods, you have to enjoy them." I also gave a list of some simple materials.

I now find that the composition of the audience is not what I expected, but I shall in fact keep approximately to my original plan.† I still intend to break off from time to time to give you an opportunity to experience the different stages to which I am going to refer. We have some string and some scissors and I think we have some squared paper.

Many people ask these days, what does discovery mean? I have tried not using the word discovery altogether; I use the word "investigation", because discovery has been so much misunderstood in the past, not only by laymen, but also by teachers. Discovery does not necessarily mean running around with a trundle wheel or a tape-measure, in fact I would love to have a bonfire of all trundle wheels and to hide tape-measures for a year. At the end of that time our children would know a great deal more mathematics than they do at the moment. (Of course there *are* times when we do need a measure.)

During the past 10 years many teachers have done a great deal of admirable experiment. It is now time to settle down and to plan our experiments. I believe we need some sort of a plan whatever we are doing with children. If, for example, we take the topic, length (since I mentioned tape-measures and trundle wheels), we should

† *Editor.* Due to illness I was unable to act as Chairman. Uninvited students took the opportunity to listen to Miss Biggs.

try to plan experiences to cover the different stages in a child's learning. One discovery is not enough for most children; for some children it is, but for some it is not. Every starting-point should be more demanding than the preceding one.

I would not want to limit a teacher in what she is doing in any way at all. What I am suggesting is that we consider the starting-points at each stage. Teachers need not use these, but they will illustrate the different stages and indicate that we do not want children to repeat the same experiences. We want them to meet some of the important ideas in a different context, so that to the child they seem different. On my syllabus I used to write, "revision of previous year's work". How wrong! What I needed was something entirely new. During new work we are able to discover what the children do not know and in which aspects they require further experiences.

Now, since I was asked to talk about discovery methods, after I had looked very carefully at all the material Mr. Chapman had sent to me of the work which was done last term, I decided that I ought to consider the question of discovery methods very carefully and try to delineate the different kinds of discovery.

The first perhaps is the most valuable of all and I would call it "fortuitous". This type of discovery is always initiated by a child or adult. It is not in any way teacher-directed. It happens when a child runs into school holding something which he has found and which prompts him to ask a question. It happens on in-service courses when I have given a starting-point or question and the teachers have taken no notice at all, but have done something which they themselves have suddenly thought of. And this is perhaps the most highly motivated type of discovery there is. I am going to give you an example of that in a minute.

The second I have called "free and exploratory" with materials provided. This and every other type of discovery is structured in some way or another. Even when we are merely providing material we are structuring the situation. No questions are asked, but materials are made available. A teacher often works in this way in order to find out what questions she should ask the children. If a teacher likes to work from questions on a work card, then the best way of going about this is to see what the children would do with the material the teacher has in mind. So there would be no question, just material, and this might well be placed on a table at one side of the classroom so that the teacher could watch the children in an unobtrusive way.

The third kind I have called "guided discovery", and here the starting question and probably materials as well will be provided by the teacher. Of course a starting question may merely be, "What can you do with this?", or "What can you find out about this?". These would be very open situations indeed. Or it might be more directed, "What might happen if ...?"

The fourth kind of discovery I call "directed" as distinct from guided. In this case the teacher directs the discovery throughout by asking questions to suit the children she has in front of her, and she will ask these questions whenever the children require it. She might go from group to group asking questions as necessary to help the children to take a further step.

The fifth kind I have called "programmed discovery" and I hope you see the distinction there. There are a number of work cards on the market (I would put many work cards into the programmed section) and many of them begin, "You will need: ... (I) Do so and so. (″) Do so and so, ..., etc.", from the beginning to the end. I have seen a teacher make up a work card of this kind. No word was spoken, but each child queued up and the teacher looked at his work. One boy had nine ticks and the teacher said, "Find the next card". I knew this teacher very well and I asked him if I could talk to the boy. I found that the boy had discovered nothing whatsoever. For discovery it is normally essential for there to be discussion. This is one reason why in classrooms children are arranged in groups. Another is that the teacher could not possibly get round to individuals and give each one the necessary discussion they need; moreover, children learn a good deal from their peers. You can see that this final kind of discovery is just like programmed learning. The planning has not been made with a specific child or group of children in mind. The questions are those which the teacher made up in a logical order.

Now I will give you time for discussion later, and if necessary we will come back to this point, but I hope that you can see that there are really differences between one type of discovery and another. I would like to add another point here, because I think this is very important. When a teacher is introducing children to active learning she finds that to begin with she has to give much more guidance, and maybe direction. But, however great the temptation, the child should not be told, because if we tell the child we rob him of self-respect; we rob him of the opportunity to say, "I took that last step myself". This is why investigation is so important, not just at the primary stage, but also at the secondary stage, at Colleges of Education, at university, yes, and beyond; the more we do for ourselves, the more involved we are.

I am not saying that we can learn everything by discovery, although children at the primary stage can learn a good deal of their work by discovery, but surely the most important thing we can do for human beings of any age is to give them this opportunity to think for themselves. After all these years I still believe this, perhaps even more firmly than at the beginning. All this is fresh in my mind because, at the beginning of November 1969, I went to the United States where I had been asked to work in an elementary school of age range 4 to 12+. I thought I had been asked to help the teachers with mathematics!

I arrived with my luggage lost, at a dinner party. I went in my snowboots and in the clothes worn for the journey, feeling extremely travel-stained. The first thing I was told by Dr. Max Beberman, the Director, was, "My experience in Great Britain has suggested to me that the right way of working is not through subjects, but through an integrated programme. Will you please introduce these teachers to an integrated programme". You can imagine what I went through during those three hours of that long-drawn-out dinner, when I was trying to think of what I should do the next day! It was most unfortunate that the teachers were not there, because if only I could have got to know them early, I might well have known some of the problems immediately. I am going to tell you some of the problems because you can learn from my mistakes. In all this I was more of a learner than either the teachers or the children.

To begin with I should explain that I had already met and worked with some of the teachers for a brief two days in the middle of their summer vacation. Nevertheless, when I arrived, I found that they were hostile, because I had been built up as a British expert. Moreover, there were eight visitors from all over the United States, very distinguished educationalists, who had come to see developments and were prepared to follow me from class to class.

The first thing I had to establish was that my job was to be a teacher's aid and that I was no expert. (This latter was not difficult!) I hoped that towards the end of the week when I had mopped up the water when it was spilt and done all the tasks which a teacher's aid would do, the teachers would accept me and be willing for me to take their ideas and help to extend these a little so that they were carried over subject barriers.

These teachers had never taught children in groups, their classes were always in rows in front of them. So you can imagine the problems which I had to face. I am going to cut this long story short, because I want to give you two examples of how I dealt with the children.

I want to emphasise that at no time did I say "Let us do this". In the previous week, for the first time, a British teacher had worked at the school. This teacher was very experienced in working to an integrated programme. After introducing the idea of an integrated approach, he had taken three volunteer classes for three topics that he had worked out for himself. Now since this resistance to a British expert had built up, it was impossible for me to work in this way. I wanted to start with any topic suggested by the teachers themselves, and to help them to develop this topic.

In one class I found that the children had been doing some surveying. These children were only 8 years old and I was sure that they had not a clue about surveying. They had no idea, to begin with, about scale. The surveying formed part of a Social Studies Unit, and the teacher was solemnly going through it lesson by lesson. I felt we had to start at the very beginning of scale. (I tried to work long periods with each teacher so that they would understand that I was really concerned about their particular children.)

At the teacher's suggestion we started with the children's own measurements. How much I had to learn! I suggested that they might draw round themselves; I thought that would give them their height. But of course they drew themselves with the feet hanging down, so we did not get their heights at all and we had to start all over again! The children were very noisy. This teacher did not want different groups to be given different things to do. This would have been much quieter. Moreover, she added, "What have we got out of this? What mathematics? What English? You know I shall be behind with what I am doing. This is too slow. Where are we going to next?" And I am sorry to say she communicated her anxiety to me, so that whenever I was working with those children, I was working under this disadvantage and I was not relaxed.

Later when I came to do some work with shapes with the class, instead of giving them the opportunity to build with shapes, as they wanted (all the things we had planned could have been learned from such a situation), I started structuring at once. You will remember that Dienes insists that a play stage comes before structuring. I felt at the end that these children had learned very little and did not enjoy their experiences either. Yet some of them had enjoyed the previous work drawing round themselves, painting their own outline and measuring their height and waist in paper and string. Eventually some children made a quarter-size model of themselves in three dimensions.

By contrast the 7-year-olds took things in their own hands and from the very beginning measuring was a great delight to them. This may have been caused by the age difference; perhaps we were doing something rather easy for the 8-year-olds. Or it may have been simply that the first teacher was over-anxious. Certainly the two situations were entirely different. The second group learned so much more than the first and yet in a very different way.

At the beginning the teacher of the 7-year-olds had said, "I want to divide my classroom into four sections; I do not think I can manage more than four". She also insisted that one corner of the room should be a quiet corner and I heartily agreed. The children who had never painted on large sheets of paper before, painted some beautiful pictures to decorate the inside and outside walls. They used the teacher's desk and a bookcase as room dividers, and moved in a round table and a comfortable chair made by the teacher in a cardboard carpentry session. I had visited these children earlier to talk to them about shapes. (It is hard to teach an American child what a cube is, because both their sugar cubes and ice cubes are flat. I hope you know the reason why it is more efficient to have flat sugar cuboids than cubes or that you will discover this for yourselves.)

These children were interested and very friendly. I asked them if they would bring collections of boxes and we managed to acquire all kinds of shapes (which they carefully sorted) except a cube. You know that it is the simplest things which give one the greatest pleasure and these children were no exception. They had a box of square section but which was not a cube; they found two different ways of making this into a cube. Then they pasted paper on the cube until it was covered. By the time they had done this and painted it, the cube was so strong we could use it for a ball. And so these children were interested in shapes from the beginning.

I suggested that they should make any model they liked. I wonder if you can guess what two boys made first of all? I can still see Charles, a very small 7-year-old, looking like a little old man, flushed with excitement. Since they had many cylinders, they made a telescope. When it was finished they immediately went to the window. "But it doesn't bring the moon any nearer, does it?" said Charles. I asked them what we could do to bring the moon nearer. One of the boys said, "I know what we need, it begins with 'L', and it's lens". So I took off my glasses and told them to try these lenses. They could not believe their good fortune! "Your lens brings the moon nearer", they shouted. "You have not started to look. My glasses could not possibly bring the moon nearer, because I am short-sighted and mine do just the reverse", I told them. You can imagine the pleasure and enjoyment we had. We experimented with various lenses and did a great deal of work in science and with shapes in particular. Every child could tell me how to build smaller cubes into bigger cubes. I had tried in vain to do this with the 8-year-olds, and yet these children, through their play, had learned far more. Another pair made a periscope. They said, "Now we want some mirrors". They learned a good deal about reflections and the arrangement of mirrors in a periscope.

You will be amused to hear that a girl made a burglar-proof jewellery box. These children had never worked in this way before and yet they were so imaginative, so willing to talk, and later eager to write about what they had done. In fact a great deal was achieved. Another source of amusement was that my two groups were always the noisiest. The teacher's groups were very quiet but it took me some time to get my groups settled down. But once the children became engrossed in their work the noise subsided.

From now on, I want to give you experiences of the different kinds of discovery. I would like you all to have a sheet of paper. I am going to start you off and to ask you what you can discover. A colleague of mine, an art colleague who always disliked mathematics, often helps me on a mathematics course. Each time she helps she takes part in the shapes session, because she enjoys this so much. Every time she makes a completely original discovery. I am going to give you her starting-point and I am going to ask you, "What can you discover?" If you do not discover very much I shall ask you to take it away—the idea I mean—and think about it because there is a great deal of mathematics in it.

The starting-point is to fold the sheet of paper in half in an original way—so find a way of doing this. Now cut it. What do you have to do to show that one piece fits the other exactly? It is always interesting to watch adults and children do this for the first time. They always make unnecessary turns. Here are two different solutions (see Figs. 1 and 2).

Notice that a similar example can be obtained by bringing opposite corners together. What is it that these two have in common? Could you find other solutions? (There are many more.) But what have all the folds in common? Yes, as soon as you make another fold you begin to suspect. In fact this is a lovely piece of mathematical continuity. If you try making fold after fold, where will one of the folds

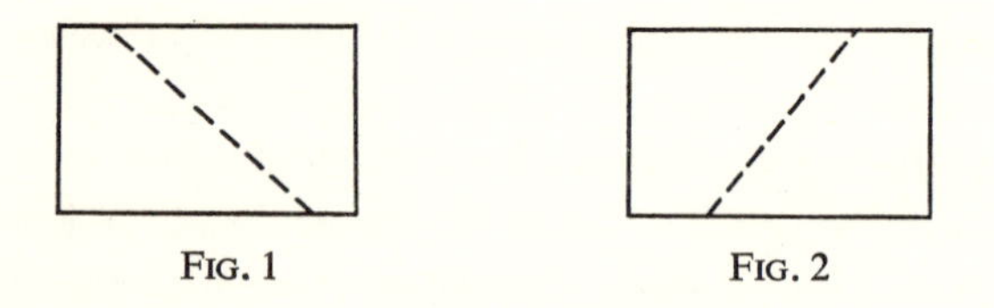

FIG. 1 FIG. 2

be? Yes, across the diagonal, and all the folds will be between one diagonal and the next diagonal in a beautiful continuous way that is characteristic of mathematics. But this is only the beginning of the discovery. The next thing which was done was that each pair of opposite corners was brought together and a fold was made. There are a number of mathematical relations to be found here.

I want to give you another problem to do. For you this is stage 2, for this colleague of mine it was a discovery, and her own discovery. I had given her cubes but she turned away from the cubes and said, "I am going to find out something with this piece of paper". So the paper was her own choice for a starting-point. (Discovery Type 1.)

You have all found that the two pieces left over are rectangles. Now take one of these rectangles and repeat the folding so that each pair of opposite corners is brought together in turn. The interesting thing is that you can then discover a mathematical relationship between the dimensions of the original rectangle and the new one. This can be repeated over and over again. My colleague was encouraged by her success and thought that she would start with other shapes. She used first a square and then a parallelogram. She discovered a delightful result with the parallelogram and endeavoured to relate this to that for the rectangle, taking the rectangle as the extreme case of the parallelogram.

What intrigues me is that using simple materials there are so many ratios and proportions that can be found immediately. At first the discovery is intuitive, but then it must be justified mathematically. This is another important fact about discovery, particularly with young children. Often they make intuitive leaps and usually their intuition is quite sound. Sometimes it is possible to say to them, "Convince me". Then they have to search for a mathematical reason. At a later stage I would say, "Search for a mathematical proof".

I am going to give you another problem to start you thinking. It is an interesting one so I want to give you the background to it. It was a G.C.E. "O"-level question set for an enterprising school. The pupils were allowed four hours to do only one question. This was the question:

"Investigate the areas of triangles with perimeter 24 inches".

I am going to suggest that you take a piece of string. It does not matter what the perimeter is. Work with a partner and investigate all the possibilities. I know of at least six methods of approaching the problem. May I tell you a story about this particular problem?

I was working with some graduates at Harvard. They were all mathematics graduates, and some of them had come from the Massachusetts Institute of Technology, where I was assured they had some of the best mathematicians in the U.S.A. I was asked to talk about children's work, but I do not like talking all the time because I believe this is not convincing. We must experience mathematics, it is no good talking about it. As someone once said, "Mathematics is like kissing; the only way you can discover its delights, is by trying it". I think there is a lot of truth in this. I gave the students this problem as a starting-point. I also included other problems initiated and solved by primary school children. I told the students that if they wanted materials, string, cubes and squared paper were available. (These were the things I asked you to bring!)

It was immediately apparent that some students always turned to materials while others did not. In this particular problem, at the end of three-quarters of an hour, those in the second group were still struggling, whereas those who had taken a piece of string had, within twenty minutes, found the complete set of solutions. Afterwards I discovered from the Professor that the first group had been a special experimental group and had been treated in quite a different way. In the United States a good deal of the university work consists of lecturing and nothing else. I know because I have attended lectures! I once asked a student next to me why he came to the lectures since we neither of us could see the screen and the student had an opened copy of the lecturer's own textbook from which the lecturer was writing on the overhead projector. Of course, he said what I was afraid he was going to say, "Because I need to get my attendance mark".

This particular group of students had been given two types of activities. All the work they did was problem work; there were group activities and also single activities for individuals. These students did not attend lectures at any time during their training. When we came to check afterwards, all those in the "string" group were from this experimental group, while the others were those who had received the usual lecture programme. I am glad to say that all mathematics students at Harvard School of Education are now given the experimental course.

Before you begin this problem I want to show you one experimental piece of work. The string reminded me of it. The teacher deliberately set out to find out what eleven of her 6-year-old children could do with scale. The first thing she asked this group to do was to put themselves in order of height. There were eleven children. Can you imagine the chaos there was when these children tried to put themselves in order of height? Eventually it was done and the teacher said to them, "I want you to draw what you have just done: yourselves arranged in order standing on the floor". [You know that young children ignore a base line, don't you? It is only when the teacher puts it in that you get a base line and when children are released from the teacher, they forget it again. Until they have gone through certain stages, children are not ready for a base line. It is only when

FIG. 3

they use large numbers that they need it. In general children choose to arrange their squares or their drawings in a rather haphazard way. They are not orderly by nature—except the rare one.]

Here are a few examples of what they did. There is the floor and sensibly the child has used the bottom of the paper for the floor. She started grading her drawings quite well until she reached the middle one. Then she realised that the drawings were going off the page so she started again. But she has got two sequences. Look at the next one in which the children are practising levitation! This child certainly has not drawn eleven figures. I do not think she can count up to eleven. The third child has drawn eleven figures, but the line comes down; it looks as though the children are skipping. The fourth drawing shows no ordering. The child has drawn a line at the bottom, but the figures are not in any order. The only child who succeeded in this difficult task was the one who turned them all to coloured rods (see Fig. 3).

Then the teacher said to the children, "You cannot always be your own graph standing up against the wall; what would you do so that we have a record of your heights?" They drew round the tallest and each child marked his height on the outline, adding their names. Then they matched and cut their heights in string. The teacher gave them a piece of paper and said, "I want you to put your string heights on this". You can guess the children's response for the paper was not nearly large enough. They said, "That isn't enough paper, we'll go and get another piece". The teacher replied, "I am sorry but there is no more yellow paper and you cannot mount your heights on paper of two colours can you?" So the children had to think of another way. What did they do to the string? It was too long, so what would you do? Halve it. Yes, in fact fold it into two. There it is. And you see that the children were getting their first experience of scale.

So often we rush children into using scale. We think they know why we use a small square instead of a big square. Children have great difficulty with this concept even at the age of 8. One point of interest arose. Mark discovered that on the outline he seemed to be taller than Susan, while on the half-scale "graph" he appeared to be shorter than Susan. This disturbed him very much and he had to go back and do his measurements all over again.

You notice that these children have learned a great deal without actually measuring. They were making direct comparisons, halving and so on. The children found out for themselves what they had to do. The teacher did not tell them, but she set the scene and she supplied the materials. Perhaps I ought to call this guided discovery because at each stage the teacher asked the next question the children needed to help them to progress.

> [Interval for attempts at the triangle problem. This was followed by discussion.]

You have experienced ideas of continuity, limits and found various relations. If for example, you fix the base, what is going to happen to the path of the vertex? Yes, you can soon establish that it is an ellipse, except in one particular case. And what are the extreme positions going to be? Can you imagine it? Yes, the area is going to be zero at each end. What is the maximum area going to be? Well, I do not want to take up too much time because I think this is a problem that you need to think about for yourselves, but I would like to tell you that the group I started working with began, I thought, in a delightful way with equilateral triangles. Is this because the equilateral triangle is symmetrical? I am quite sure that it is the same with children, symmetry has perceptual appeal. And when I asked what should we study next, the answer came, "Isosceles triangles". So the next to be investigated were isosceles triangles.

Here again you get two extremes. Often adults or older children make an intuitive leap. How many of you found out what happens, for example, if you fix the base? Where did you find maximum area if you did fix the base? And which shape has the greatest area of all? Which triangular shape? If you do not know, would you like to make an intuitive guess? All right, the equilateral triangle.

Now at one extreme you can prove this by calculus. I have been struggling for some time to find a situation which would help older primary children who had discovered that the isosceles triangle is the largest, to find out for themselves that the equilateral triangle has the greatest area. I want you to find a proof without using the calculus. It is not easy, but there is a delightful method of approximations, starting with the fact that with any one base, the isosceles triangle has the largest area. I found a teacher using this method and you may want to use this yourselves.

This problem could be tackled at many different levels. For example, slower secondary pupils can draw the various triangles in colour. This time last year I was in Malaysia working with teachers. Throughout, their work was more attractive than any other work I had seen. They wear very attractive and colourful clothes too, and all the work they did reflected this delight they had in colour. I think that we should think of this continually in mathematics, it should be a subject of delight. I am not saying that all children need the same experience, in fact I have tried to show you from this question that sometimes people need very different experiences. It is not always the brightest pupils who have no need to use materials. Before I finish I shall give you an example of a boy who made a remarkable discovery; he would not have done this if his teacher had not given him initial experience when the boy was 10 years old, in learning mathematics through first-hand investigations.

Now before I go on, would you like to ask me any questions?

Student: "When we talk about discovery at secondary level, is there time, in terms of pupil's time, and also cost of money in time, for all the necessary work to be done by discovery and investigational methods?"

Miss Biggs: "Could I perhaps answer you this way. This important question is always being raised, as you know. There are not a large number of secondary schools working in this way, but I want to tell you about one teacher's experience. She was trained in this institute about 9 or 10 years ago. She began 18 months ago to introduce this way of working in a Grammar school. She introduced it throughout the school. When you first begin to teach in this way you may find it easier to begin with a single group. This will not only train your pupils, but it will also accustom you to this way of working. Now this teacher divided her classes into six groups because she had 35-minute periods. She planned to give 5 minutes to each group so she could not manage more than six groups. Towards the end of one term, one group was going to have examinations, so she decided to put all her classes back into the traditional situation. She found she had a mutiny on her hands! Those who were going to sit examinations brought along their syllabuses and said, 'Do you realise that we have covered more of the syllabus this term than ever before? We realise how much we have learnt from each other'.

"But let me also confess that I know from bitter experience that this active method of learning seems interminably slow to begin with and until you get it established, a little worrying. Later, however, I found that I could go right outside the prescribed syllabus and we could still cover our examination syllabus without difficulty.

"What is the reason for this? First let me emphasise my three aims for the sound learning of mathematics. The first and most important is to give pupils the opportunity to think for themselves. Secondly, we must give pupils the chance to experience the order and pattern which are the essence of mathematics and which are to be found in both the natural and man-made environment. Thirdly, we must give them the skills. There are a certain number facts that every one of your children should know. Moreover, they need practice too. The trouble is that too often in the past we were given practice at the wrong time. It was given before we had the necessary number knowledge. In those days it was thought that mechanical practice would result in learning. In fact practice, before children are ready for it, usually drives them to use inefficient methods. For example, if a child does not know 8×7 when he needs this, he may start at the beginning if he has been brought up by traditional methods. If he has not and he knows 2×7, he may go on doubling; he may say 2×7, 4×7, 8×7, a much better method because the child is really looking in a different way at mathematics."

Let us consider the number language children require. Are there any infant teachers here? None at all. Then I will give you an outline of mathematical experiences provided in infant schools.

The children will have various experiences in length, time, capacity, volume, area, money, number and shapes. In addition, infant teachers are still striving to get children to know all their number facts up to 20. Now there are some children (not a great many) who are able to go up to 20 and far beyond. If a child is at home with numbers and enjoys them, let him go just as far as he likes as long as he is enjoying himself. But there are many children who arrive at Junior school knowing no number facts at all, or very few. I think the reason is that Infant teachers are striving to get so much knowledge into the children. What do I mean by number knowledge? I mean that if I ask a child to add 5 and 3, she will say 8 immediately. This does not happen in the first year except with a very few children. You see children have to begin by learning the counting numbers and what they are. It is only when you begin teaching young children that you come face to face with this difficulty. For example, if I ask a young child how many animals he has and he begins to count, he can only do this successfully if he has things to handle, because, of course, counting is a one-to-one correspondence between each object and the appropriate number name. When, say, the child has counted a set of five objects and I ask him to show me five, in the early stages he simply shows me the last one he has counted. He does not know at this stage the meaning of the cardinal number of a set. He does not realise that first he matches each object to a number name, and that when he runs out of objects, the set of objects is a matched set with cardinal number the last number he has used.

When, therefore, I ask the child to show me five, I expect him to pick up the entire set, not merely to show me the fifth one he has counted. This is not an easy idea for children so I want you to consider what I am saying against this background of experience. I believe that as far as number knowledge is concerned Infant teachers should first aim at a thorough knowledge of addition and subtraction facts to ten. There are a few other facts I want to discuss, but if children know these thoroughly before they are transferred to the Junior school, I think that children would have a much better foundation and they would feel more confident.

But there are, of course, other facts which children need to know. Do you all know what I mean when I say an operation table of addition for the numbers 1 to 10?

10	11	12	13	14	15	16	17	18	19	20
9	(10)	11	12	13	14	15	16	17	18	19
8	9	(10)	11	12	13	14	15	16	17	18
7	8	9	(10)	11	12	13	14	15	16	17
6	7	8	9	(10)	11	12	13	14	15	16
5	6	7	8	9	(10)	11	12	13	14	15
4	5	6	7	8	9	(10)	11	12	13	14
3	4	5	6	7	8	9	(10)	11	12	13
2	3	4	5	6	7	8	9	(10)	11	12
1	2	3	4	5	6	7	8	9	(10)	11
(+)	1	2	3	4	5	6	7	8	9	10

FIG. 4

Figure 4 will show you what I mean. The tens are marked in because we are concerned with the numbers whose sum is up to ten. What do you notice about the tens? They form an interesting pattern. Why do they go down in equal steps? What is the mathematical relationship? Suppose you think of the bottom row as x numbers and the left-hand column as y numbers. Yes,

$$x + y = 10$$

because the sum of the two numbers is ten.

Do children really need to know all these facts? Let us look at it more closely. Look at the leading diagonal. What is it? I will read it to you. It goes 2, 4, 6, 8, 10, etc. What are these numbers? Even numbers, yes, what else? Yes, its the 2 times table. Infants will be able to double numbers far in excess of 10. A 6-year-old said to me the other day, "Give me a number and I'll double it for you". I gave him 37 (he was calculating in his head). He said, "Two thirties that's sixty, two sevens that's fourteen, seventy-four". He continued, "Don't tell my teacher". I asked why not since his method was such a good one. The boy answered, "She makes me write it down and I don't understand her method, so I do it in my head and then I write it down her way and I always get my mark, so don't tell her". But you see this mental calculation shows a grasp of numbers which is just what we need to ensure. There is no point in written calculation when a child can calculate so efficiently in his head.

To continue, what do you notice about the numbers on either side of the leading diagonal? Yes, they are symmetrical. Why? If, for example, we have 5 add 3, in our addition table, we get the same result as 3 add 5. Can you check this from the table given in Fig. 4? Yes, we have

$$5 + 3 = 3 + 5 = 8.$$

Thus we can observe this very important law which holds for addition. It is not only necessary for children to discover this fact for themselves, it is also important for teachers to see that they use this law. It is so much easier for a child to add 2 to 5 than it is for him to add 5 to 2. So you see that although I have said, "number facts up to ten", I am thinking of other facts too.

There is one other set of facts which I would expect some infants to know. This is adding 10 to numbers from 1 to 10. There is a stage when children adding 10 and 7 will count up in ones to get 17. One day the teacher thinks she has won the battle at last and that whatever number she gives the child to add to 10 he will give the answer immediately. Look at the problem caused by our language. We say fourteen and seventeen and yet twenty-four and twenty-seven. This stage requires effort on the part of the teacher and children. Moreover, one day the children can give the answer immediately, but the next day they have forgotten all about it again! When the children are transferred to the Junior school (and even sometimes on transfer to the secondary school) we find that they have forgotten these facts, since, temporarily at least, they feel insecure.

How are we going to find out whether children know these facts? There is only one way. Every infant teacher gives each child an opportunity to read to her most days. Why not do this with number sometimes? We can only learn about a child's number knowledge by taking him alone. To begin with it took me 10 minutes to find out how much number knowledge a child possessed. Now I can do it in 1 or 2 minutes. (At a later stage children will also need to know the repeating pattern 9 and 7, 19 and 7, 29 and 7, ...)

Now I want to take a leap and to look at the other end of the Junior school. In other words, I want to look at the child entering the secondary school, because these are the children with whom you in this audience are going to be concerned. Are there any of you going into primary schools? Only one. Everyone else is going to secondary schools, so let us think about this stage.

Again I think there is a minimum number knowledge which we should aim at and which I hope teachers will achieve for 90 per cent of the children. Out of that 90 per cent there may be a third who will know as much as, hopefully, you do. Do you know all the multiplication tables? If I said 8×7 would you give the correct answer immediately? Let us try it out. I am disillusioned because I often run in-service courses for teachers. Now that we use the metric system in arithmetic, the score is, I regret to say, under 50 per cent. I find the same result for both secondary and primary teachers. So try to see that your children know a minimum. What is this minimum? Well first of all they have to have the number facts up to 10 by immediate recall. Then they need to extend this knowledge up to 20. I believe that the slower children can know these facts at two levels whereas quick children know them thoroughly as you and I did. (Remember that you belong to a small percentage.) Slow children need a target. This is what I think is probably the minimum. First immediate recall of knowledge of addition and subtraction number facts up to 10. Once we can add 10 to a number we can easily add 9. Let us add 9. Once you have added 10, then by quick recall you can add 9. I hope the child is going to say, "Of course I know that 9 and 7 are 16, because 10 and 7 are 17 and it is one less". I would rather a child does this because he is thinking about the inner relations of number.

There is yet another set of numbers with which we can help children. I call these the near-doubles. It is interesting to see that these numbers come on either side of the doubling line (or multiplication by 2). Let us take 13; 13 comes by adding 7 and 6. That is another fact which children obtain through quick recall. How are these children going to do that? I hope they are going to say, "I know that it is 13, because it is 2×7 less 1, or it is 2×6 plus 1". I hope for both those reasons. Here you see the child investigating, exploring and reasoning for himself.

You will see that only six number facts between 10 and 20 remain, since all the others have been dealt with in one way or another. You will all have come across the number line when I add, as essential number facts, extensions such as 9 and 7, 19 and 7, 29 and 7. You know what I am talking about.

I believe that we should not give children written practice until they have the necessary number knowledge that I have mentioned. What do you consider would be a minimum for multiplication tables? Some of the children, like yourselves, will know them all. What is the minimum? I think there are three that they must know. They must know their 2 times, because then they can get 4 and 8 times; the slow ones by quick recall. They must know 3 times, but I would select another one rather than 5 times. My choice would be the 10 times. Ten, everyone must know, and this brings me to my next point.

Children should be given opportunities to discover their own methods for the reasonable calculations they meet. In other words, they should invent their own methods of long multiplication and long division. Once they have invented their own methods we may have to help them, by guided or directed discovery, whichever is appropriate to the child, to make those methods as efficient as possible. I am going to give you two examples. I had enjoyed a wonderful day; the school I was in had for some time been using methods of investigation for the children's learning. I asked the children many questions and at the end they always told me what to do. They could not always carry out the calculation, but they always knew what they had to do. How wonderful it is when children have no fear of problems! Towards the end of the day I was with a class of 10-year-olds and we had been talking about large numbers. I asked them what a thousand looked like. On the floor there were nine-inch tiles, so I asked them whether they thought there were a thousand tiles on the rectangular floor. (I had no idea, of course.) The children counted the tiles one way—there were 36—and then they counted them the other way—37. To my amazement the children continued counting in ones. I pointed out that counting would take too long and asked them to think of another method. There was no response at all. Then I asked, "What do you have to do to the numbers to obtain the answer?" They shouted, "Multiply them". But still they still did nothing. Now I thought that these children would be perfectly capable of multiplying two numbers together. So I asked them if they knew 30 × 30. "900", they shouted. One boy added: "And 6 × 7 is 42 so that gives us 942." For a moment I wondered why I did not know this method! It sounded so simple! Then I pulled myself together and suggested that each child should draw a picture for himself. I emphasised multiplication by 10 so you can guess how they divided the room. See Figs. 5 and 6.

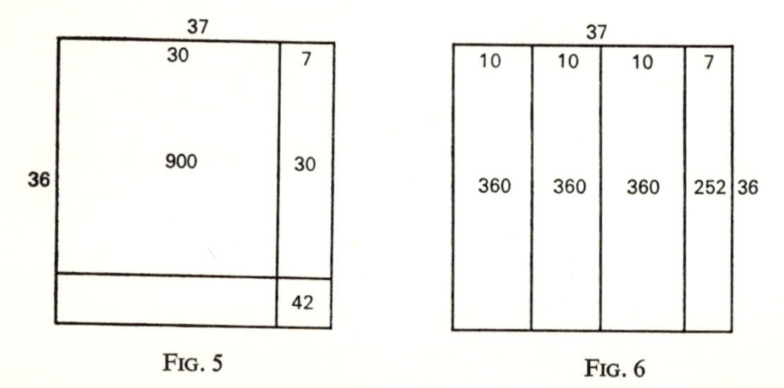

FIG. 5

FIG. 6

Both diagrams can lead to the traditional method of multiplication in which we have 30 multiplied by 30 plus 7 and 6 multiplied by 30 plus 7. I found some slow secondary pupils in the United States using this method illustrated by Fig. 5. How much better that they should use a method which they could master and which they could always use, than a method which they didn't understand at all!

I want to give you another question, this time leading to division. This will be a structured discovery! I was in the first metric lift I have ever seen. A notice read: Maximum load 2000 kilogrammes. Standing next to me was Mr. Jones who had just been on a course with me. I asked him how much he weighed and he replied 85 kg. So the question I want to ask you and Mr. Jones is, "How many Mr. Jones could we safely carry in the lift?" How would a child answer a question of this kind if he had not been taught the long-division method? What is his first way of doing it? Yes, subtracting 85 over and over again. When a child has come to the end of subtracting 85 kilogrammes at a time, what is the question I am going to ask him to help him to refine his method? This is an important stage. In fact I would ask the children if they could subtract easy groups (or multiples) of 85, instead of taking away one 85 at a time. I hope they would suggest multiples of 10. I should certainly see that children were able to multiply any whole number by 10. I have noticed that when children invent this method for themselves they go straight on to subtract another 10 without difficulty.

2000 kg	
850	10 Mr. Jones
1150	
850	10 Mr. Jones
300	
170	2 Mr. Jones
130	
85	1 Mr. Jones
45	23 Mr. Jones

Often children may continue to subtract 85 one at a time or 170 and then 85. The important thing is that every step has been taken by the children themselves. The bright children will refine the method still further, and subtract twenty 85's.

Other methods are sometimes unmathematical. When using the traditional method we say 85 into 2, 85 into 20, 85 into 200. What nonsense! Now you could justify this method as long as you do not say 85 into 2. You must say 85 into 2 thousand, 85 into 20 hundred, 85 into 200 tens. But in the first instance children will not invent this method.

There are very few children who arrive at the secondary stage able to do long division. In an area where there is still an 11-plus examination, only 4 per cent of the children performed the long division correctly. So we have nothing to lose! It is an interesting historical fact that about ninety years ago the method I first described was taught in elementary schools, because it was thought that was the method children would understand. Today we have gone back to the other method!

I have given you but a glimpse of all the possibilities. I have children's work here on shapes and a variety of relations which children are able to discover through first-hand experience. But there is one piece of work which I want to describe because it is very important for you as secondary teachers. You all know what this graph represents:

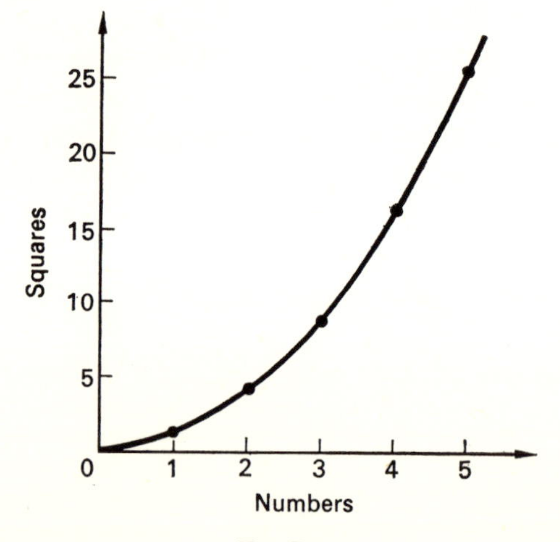

FIG. 7

It was made by 10-year-olds. It is a graph of the squares of numbers. The children first built a sequence of squares using identical unit squares. They were delighted to find that square numbers formed squares when they built them with unit squares and that cube numbers were cubes.

Now the 10-year-old boy whose work I am about to describe had an I.Q. of about 110. This boy had only been using methods of investigation for about two months. He brought the graph of the area of squares to his teacher and said, "I do hope you don't mind, I don't want to do any more of this work", indicating the attractive materials around the room. This boy had drawn ordinates on the graph. He said, "I can see that the area under the curve is changing and I want to find the pattern". He also posed another question on gradients. He had been working with a truck running down a slope. He had asked the teacher, "If I prop the slope on one brick and time the truck from top to bottom and then I prop the slope on two bricks and time the truck from top to bottom, will the time be halved the second time?" The teacher suggested that Peter and his two friends should experiment to find the answer to this question. So Peter's second question concerned the pattern of gradients at various points on the curve of the squares. His teacher asked, "But how do you know that there is a pattern?" Peter replied with great conviction, "In mathematics there is always a pattern, you've only got to look for it". It took Peter and his two friends six months to discover the patterns. But no one knows how long it took Archimedes! Peter had to calculate areas by counting squares, which is probably the method used by Archimedes. At the end of the six months Peter had discovered the calculus, despite the fact that no one at home could give him help.

I would urge you all, for I believe we owe it to every pupil, to give the children the chance of investigating a problem before we collect together and then summarise the pupils' attempts. I am not saying that we should not have class discussions, but learning at different levels through a wide variety of investigations is the soundest way of learning.

I believe this method is the best way to give our pupils real excitement in mathematics. I believe too, that it is only when we give our children a chance to think for themselves that they realise their full potential. You will remember from your own experiences, that for a good deal of time at school you were working at about 50 per cent capacity. And I believe that sometimes our adolescents rebel because we do not let them think for themselves.

There is a book that I recommend to you all. It is called *The School I Would Like*, edited by Edward Blishen. It consists of a collection of essays by boys and girls aged 12 to 17. When you read this book you will know what some secondary children feel about teaching they receive and you will realise that they really should have an opportunity to think and to decide for themselves.

So I give all of those of you who will be going out to teach in secondary schools my very good wishes. I hope that you will find some way, even if it is very tentative to begin with, of introducing learning by investigation. Do not be discouraged at any time. There is no other profession as enjoyable as teaching.

A Note on Discovery

J. F. DEANS

Discovery

As a preliminary to a discussion of discovery as a way of learning mathematics let us consider what is implicit in an act of discovery. They way in which the word is used is commonplace enough: Columbus discovered America; Marie Curie discovered radium; George Smith discovered a route which avoids the Exeter by-pass. It would seem that we discover something that is "there" in some sense, and in addition to objects and locations this might be some physical, mathematical† or other property as, for instance, Boyle's law for gases or that $\sqrt{2}$ is irrational. A discovery is not necessarily unique; someone else seems to have discovered America which affects the question of priority but does not invalidate Columbus's claim to a discovery. Simultaneous but independent discoveries have not been uncommon in science and mathematics. One individual might even discover the same thing on more than occasion if the original discovery has been forgotten in the interim, as for instance how to make a convex hexagon from the tangram pieces. Something new, however, must be brought to light, at any rate for the particular individual at a particular time. Our discovery may or may not be purposeful; we may find out what we set out to, but on the other hand we have all made accidental discoveries.

† In what sense, if any, there exist theorems undiscovered as yet by anyone, given a system of axioms, is a philosophical question. We shall not attempt an answer here.

Discovery differs from invention or artistic creation although discoveries will certainly be made in the course of these. Watt did not *discover* the steam engine, but in the process of inventing it, he discovered something about the behaviour and uses of steam. That discovery is not an inevitable outcome of mental or physical search, however assiduous, need hardly be said.

How much of the effort must be yours for what you find out to rank as your own discovery? One could imagine the provision of a series of clues leading towards the solution of the tangram problem mentioned above. These might take all the difficulty out of the puzzle for a mathematically inclined adult, yet might leave quite a challenge for a child. Gibby's article on programmed learning refers to the difficulty of making the gradation of the units testing enough to maintain interest, yet not so difficult as to slow down progress unduly. In structuring a situation so that mathematical learning might take place, in providing guidelines, clues or leading questions, we have a task differing from the construction of a linear programme in that we must allow freedom for the child's thought to range widely, but we have a similar problem in providing the correct gradation of experiences. Friedlander† suggests an alliance between discovery and programmed learning: "a prerequisite for fruitful, intuitive discoveries might well be linear programmes on how to make them". However, one would feel the "discovery label" to be appropriate only if the contribution on the child's part is substantial. If great benefits do in fact accrue from a child's *discovery* we must avoid over-structuring the situation so that it begins to approximate to a conventional programmed learning sequence. On the other hand, too little guidance can lead to the child's aimless wandering.

According to Piaget new experiences are met on the child's part by either assimilation or accommodation. If he has available a schema with which the new experience ties in he can assimilate it. If, on the other hand, the element of novelty is too great for assimila-

† B. Z. FRIEDLANDER, "A psychologist's second thoughts on concepts, curiosity and discovery in teaching and learning", *Harvard Educational Review*, Winter 1965.

tion to take place he will have to accommodate to the new situation by modifying existing and relevant schemata. It is through such alteration of existing structures that new ones emerge and the child's conceptual equipment grows. But only too frequently in mathematics teaching children have been called upon to make accommodations beyond their capacities. When a new topic, or some new aspect of a topic already dealt with, is introduced, it is essential that the child's existing knowledge is adequate, and is so organised, that accommodation can take place. If the jump required is too great the child must strive to avoid utter failure by some other means, and rote learning is resorted to. But this is a dead-end strategy. There is a place for rote learning, but it is peripheral to the main process; what is learned by rote is sterile, and cannot give rise to new growth. E. A. Lunzer, in his *Recent Studies in Britain Based on the Work of Jean Piaget*, sums the matter up "... the more satisfactory means of advancing learning should be that which (a) brings the child face to face with the inadequacy of existing schemata; (b) offers him the means and the guidance for making new accommodations; and (c) encourages their integration with previous schemata".†

These points have relevance for the devising of schemes through which a child is led to learn by discovery. J. G. Wallace, in *Concept Growth and the Education of the Child*, describes an inquiry conducted by S. H. Banks‡ in which secondary modern school pupils discovered scientific principles through experiments. Great interest was shown by the children when they were actively engaged in verifying their own hypotheses. Lunzer§ links this inquiry with one of Smedslund's in which techniques of teaching transitivity of weight were compared. The first group had repeated experience of balancing three objects in pairs, A against B ($A > B$) and B against C ($B > C$), being then required to deduce the relation between the extremes, which they were allowed to verify. The second group was required to arrange sets of three objects in ascending order of weight, without being

† Page 48.
‡ Page 157.
§ *Op. cit.*, p. 51.

directed in their procedure as was the first group. As Lunzer puts it, "... the question of transitivity was forced on these children in the context of a genuine problem", and this group showed a gain in understanding of transitivity of weight, while the first group did not. He concludes, "... it is by confronting the child with phenomena which are *unexpected* in terms of his existing schemata and allowing him the opportunity of trying out modifications for himself that fundamental reorientations may be brought about".

The effect of surprise, of a certain degree of complexity, would seem to be catalytic. The teacher's quandary is what Hunt has called the "problem of the match". The situation with which the child is presented must exhibit recognisable links with the familiar, calling on schemata already formed. For these schemata to develop, however, accommodation must be demanded, and while the accommodation required must not be too great, so as to cause breakdown and regression, the element of novelty, of the unexpected, is the spur to progress. An approach to mathematics teaching which utilised discovery, particularly in connection with the first meeting of some new principle, would seem well placed to capitalise on the effect of the unexpected.

If the mathematics learning resulting from discovery is superior in some way to that achieved through other methods (and this remains to be proved), it would seem to be due to two factors:

(i) understanding is idiosyncratic and the path to new insights will necessarily be idiosyncratic if it is the individual's own discovery;
(ii) the affective boost of the "Aha!" experience, the paradigm of which is Archimedes' shout of "Eureka!" and his associated behaviour. The Gestalt school of psychologists have drawn attention to the feeling of satisfaction which accompanies the achievement of closure.

Now the idiosyncracy of the discoverer's route and the vividness of his "aha!" feeling would seem likely to be inversely proportional to the number and significance of the clues provided. We do not enjoy a detective novel with a too obvious solution. Some educators mistrust discovery that is too closely guided and feel the need to distinguish it from "genuine" or "real" discovery. David Wheeler,† writing of structural apparatus, makes a demand, "... for a material which can be used in such a way that children (genuinely) discover the essential concepts and processes".

Of course totally unguided discovery would be very unproductive and no one would advocate such a thing. One would hardly leave a child in a field at the age of 5 and come back 10 years later, having of course catered for his bodily needs, to see what he had discovered. In making some provision for the child's activities his course is already influenced. Any structural apparatus merely by being structured leads the child in certain directions; apparatus like Cuisenaire into different types of situations than an apparatus like Dienes' Multi-based Arithmetic Blocks. The teacher provides the environment and the equipment and in so doing guides the child's thinking and investigations direction. Discovery is not to be qualified only by the adjectives "genuine" and, presumably, "spurious", but according to the degree, greater or lesser, of guidance provided.

W. J. Glennon's‡ survey of current thinking on mathematics teaching puts discovery in perspective among other theories. He outlines three theories of the purpose of the curriculum and three concerning the method of teaching, of which discovery is one. The first of the theories of the curriculum is the *psychological*, which minimises systematic instruction in favour of the expressed interests of the child. The extreme form of the theory is illustrated by A. S. Neill's only reference to primary school mathematics in his book *Summerhill: a Radical Approach to Child Rearing*: "Whether a school has or has not a special method for teaching long division is of no significance, for long division is of no importance except for those who *want* to learn it. And the child who *wants* to learn long division *will* learn it no matter how it is taught."

† D. Wheeler, "Structural materials in the primary school", *Mathematics Teaching*, Spring 1963.

‡ *The Arithmetic Teacher*, Feb. 1965.

For the *sociological* theory of the curriculum the criterion is adult usage. The list of our basic requirements is not a lengthy one and the resulting syllabus is rather dull.

A renewed interest in the *logical* theory of the curriculum has come about as a result of the evident scientific and technological needs of today. There has been a call for more, and more rigorous, mathematics in our schools. Glennon characterises recent experimental programmes according to their degrees of commitment to this theory as conservatively modern, ultra-modern and "pie-in-the-sky" modern.

The *didactic*-theory of method has a persuasive advocate in D. P. Ausubel, a convenient source of whose views is *Readings in the Psychology of Cognition* edited by Anderson and Ausubel. Most teaching, he asserts, has been and needs to be done by telling, but this process need be neither authoritarian nor rote. Ausubel distinguishes the rote-meaningful dimension from the reception–discovery dimension; reception does not necessarily involve rote learning, while discovery need not be meaningful. The first of these assertions will be readily conceded; the second may be illustrated by an anecdote from Friedlander's article. A teacher had written on the board a number of words beginning with "c" hoping to lead the children to see when the phonetic value was soft as in "cent". A boy "discovered" that the criterion was whether the "c" was a capital or small letter, although the size of the letters as written was virtually identical. "His great insight", says Friedlander, "gave him the full Brunerian glow of satisfaction". The didactic method was further held to be an efficient method of transmitting large bodies of knowledge.

Psychotherapy as a theory of method has as its principal aim the improving of the learner's relations with others and the development of his self-discovery. Growth in the learning of the classroom's traditional subjects is regarded as much less important.

Discovery is the other theory of method that Glennon discusses. Its link with the psychological theory of the curriculum is obvious, but it need by no means be absent from a development along sociological or logical lines.

It might be mentioned that Professor Glennon hopes for a synthesis of the three theories of the curriculum going hand in hand with a similar synthesis of the method theories, with an avoidance of the errors embodied in any one of the views when carried to an extreme.

To return to discovery. The method is not new; under the title "the heuristic method" (a label stemming from the same Greek word as Eureka!) it has had a long history, particularly in science teaching. The name of Dewey is prominent among those who have favoured this approach. An American book on *The Teaching of Mathematics in Secondary Schools*, by Schultze, published in 1912, lists some advantages and disadvantages of the method:† Students think for themselves and are not merely listening for information. They acquire a real understanding of the subject. Their interest and the resultant willingness to work are greater when they are taught heuristically than when they are taught by the information method. Teachers are in complete touch with their classes. Home study is not nearly so heavy or tedious as when the informational methods are used. On the other hand, the heuristic method is slow especially in the beginning. It is sometimes difficult to make students discover certain facts. The method does not work well in the hands of every teacher and the method is difficult for the teacher for he cannot simply follow a textbook. Half a century later these points are still pertinent for a discussion of the discovery method in the classroom.

The name most frequently associated with the method today is that of Jerome Bruner. His contributions in Anderson and Ausubels' book are well worth reading as is his *Process of Education*. The latter records his account of a conference of educators, mostly scientists and mathematicians, at Woods Hole, Massachusetts, in 1959. "The discovery method", he says, "epitomized the conference".

He defines discovery as "a matter of rearranging or transforming evidence in such a way that one is enabled to go beyond the evidence

† Quoted by A. L. HESSE, "Discovering discovery", *The Arithmetic Teacher*, Apr. 1968.

so assembled to new insights".† (One is reminded of the Gestalt view of problem solving to be found in Wertheimer's *Productive Thinking*.) He is enthusiastic about the potentiality of discovery as a teaching method, but he is well aware that it is not the whole story, and also that its employment does not automatically ensure successful learning. "The history of science", he points out, "is studded with examples of men 'finding out' something and not knowing it".‡ "Discovery, like surprise, favours the well prepared mind".§

He discusses four benefits deriving from discovery:

(i) increase in intellectual potency,
(ii) the shift from extrinsic to intrinsic rewards,
(iii) learning the heuristics of discovery,
(iv) the aid to memory processing.

Under the first heading he distinguishes between two approaches to problem solving, the purposeful, organised approach, and that which gathers information at random. A child's strategy in a twenty-questions game illustrates these tendencies. He put forward the hypothesis that emphasis on discovery learning leads a child towards the constructive, organised approach, but he admits that there are as yet no supporting research results.

Discovery itself can be rewarding. Curiosity, the fun of finding out, seems to have been an incentive lower down the evolutionary scale with Harlow's monkeys. Certainly in classroom sessions given over to discovery the need for the traditional sticks and carrots is little in evidence.

The third heading is related to the first, perhaps somewhat as tactics is to strategy, comprising skills of a more particular nature. "It is my hunch that it is only through the *exercise* of problem solving and the effort of discovery that one learns the working heuristic of discovery". Bruner sees engaging in inquiry as the only

† "The Act of Discovery", *Harvard Educ. Rev.* **31** (1961) and reprinted in Anderson and Ausubel (eds.), *Readings in Psychology of Cognition.*

‡ *Loc. cit.* [The Act of Discovery].

§ *Loc. cit.* [The Act of Discovery].

means of improving in technique, and would seem to discount the usefulness of books such as Polya's *How to Solve It.* Certainly when problem solvers of the first rank, such as Poincaré, describe their techniques, experience and intuition take pride of place over any system of rules.

In the related field of invention, W. J. J. Gordon's book *Synectics* gives an entertaining and thought-provoking account of some recently developed techniques which have found application in the scientific and commercial fields. These push intuition a stage further in attempting to draw on the mind's unconscious resources. Of synectics theory Gordon says in his introduction,† "It is an operational theory for the conscious use of the preconscious psychological mechanisms in man's creative activity. The purpose of developing such a theory is to increase the probability of success in problem-stating situations". Although the examples cited in the book are of industrial applications, an investigation into the employment of the techniques in schools and colleges is mentioned. When he puts forward the hypothesis that "in creative process the emotional component is more important than the intellectual, the irrational more important than the rational",‡ he has not poetry or painting in mind, but the invention of hydraulic jacks and vapour proof closures for space suits. He continues: "it is these emotional, irrational elements which can and must be understood in order to increase the probability of success in a problem-solving situation". With such an emphasis on the emotional and the irrational, this theory may seem a far cry from the teaching of mathematics, and even directly antithetical to it; but Gordon is not rejecting the rational, as his transcripts of problem-solving group sessions show. He is trying to harness those elusive processes, inaccessible to introspection, to which the autobiographical writings of so many mathematicians and scientists bear witness.

The members of a group which is concentrating on a particular problem area are encouraged to employ fantastic metaphors to try

† Page 3.
‡ Page 5.

to make the problem accessible. Several illustrations are given in the book of one member's extraordinary analogy sparking off a train of thought in another, and so leading to an elucidation of the problem. As far as education is concerned, these investigations would seem to be in their infancy; their application would require a certain degree of sophistication in group members, who might be drawn from the upper secondary forms or from college students. Few of us might feel drawn as yet to explore the possibilities of the methods which Gordon describes, but they have been found fruitful in some fields, and may yet open up a new dimension in educational technique.

Under Bruner's fourth heading, the aid to memory processing, the superior recall of ideas which are linked and not isolated is noted. Most rote learning involves a collection of facts which although perhaps similar are not linked to each other; the recall of one does not in general help in recalling another. Memory-improvement systems, for which one sometimes sees advertisements, and of which Ian Hunter's article in *Science News* No. 39 gives a useful survey, utilise the potency of a link between one idea and the next in a set to be memorised. The linking mechanism may be prescribed in detail by the system, or a mechanism may be described in general terms, the user devising the details to suit himself. The latter would seem the more likely to produce a satisfactory mnemonic system. An individual's learned schemata generally exhibit idiosyncratic linkages and the personal nature of discoveries would aid future recall in that they would fit into the existing schemata in a natural way.

Anecdotes concerning isolated but significant discoveries by individual children abound. The Schools Council's Curriculum Bulletin No. 1, *Mathematics in Primary Schools*, largely the work of Miss E. E. Biggs, is a good source. The Plowden Report quotes the following example:

> Some ten year old children had collected a number of bird and animal skulls and became interested in comparing the capacities of the brain cavities. They had to think out an efficient method of measuring them, and then construct some cubic receptacle for measuring the dry sand that they had poured into the cavities. The cubic inch that they had used for the cat's and

the rabbit's skulls proved to be too large for the bird's and the interesting discovery that a cubic quarter inch was not the same as the quarter of a cubic inch was not likely ever again to be forgotten.†

Miss E. E. Biggs has for some time been going up and down the country organising courses to popularise this kind of approach. In the introduction to the Curriculum Bulletin No. 1, she writes that "it contains a summary of intensive work in the learning of mathematics by discovery methods carried out with children and teachers during the past six years". The Nuffield Project in its first progress report sees its work as "... helping teachers to find out about, discuss among themselves, try out and apply—in their own way and in the classrooms—what is often called the 'discovery' approach to mathematics teaching". Both of these important sources of influence on classroom teaching, at least up to Middle School level, see themselves as propagating the discovery method. In some of the Curriculum Bulletin's examples and in some to be found in the Nuffield literature, the element of discovery does not seem prominent and "activity" would better describe these. The set of all discoveries would presumably be a subset of the set of all activities, and while the discoveries are undoubtedly important, those activities which do not involve discovery are not without value. The "real life" school of thought from Brideoake and Groves onwards has much to offer in the realm of productive activity. To gain some of the insights Piaget shows to be lacking in young children they would seem to need repeated physical experience of certain manipulations. When a discovery has been made by a child, related activity should help to consolidate his findings.

There is widespread optimism about the use of the discovery method. The Plowden Report says: "There is ample evidence that many of the claims made for the new approach are well founded".‡ There is in fact no such thing as "ample evidence"; the opinions of teachers who enjoy using the discovery method, the effect of which is heightened by a too easy dismissal of detractors as stick-in-the-muds, are no substitute for research findings. In Britain these are

† Paragraph 654.
‡ Paragraph 656.

lacking, and Ausubel's† summary of American findings points out the difficulties involved in such research and the temptation for enthusiasts to come to prematurely optimistic conclusions:

> Careful examination ... yields these three disheartening conclusions: (a) that most of the articles most commonly cited in the literature as reporting results supportive of discovery techniques actually report no research findings whatsoever; (b) that most of the reasonably well controlled studies report negative findings and (c) that most studies reporting positive findings either fail to control other significant variables or employ questionable techniques of statistical analysis. Thus, actual examination of the research literature allegedly supportive of learning by discovery reveals that valid evidence of this nature is virtually non-existent.

Max Beberman's U.I.C.S.M. project, described in *An Emerging Program of Secondary School Mathematics* stresses unambiguous language and discovery as the two essentials for understanding mathematics. Davis' Madison project eschews exposition; he would like to see discovery ubiquitous in the mathematics curriculum. Most other British and American projects find a place for discovery in mathematics learning. One cannot doubt that the approach is of value. The problem is to determine for which topics, for which areas of mathematical learning it is appropriate and in what ratio to other methods in the various age groups. The child's need for repetition and practice in learning mathematics is not always given proper emphasis by advocates of discovery. This is an aspect of mathematics learning with which various systems of structural apparatus are well equipped to deal. Much more research is needed, with realistic teaching of control groups. These must not be Aunt Sallies while the experimental groups are taught under what J. B. Biggs has called "greenhouse conditions". Lee J. Cronbach's assessment is sound:

> We need experiments that carefully control the time allocation to discovery, to know how much slower it is. And we need long-extended, carefully observed educational studies. I think we will find that a rich mixture of "discovery-to-presentation" (telling) is best to get the learner started, but that after he is well on the road, a leaner mixture will make for faster progress and greater economy.

† "Learning by discovery: rationale and mystique", *Bulletin of the National Association of Secondary School Principals*, Dec. 1961. Quoted in Glennon, *loc. cit.*

Further Reading

In addition to the books and articles mentioned, the following will be of interest:

Arithmetic in Action, Brideoake and Groves.
The Psychology of Invention in the Mathematical Field, Hadamard.
The Child's Conception of Number, Piaget.
Children Discover Arithmetic, Stern.
"The growth of mathematical concepts in children through experience", Dienes (in *Educational Research*, Nov. 1959).
"Mnemonic systems and devices" in *Science News* No. 39.
"Mathematical creation" in *The World of Mathematics*, Vol. 4, Newman.

Structural Apparatus

J. F. Deans

Cubes, cylinders, tablets and bars of wood or plastic, some coloured and some plain, have for so long been a feature of the primary school mathematics scene that a detailed description of the many varieties of structural apparatus available at the present time may be dispensed with, even if space would allow such a thing. Sources of information about individual apparatuses are given in the brief bibliography, but useful summaries are to be found in J. D. Williams' series of articles in *Educational Research*, February and June issues 1961, and in the same author's section on the teaching of number in *Primary Education* edited by Peterson. J. B. Biggs' survey of the methods used in teaching mathematics in primary schools† showed that even in 1959 a substantial number were making use of structural apparatus; a similar undertaking today would certainly show an increase. However, the British educational system allows such freedom to heads of schools that one can have no confidence in forming a precise picture as to what must be taking place in a school which uses, say, the Cuisenaire apparatus. In 1960 W. A. Brownell, making his "Observations of instruction in lower grade arithmetic in English and Scottish schools",‡ found that where the Cuisenaire rods were used there was great variation in the extent and manner of their use. He did not find it easy to discover schools which were entirely traditional or entirely non-traditional; almost all used mixed methods and these would vary within a single school. This was a surprise to an American visitor, but it is commonplace to us; such a situation,

† *Educational Research*, Feb. 1961.
‡ *The Arithmetic Teacher*, Apr. 1960.

however, makes for obvious difficulties in the conducting of research, especially as on the appointment of a new head, or even without such an upheaval, the method of teaching followed in a school might change considerably. If the extent and manner of use of an apparatus for which there are detailed instructions varies, how much more variation might we expect where instructions consist only of a leaflet of suggestions. Further, if the use of an apparatus varies widely even where the inventor's instructions are detailed, then we cannot expect to be too specific about the achievement of the apparatus in the classroom.

The systematic use of physical means of representation as an aid to calculation goes back to the abacus, but even the kind of apparatus already referred to is not of recent origin. Pestalozzi used numbered lengths of wood; Froebel, among other number apparatus, used "boats" to hold a number of cubes rather as does the Unifix apparatus today. Tillich's set of ten segmented blocks, first used at the beginning of the nineteenth century, were uncoloured, of the size of the Cuisenaire pieces, but resembling the "Longs" of Dienes' Multi-base Arithmetic Blocks. Maria Montessori devised a variety of number apparatus. Some kinds, like her sandpaper numerals, were not structural, but others foreshadowed modern developments. Young children working under the Montessori system had rods which represented the numbers 1–10; the unit was not a cube but a 10-centimetre length with a 2·5-centimetre square section. Alternate segments of a number rod were coloured red and blue. Other representations of the numbers 1–10 were smaller in size for the use by the older children who could easily manipulate them and at the same time, being smaller, could the more easily represent large numbers. These consisted of sets of beads on wires of graduated lengths and there were two distinct sets of these, one in which beads on a wire were coloured alternately black and white, and another in which a different colour was allocated to each of the ten number bars. This second set of bars, although coloured, was known by the number names as in the more recent Stern apparatus where coloured wooden bars are used, and not by the colour names as is the case with Cuisenaire apparatus. However, Montessori noted that children who were asked to add two numbers sometimes reported picturing the two coloured lengths end to end and the appropriately coloured "answer" length beside it.

In addition to the bead bars described, there were bead squares and cubes corresponding to each bar. For instance, unit bars and four-bead bars could be used in conjunction with four-by-four bead squares and four-by-four-by-four cubes. The familiar parallel with this apparatus today is the set of Dienes Multi-base Arithmetic Blocks with its Units, Longs, Flats and Blocks. Through the set based on them, notation and the processes of computation were taught. Montessori introduced a cardboard thousand-cube as being cheaper than the rather expensive bead cube, but noted that with its six faces, each marked to show a hundred small squares, it was sometimes mistaken for a representation of six hundred. The Montessori apparatus is, of course, used today in Montessori schools, but after a brief vogue in the fifties it is seldom seen in schools outside that movement. Those who regret this may reflect that the compliment has been paid her which other successful educational innovators have also received, that many of the methods which she devised for the teaching of arithmetic are in everyday use in our schools although neither acknowledged nor even recognised as hers.

Among other parallels with more recent structural apparatus systems we notice the multiple embodiment Montessori provided for the representation of the numbers 1–10. Dienes, in his book *Building up Mathematics* (and elsewhere), insists on this as a basic principle of mathematics teaching. "To allow as much scope as possible for individual variations in concept formation, as well as to induce children to gather the mathematical essence of an abstraction the same conceptual structure should be presented in the form of as many perceptual equivalents as possible".† The importance of multiple embodiment—"the more the merrier principle" as Dienes calls it—can be seen clearly in the regrettable consequences of ignoring it.

† *Op. cit.*, p. 44.

One has met the child who has learned to do simple sums using apples as a concrete embodiment, and can work out, even in the absence of apples such problems as, "If you have five apples and eat two, how many you left?", but posed the same problem in terms of oranges cannot tackle it. We recall Eileen Churchill's story of the little girl who had painted ducks on a pond, with four roughly forming a square and one above, and who would not acknowledge that there were five "because there is not one in the middle". Cases are not infrequent of children who have been well drilled in finding the areas of rectangles, and have the formula $A = L \times B$ at their finger-tips, but can assert that a triangle, having no $L \times B$, has no area. The notion that three is light green is not unknown.

Multiple embodiment, as one would expect, is the dominant feature in Dienes' M.A.B. and A.E.M. (Algebraic Experience Material). With the M.A.B., place value and the four rules are taught through the manipulation of sets of blocks representing numbers in different bases. Algebraic concepts are presented in his A.E.M. set through a "Leicestershire"-type balance, pegboard and pegs, plywood squares and rectangles and by other geometrical means.

Structural apparatuses (the term was originated by Catherine Stern), may be fruitfully studied by using a series of dichotomies. We shall consider some of these:

(i) *Analogical/Abstractive.* J. B. Biggs has used these terms, and also refers to the first group as uni-models and the second as multi-models. An *analogical* apparatus utilises essentially a single way of representing the numbers 1 to 10 (or 1 to 12 in the case of Colour Factor rods); they represent a single analogy with the set of integers, and the operations with integers can be paralleled by manipulations with the pieces. Comprising a *single* embodiment, care must be taken to provide other number experiences. The ordinal aspect of number is much over-emphasised in the Cuisenaire scheme, whereas the Stern apparatus provides opportunities for experience of ordinal and cardinal properties to alternate with each other so that a proper balance is maintained. Although incorporating a variety of apparatus, Stern's scheme might more appropriately be placed in the analogical camp. In *abstractive* systems a given concept is presented through a number of representations, the expectation being that the concept will be grasped and the medium (so much "noise") by its very variation will not be seen as essential. Dienes' apparatuses are paradigms of abstractive systems.

(ii) *Logical/Psychological.* In the Cuisenaire system as set forth in Cuisenaire and Gattegno's *Numbers in Colour* and in Gattegno's textbook series the *logical* development of arithmetic is pursued without reference to the child's experience of the world or his inclinations and interests. Coverage is thorough, with perhaps a little over-emphasis on those topics which the rods do rather well such as the multiplication of a fraction by a whole number, and the logic cannot be faulted. But the presentation is dry and in this respect compares poorly with almost any other arithmetic textbook on the market; multiplication and division of fractions as dealt with through the rods seem pedagogically impracticable.

A good coverage of elementary arithmetic is achieved also by the Stern apparatus which can be placed in the *psychological* category, not because of its evident dependence on the Gestalt school of thought (Stern worked for some time with Max Wertheimer) but because of the attention paid to the facts of child development. The size of the pieces is adapted to the manipulative abilities of young children—the pieces used in America are even larger than those available on the home market—there is a variety of trays, boxes and boards which make the building up of the various configurations of blocks and cubes easy and in themselves avoid a presentational monotony. The non-structural provision for recording is intended to help the child to progress arithmetically without being held up by lack of manipulative ability. That the "story" of ten, the series of number bonds with ten as the total, is dealt with before the "stories" of smaller numbers is due to the fact that it worked better that way round in practice.

In his book *Concept Formation and Personality* and elsewhere, Dienes has outlined his theory of learning and his various apparatuses have been designed and their method of use planned with this as a guide. They fall into the psychological category. One meets in Dienes deviations from the expected order of development, such as the treatment of division before multiplication. "It is usually found quite absorbing by children", he says in his manual, "and is much easier than multiplication".

Margaret Lowenfeld's Poleidoblocs provide another example of an apparatus behind which lies a well-formed view of child development and concept formation. For the use of these attractive and well-made aids, the coloured G set of geometrical solids and the plain A set of cuboids, she offers advice but no detailed progression of work. She invites teachers to communicate their experiences and observations of children's use of the apparatus to her. One sees it used in schools by the youngest children for free play in the role of building blocks, its employment becoming more structured with increasing age until the upper juniors are using the apparatus for fraction manipulation and the mensuration of cylinders and cones, pyramids and prisms.

The logical/psychological distinction, as well as another interesting one between (iii) *open* and *closed* systems is brought out in a chapter in Doris Lee's *A Background to Mathematical Development*.

In an open system, while there is full coverage of the arithmetic syllabus, the child's mathematical experience outside of work with the sets of blocks and so on are called upon. Such experiences are recognised as having a complementary value to the more formal progression of work with the blocks. In a closed system the apparatus is seen by its devisers as sufficient in itself to supply all that is required for the child's arithmetical, or even mathematical development.

(iv) *The place of colour, central or incidental?* Dienes' M.A.B. is uncoloured for the obvious reason that the part the pieces play is determined, within a given base only by their shapes: Unit, Flat, Long or Block, and these are readily distinguishable. In the Avon apparatus, red dots and blue numerals are used, but no significance is attached to these colours which merely enhance the appearance of the pieces. Stern uses colour to enable the children to pick out easily from a pile of blocks the ones required; the colour names are never used by the teacher, and in practice are rarely used by the children. It might, at first sight, seem unlikely that in working with coloured blocks the use of colour names could be avoided, but in Stern's system, once a number name is associated with a particular block it is henceforth described by that name. In the progression of experiences with the blocks before the introduction of numbers, a block involved in the work at a particular moment is always present, "Where does this block go"?, is identified through its position in a progression, or through the answer to the question "Which block fits this groove?"

Colour in the Cuisenaire and Colour Factor systems, on the other hand, is essential. How far its use is a help to the child and how far a stumbling-block is not easy to determine. Many a teacher has asked herself whether the colour equations such as $r + y = b$ are a waste of time. Montessori's experience shows that colour linkages may be used naturally as a memory aid by children, but is the effort involved in harnessing this tendency worth while? Are there children whom such an approach does not suit? Then there is the (surely undesirable) tendency to associate the number and the colour so closely that either may be substituted for the other. "Miss, you've got a five dress on today" is not apocryphal. In a film of the Cuisenaire apparatus in use, a child suggests "Blue" as a solution to the division of 27 by 3; Dr. Gattegno who conducts the session is not perturbed. Professor Lee, in the section of the book mentioned above, concludes: "Clearly the colours are of no importance whatever, except perhaps for incidental ease of picking up the right rod on any particular occasion". W. A. Brownell, in a piece of research to be discussed at some length below, finds that arithmetic may be well taught using Cuisenaire rods, but advocates that the colour names should not be used: "... nothing is gained—and much may be lost—by having to refer to the orange rod and to the black rod instead of more precisely to the ten rod and the seven rod respectively".

Another dimension worth exploring as a distinguishing feature among apparatuses is:

(v) *the consideration of each number as a single entity and, alternatively, as a collection of units.*

In the approach employed in the traditional classroom the aids available to help in calculation, counters, straws, fingers, are of the latter type. Gattegno discourages recourse to the counting of units in the use of the Cuisenaire apparatus. The ordinal aspect of number is emphasised at the expense of the cardinal. In the case of most other apparatuses, however, it is evident that their devisers intend children to be constantly aware of both the cardinal and ordinal properties. A block of the Stern apparatus, so many units in length, is lightly segmented so that if the number associated with it is forgotten the units composing it might be counted. In work with her Pattern Board units are placed in recesses in each board, and the odd or even character of the numbers 1 to 10 is exhibited. In work with her Counting Board a series of grooves accommodates the ten blocks, but the grooves are boldly marked to show the decomposition into units, and in associating number names with the blocks, unit cubes are counted into the grooves. In B. R. Jones's Avon apparatus the unit is an inch square and is marked with a large red dot. The pieces to represent the numbers 1 to 10 are cut out of plywood to exhibit the odd/even pattern. They can be arranged ordinally, in staircase form, and while the pieces are used in calculations the numbers they represent are probably viewed by the children as single entities; but their cardinal aspect is prominent and the counting of units could easily be resorted to. The unifix unit cubes can be "popped" together to make blocks which can then be ordered, compared and combined through a variety of supplementary apparatus in several respects resembling that of Stern. Shaw's Structa is another apparatus where unit pieces are counted out and then combined, to form in this case cylinders which represent numbers, but in which the component units are always evident.

Brownell's experiences when seeking "Cuisenaire schools" remind us that once the teacher has obtained his set of apparatus, he can use it more or less as he chooses. Some schools welcome the Dienes ten-box and reject the rest of his M.A.B.; one comes across Cuisenaire rods with numerals inscribed on them. Whether, then, (vi) *an apparatus is first introduced to children through free play or by exercises directed by the teacher* may be decided by each individual, but each system advises one or the other approach. The user of the Cuisenaire apparatus is advised that a period of free play extending for the youngest children over several weeks should precede directed play. During this time the discovery that all rods of the same length have the same colour, and vice versa, will be made. The regular gradation of the rods might be discovered, but the construction of the staircase is in fact introduced at the beginning of the directed play. Colour Factor recommends free play at first; the staircase has a diagnostic significance, its appearance during a child's play indicating that he is ready for a guided exploration of more formal properties of the system. Although several of the components of Stern's material are suitable for building towers and pyramids and for a variety of play situations, she recommends that it should be introduced to the child through an exercise on the ordinal relationships of the blocks using the Counting Board. Each new exercise is ideally taken by the teacher with an individual child, who then works with a group which has been introduced to it. At any one time each child has a repertoire of exercises at which he works, discarding the earlier ones as he meets new ones. It may be interesting to note the place of the staircase in Stern's system. The grooves of the Counting Board aid the placing of the blocks in order in the earliest exercise, but the staircase proper is first met in the context of the Unit Box, a square tray with a side ten units in length. The teacher places the first three blocks in the empty tray one by one, then passes it to the child in the expectation that he will perceive the pattern which is unfolding. Stern, in a number of places, uses this technique of starting to construct a pattern, and then without any verbal instruction to the child, passing it on to him to complete.

Some apparatuses do not lend themselves to free play. Exercises with Structa are directed by the teacher from the start; but the small unit cylinders are not adaptable to building play, although free pattern making can take place on a perforated board. Similar observations apply to the Unifix apparatus.

Among products which are offered for sale in a competitive market it is natural, but none the less regrettable, that there is a tendency to claim too much. Gattegno and his disciples are to the fore in this. In *Numbers in Colour*, Cuisenaire and Gattegno write of their scheme: "It is a *complete answer* to the mathematical and psychological problems set by the teaching of number as seen in arithmetic".† Their italics assure us that they mean it. J. V. Trivett echoes this: "The Cuisenaire–Gattegno approach presents a unified, complete learning programme in mathematics from the early number work to sixth form studies".‡ It is surely not too difficult to think of some aspects of mathematics which might escape the net with profit to both teacher and pupil.

In the face of such claims what does research reveal? The situation for structural apparatus may be described in terms not too dissimilar from those employed by Ausubel reviewing research into discovery methods (see p. 252). Let us be clear, however, that even if this is so it by no means indicates that there is no place for such apparatus, but it seems certain that it should not occupy the whole of the stage, and probably not even the centre of it. Large-scale research in Britain is rare and as one of the few projects to involve enough children to command some respect J. B. Biggs' survey described in *Mathematics and the Conditions of Learning* deserves attention. In a section§ summarising his findings he makes a number of references to structural apparatus of which these are the chief:

> Using conventional problem and mechanical tests as criteria, there is no evidence that the use of the uni-model structural materials, such as the Cuisenaire or Stern materials, will produce results with average children that differ from those obtained under traditional methods, in similar school conditions.

† Page vi.
‡ *Technical Education*, Feb. 1960.
§ Page 266.

> In the remedial situation uni-model ... methods ... were better than traditional methods.
>
> Multi-model methods produced higher mechanical and concept scores and more favourable emotional attitudes to school and to arithmetic in particular than did traditional methods.
>
> Multi-model methods, while they benefit all children, favour especially the dull and backward. This was in marked contradistinction to the uni-model pattern.

Since, of the structural apparatuses available, the Cuisenaire is the most widely used, it is natural that the bulk of research into the use of structural material should include reference to it. Brownell's project of 1962, described in "Arithmetical abstractions—the movement towards conceptual maturity under differing systems of instruction" (University of California Press, 1967), investigated the results of teaching arithmetic through the Cuisenaire rods, the Dienes M.A.B. and the conventional (traditional) approach. His work involved 1400 children in English and Scottish schools, who at the time of testing were completing their third year of schooling. Most research projects which set out to compare the merits of various approaches to the teaching of arithmetic rely on the results of mechanical and problem tests. Brownell's findings are particularly interesting in that he used observation and interview techniques; each individual child was interviewed for about an hour on his knowledge of number combinations, his ability to perform computations and to solve problems. A record was made of each child's performance, and of what he said voluntarily and in reply to questions. The interview approach was adopted to discover how each child obtained his results, and what he understood; the experimenters sought to chart progress towards abstractness and maturity of arithmetical concepts. Necessarily the number of calculations which could be thoroughly explored in the time with each child was small. There were two number combinations for each of the four rules, and the child was asked how he had arrived at his answers. One mechanical calculation, e.g. 92−45, was included in each operation, and it was noted how well the child could explain what he was doing. There were twelve problems, three in each operation, of which the following is an example: "Tom had 18 books to put away in the cupboard. He put them in equal piles of 6 books each. How many piles of books did he make?"

Although the Cuisenaire and conventional approaches were pursued in the classroom in a more or less thoroughgoing manner, the Dienes approach, Brownell decided, must be accounted a hybrid, since the more purely mathematical programme of Dienes was augmented on Sealey's advice by some of the more traditional early primary mathematical activity. In the sample of Scottish schools only the conventional and the Cuisenaire approaches were investigated, and the Cuisenaire approach gave significantly better results than the conventional. However, in the English schools the conventional approach was significantly ahead of Dienes, which was not much ahead of the Cuisenaire.

To explain this seemingly anomalous result Brownell points out, as has been noted, that the label "conventional" or "Cuisenaire" does not guarantee conformity to a precise teaching programme. He also points out that the quality of instruction in all programmes ranged from excellent to poor. In Scotland something of a team spirit was in evidence among the Cuisenaire teachers, while in England a new and more positive attitude towards mathematics teaching appeared to be spreading. Brownell concludes:

> Now finally, the explanation as the writer sees it: the Cuisenaire programme was better taught in Scottish schools than in the English and the conventional programme was better taught in the English schools than in the Scottish schools. In a word, in the schools of each country the better taught programme, whether Cuisenaire or conventional, was the more effective of the two in promoting conceptual maturity.

The quotation from J. B. Biggs' *Mathematics and the Conditions of Learning* made above differs little from an earlier summary of his findings published by the N.F.E.R. as an information pamphlet. Having compared the use of the Cuisenaire and Stern materials with the traditional approach he adds: "Of more importance than the method, was a factor that seemed to be identifiable with teacher ability".

This conjecture, perhaps omitted from the final draft as being too subjective, supports a conclusion similar to Brownell's that, taught well, both the traditional approach and one employing structural apparatus can produce good results. It is worth noting once more that the criteria that Biggs employed were conventional problem and mechanical tests, while Brownell sought to explore understanding through individual interview. Brownell notes that weakness in several areas of arithmetical competence, particularly in number combinations and subtraction, was characteristic of the result of the Dienes programme. He suggests that it is more suitable for children at a later junior stage, and that decimal instruction could with profit precede work with other bases. Inevitably, not all that a particular approach can do well can be taken into account in an experimental assessment of this kind, but against the early introduction of fractions through the Cuisenaire rods and the non-denary base work with the Dienes M.A.B. may be set the familiarity with weights and measures which the conventional approach produces.

Reports on Cuisenaire research in Canada which had appeared over a number of years were conveniently collected in 1964 by the Canadian Council for Research in Education into one volume, *Canadian Experiences with the Cuisenaire Method.* In his introduction to the book, F. G. Robinson points to "an almost universal enthusiasm for the method among those who have tried it, and deep conviction regarding its effectiveness". However, he goes on to note that some observers had been "disturbed by the Cuisenaire appeal to *demonstration* rather than *experimentation* as a means of convincing the uninitiated. This factor, combined with the frustratingly vague language of the movement's major publications, an obvious reverence for a 'father image', and intolerance of criticism, has, in some people's minds, given a somewhat occult flavour to the movement". The projects described below are interesting as a study of experimental method, as well as for the results they have produced.

In one experiment in Vancouver, a group of twenty-eight pupils did better than the controls, but in further experiments there was no superiority, except on a special Cuisenaire test. The appearance of such a special test is not uncommon in research involving Cuisenaire material. It usually consists of items of the type:

$$(\tfrac{3}{4} \text{ of } 12) + (\tfrac{2}{5} \times 10) - (\tfrac{1}{3} \text{ of } 9) = \square,$$

for the teaching of which the rods are well adapted, but which most teachers do not consider appropriate in the early primary years. The situation which recurs in these cases is that the Cuisenaire subjects score some marks on these items while the control subjects, who have never seen this type of calculation, do not. One cannot help suspecting an element of partisanship in experimenters who make much of such a hollow victory. The teachers of Vancouver who had used the apparatus were enthusiastic and saw many benefits stemming from it. There was, however, an "apparent tendency for scores in arithmetic reasoning to decrease with greater length of time given to Cuisenaire instruction". The report noted that the Cuisenaire was recommended to be one among a number of aids; individual differences in children require this variety.

In the Burnaby evaluation the same test was given in June, at the end of the summer term, and in September, after two months' vacation, to Cuisenaire and control groups of infants who had completed their first year of schooling. The items were simple addition and subtraction bonds. As one would expect both groups obtained lower scores, but the falling off of the control group was much greater than that of the Cuisenaire group. One is reminded of the experiment on children's memory conducted by Frank and Bliss under Piaget's supervision, where a subject's ability to reproduce, after a lapse of several days, an arrangement of cubes was tested when originally the child had (i) merely looked at them, (ii) watched an adult arrange them, and (iii) actively copied the arrangement. The second group improved very little on the first, but the third was superior to both of the others. The handling of the material had been an evident aid to memory.

In an experiment conducted in Saskatchewan, while the Cuisenaire method obtained better results than the traditional approach at mechanical calculation: "In the traditional kind of problem solving it appears as though the knowledge of fundamentals has not been transferred ... the programme has not emphasised the solution of the traditional written problem". This view was endorsed by workers in Kitchener, Ontario, who noted that "... the ability of students using Cuisenaire materials to manipulate symbols far outstrip their comprehension of the practical significance of these manœuvres. This phenomenon usually manifests itself when the criterion tests involve word problems".

In the Kitchener experiment efforts were made to overcome a number of obstacles to valid results. Eight pairs of classes were matched for I.Q., for the ability of the children as rated by the teacher, and for class size. Extremes of ability were eliminated. The period to be allocated to arithmetic was set at 30 minutes per day, and all teachers involved were relieved of some of their duties for the purpose of preparation. This last provision was connected with an attempt to eliminate the Hawthorne effect. It will be recalled that this derives its name from the Hawthorne works of the General Electric Company, in Chicago, where an experiment was conducted beginning in 1927 to seek relations between conditions of work and the incidence of fatigue. Output norms were established under controlled conditions of temperature, humidity and so on. Then conditions were systematically varied; rest breaks were altered and working hours reduced, but output went up. Towards the end of the first phase of the experiment they reverted precisely to the initial conditions—and output reached 20 per cent above the norms originally established. It was clear that the affective boost produced by the mere involvement in an experiment had led to the increased production. In research into teaching methods the Hawthorne effect leads to an enhanced performance of the experimental group quite independently of any merit in the experimental teaching method. It has even been suggested that educational research is a good investment for public money whether or not it produces valid results, since the teachers involved are so beneficially affected. In the Kitchener experiment a leader was provided for each group of teachers, the Cuisenaire group and the controls. The respective groups met at regular intervals, and the non-Cuisenaire team evolved its own teaching materials. In the final tests no significant difference was found. The comment was made "Though not proven by this experiment it would seem that the generally good results obtained are due more to teacher involvement, interest and motivation than to any specific method".

In Quebec, Madelaine Goutard demonstrated the use of the rods and talked to teachers; Dr. Gattegno conducted a summer course. It is recorded that 40,000 boxes of the rods were purchased in the province. In tests it was found that Cuisenaire groups were ahead on a special Cuisenaire test, but behind on problems; there was no significant difference in mechanical arithmetical performance. The writer of the account concludes: "But whatever the information given by these studies, the best evidence of Cuisenaire achievement in Quebec will be found in the many examples quoted by Mlle Goutard in her book". An example of the regrettable but widespread readiness to be convinced by demonstration and anecdote rather than experiment.

Lucow's† research in Manitoba involved over 500 children and concluded: "The Cuisenaire method is an effective one in the teaching of grade three arithmetic. Some non-Cuisenaire methods in the hands of experienced teachers yield results that are just as good. Children should be taught by whatever method they respond to best; no teacher should limit herself to one method of instruction in the face of the abundant individual differences in children".

Passy's‡ work in New York involved 1200 children. Those using the Cuisenaire apparatus achieved significantly less at the 5 per cent level on an elementary arithmetic battery than either of two other groups following respectively a "meaningful" programme and the regular curriculum.

Among points which J. D. Williams§ brings out in his book are three which draw attention to particular points of usefulness of structural apparatus:

† *Op. cit.*, and summarised in *The Arithmetic Teacher*, Nov. 1963.

‡ *The Arithmetic Teacher*, Nov. 1963.

§ *Op. cit.*, p. 94.

> Because the material is "self-corrective" the pupil is able to and should be encouraged to learn by discovery. The pupil's manner of tackling arithmetical problems can be seen very clearly in his manipulations of the material, which can therefore be used as an instrument for finding out about his style and level of thinking about such problems. Even at a point in his development at which he is incapable of mature appreciation of certain mathematical principles, the pupil can gain a *readiness* for them from acquaintance with their concrete parallel.

For the teacher structural apparatus has attractions. Despite many claims it is not a panacea, but it can be and should be used to make mathematics more attractive and more comprehensible than it has been in the past. The temptation to over-emphasise its use to the exclusion of some other aspect of mathematics teaching must be avoided; it is all too easy to join one of the "Supporters' Clubs", to cheer the approach that you fancy and to pour scorn on findings which cast doubts on it. Mathematics teaching today has many facets, sometimes to the despair of the conscientious teacher who is trying to keep up with them all, but it has certainly become a more exciting and constructive enterprise than the pedestrian affair of a few years ago.

Suggested Reading

Shaw, "The development of the child's conception of number", *Educational Review*, June 1961.

In *Mathematics Teaching*:

1. Poleidoblocs, in No. 29, Lowenfeld.
2. Psychological Aspects of Teaching Mathematics, in No. 25, Harris.

In *Teaching Arithmetic*:

3. Thinking behind the Rods and Blocks, Lovell, in Summer 1963.
4. The Avon Apparatus, Part 1, Jones.
5. The Avon Apparatus, Part 2, Print, in Autumn 1963.
6. Unifix Apparatus, Part 1, Lockett.
7. Unifix Apparatus, Part 2, Williams, in Spring 1964.
8. Mastery in Mathematics, Stern, in Spring 1965.

Computer-assisted Instruction

W. L. BENEDICT NIXON

"COMPUTER-ASSISTED instruction" (C.A.I.) is a term which signifies the use of a computer (specifically, a programmable automatic digital computer) as an instructional tool—roughly speaking, as a "teaching machine". The term begs some questions from the educational point of view, as will be evident from comparing it with the earlier term "computer-assisted learning" (C.A.L.) which it has displaced. It is therefore convenient to adopt the acronym "C.A.I." as a non-committal label merely to identify the subject under discussion.

I. Computation

In order to appreciate the potentialities and limitations of the computer for any application in its already wide and rapidly growing range, and for C.A.I. in particular, it is a help to be clear just what a computer is and what it can be made to do. What a computer does is called *computation*; and here a preliminary misunderstanding to be cleared up is the notion that computation is the same thing as *calculation*. Certainly computers can do arithmetic, and their calculating ability is of immense practical importance; but arithmetical calculation is a rather specialised and sophisticated kind of computation, which is itself something at once more elementary and more general. Correspondingly, a computer is in essence a remarkably simple machine, with capabilities which are in themselves rather rudimentary; it is in its applications that nearly all the complexity and sophistication associated with the computer are to be found.

In the first place, a computer is a machine in which recorded information may be stored. So is a typewriter: as I type what you are now reading, what I want to convey to you is being recorded as a sequence of ink-marks on the paper in the machine. This recording is of the kind called *digital*: my typewriter has a fixed sets of keys corresponding to a fixed "alphabet" of marks from which I can select to make the recording. This includes the letters "A", "a", "B", "b", ... (which comprise what we ordinarily refer to as "the alphabet") and also "0", "1", "2", ... (which are the ordinary denotation of the term "digit"), as well as punctuation marks like ".", ",", ";", ... and symbols such as "+", "&", "£", ... By extension, the whole set may be called an *alphabet*, and each element of it a *digit* (or alternatively a "character"). In addition, the absence of a mark, produced by the space-bar and called "space", and the break between one line and the next, called "end-of-line" or "newline" and produced on an electric typewriter by a special key, both count as members of the typewriter alphabet and so as digits. All that is required of each member of a digital alphabet is that it should be clearly recognisable as such and clearly distinguishable from every one of its fellow-members. A *digital recording*, then, is a sequence (or more generally, a pattern) of occurrences of digits from an alphabet.

The "numerical digits" "0", "1", "2", ... comprise a restricted alphabet which is commonly used for recording *numerical* information. There is, however, no difference, as far as the method of recording is concerned, between typing numerical information using "0", "1", "2", ... and non-numerical information using "A", "a", "B", "b", ...—this is one aspect of the relationship between calculation and computation.

The size of a digital alphabet is relevant only to the compactness of the recordings which result from its use. Instead of a typewriter, I might use a Morse key and buzzer together with an audio tape-recorder. The Morse code specifies a recoding of each digit of the typewriter alphabet as a specific group of elements selected from a much more restricted alphabet; this has only five digits—two "marks", namely "dot" or "short" and "dash" or "long", and three "spaces", namely "short", "long" and "double-long". A Morse-code recording is therefore many times as long, in respect of the number of digit-occurrences in it, as the corresponding typewriter recording. Alternatively, had I been Chinese, I would have had an alphabet of thousands of distinct characters at my disposal, with a corresponding economy of recording.

A typewriter-like device, connected directly into or "on-line to" the computer, is the typical computer *Input*. As you type on this, the sequence of keys you strike, and therefore of the digits you select, is recorded as marks on paper for your benefit, and is also recorded, magnetically or electronically, in the *Store* of the computer itself. As a matter of fact, the full typewriter alphabet is not used within the computer: there is a systematic recoding of the typewriter digits in terms of a more restricted alphabet, as with the Morse code. The computer alphabet is in fact the smallest possible, consisting as it does of only two elements; these, which are called "bits" (short for "binary digits"), are usually represented outside the computer as "0" and "1" in print and as "nought" and "one" in English speech. Typically, each typewriter digit is represented as a distinct sequence of eight bits: for example, "01000001" for typewriter "A", "11100001" for typewriter "a", "01000010" for typewriter "B", "11100010" for typewriter "b", "00110000" for typewriter "0", "10110001" for typewriter "1", "10110010" for typewriter "2", and so on. For our present purpose, we may ignore this recoding, and regard the computer as dealing directly with the typewriter digits.

It is of supreme importance to realise that the *meaning* of a digital recording in the Store of the computer resides not in the computer, but with the person who stored it there. I might type

Baa baa black sheep – have you any wool?

and you might type

Shall I compare thee to a summer's day?

whereas the brainy baboon, tiring of the tunes from his bassoon, might type

hf&n Ug(6:? ghTRjBn(0£ + gggggk.

These are all one to the computer, each being merely a string of digits differing from the other two in respect of the digits it contains.

Unlike a typewriter, a computer is more than a merely passive recorder of the digitally expressed information entered into it; it has active capabilities too. These are ultimately reducible to the ability to generate stored digits, either absolutely, as copies of digits already stored, or conditionally, as a result of comparing pairs of stored digits—generating a stored "S", for example, if two digits are the same or a "D" if they are different. (In reality, this reduces further to, for example, the generation of a "1" if two bits are the same or of a "0" if they are not.) Two sets of stored digits—two English words, for example—may thus be compared by a series of pairwise comparisons of their constituent digits. It is, moreover, the case, which must be merely noted here, that for example the addition of two numbers is reducible to a process of digit-comparison and digit-generation applied to the stored *numerals* for the numbers to be added, a numeral being the row of digits by which a number is digitally represented. Thus everything a computer can be made to do depends on a few very elementary abilities, which are the key to the significance of the term "computation".

A typical elementary computation would be to store within the computer the text of a literary work, by typing it on the input typewriter, and then to have the computer list each distinct word in the text—to generate the author's vocabulary, in other words. A more elaborate example, which involves a modicum of calculation, would have the computer generate the vocabulary together with the number of occurrences of each word in it in the text; the result would be a stored "table" consisting of each word accompanied by the numeral for its frequency (the numeral being just another row of digits like the word itself). Of course, the results have to be got out of the computer; typically, the typewriter used for input is also the computer's *Output*, its printing mechanism being operated by the computer instead of from the keyboard. These examples exemplify the paradigm of a computer computation: starting from an initial block of stored digits, the computer generates further stored digits by means of repetitions and elaborations of its basic digit-generation and digit-comparison abilities, and outputs such of these generated digits as the user wishes to see.

Two points should be re-emphasised. The first is that the significance of what is initially input and finally output rests with the human user—the vocabulary computation just described is equally applicable to a play by Shakespeare and a baboon-text typed at random. The second is that what you and I understand as "arithmetic" is just another digit-manipulation to the computer. Both points are illustrated by comparing the arrangement of a set of numerals in numerical order with the arrangement of a set of words in dictionary order. In each case, the computation is based on a defined order for a subset of digits: "0" then "1" then "2" ... then "9" in the one case, and "A" then "B" then "C" ... then "Z" in the other. Apart from matters of detail, the two procedures are effectively the same computation.

II. Programming

How does the computer "know", for any particular computation. which digits to compare and what digits to generate? The answer is that it does not; it is the human user of the computer, for whom alone the digital goings-on inside the computer have "meaning", who knows; and he must "tell" the computer what it is to do by supplying it with instructions in the form of a *computer program* (note the spelling). Using the input typewriter, I type

```
100 PRINT " MARY HAD A LITTLE LAMB"
110 STOP
```

and the computer dutifully records in its Store the digits I have typed in the order I typed them. But now I type

RUN

The computer does more than merely record this digit-sequence; it also *recognises* it, in the same sort of way that the telephone exchange recognises a (digital) telephone number. The recognition acts as a trigger: the computer now goes through its stored record of what I typed previously and recognises this too—as a sequence of instructions it is to *obey*. The upshot is that the computer types out

MARY HAD A LITTLE LAMB

on the input/output typewriter, then stops and waits for more typing—which is just what the previously typed program tells it to do (the numbers at the left of the instructions define the order in which they are to be obeyed, which therefore need not be the order in which they are typed).

Until I type RUN, the computer merely records what I type; it is a kind of super-instruction, which directs the computer to treat what it had previously recorded as instructions to be obeyed. I could have typed

100 BRING ME MY BOW OF BURNING GOLD
110 BRING ME MY ARROWS OF DESIRE
120 STOP

which is also a set of instructions, and which would have been recorded in the store. Had I now typed RUN, however, nothing would have happened—not because the computer cannot be made to recognise BRING instructions as well as PRINT instructions, but rather because bringing bows and arrows is not computation whereas printing MARY HAD A LITTLE LAMB is. The term "program" means "a set of instructions *for a computation*". There are two points to be noticed: first, a computer can interpret digital records stored within it as instructions for computation; and second, the process of so interpreting stored records is itself a computation.

Here are some more programs (in each case, what the computer types in obedience to the program is underlined):

```
(a) 100  PRINT "5 × 8 + 3"
    110  STOP
    RUN
    5 × 8 + 3
    ---------

(b) 100  PRINT 5 × 8 + 3                [here, some calculation is
    110  STOP                            prescribed]
    RUN
    43
    --

(c) 100  LET X = 5                      [X and Y are variables
    110  LET Y = X + 3                   which are given numerical
    120  PRINT X + Y; X × Y; 3 × X − Y   values]
    130  STOP
    RUN
    13   40   7
    -----------

(d) 100  LET X$ = "88888"
    110  LET Y$ = "* * *"
    120  PRINT "P"; Y$; "Q"; X$; "R"; Y$; "S"
    130  STOP
    RUN
    P* * *Q88888R* * *S
    -------------------
```

In this last example, X$ and Y$ are variables which have values, as are X and Y in example (c); unlike X and Y, however, they have values which are merely digit-rows, which even if they look like numerals for numbers (such as the value of X$, which is the digit-row 88888) are not so interpreted by the computer. It will be clear how the "quote" symbol ("") is used to distinguish between mere digit-rows and digit-rows which are to be taken as signifying numbers for calculation; also, the symbol "$" is used to distinguish between the two kinds of variable.

So far, all the *operands* of each computation—that is, the digit-rows, whether or not they are numerical, on which the operations specified by the instructions are to be performed—are given in the program itself. This is no longer the case in the next example:

```
100 PRINT "WHAT IS YOUR SURNAME";
110 INPUT N$
120 PRINT "YOUR SURNAME IS"; N$
130 STOP.
```

The effect of the instruction INPUT N$ is that the computer types a question-mark, then stops and waits for something to be typed; whatever is in fact typed becomes the value of the digit-row variable N$. Thus after I type RUN, the following dialogue between the computer and me will take place:

```
WHAT IS YOUR SURNAME? NIXON
YOUR SURNAME IS NIXON
```

Whatever I type after the question-mark, say MAO-TSE-TUNG or even 5 × 8 + 3, will be accepted. In the following example, the computer is instructed to examine what is typed and to discriminate between certain possible answers:

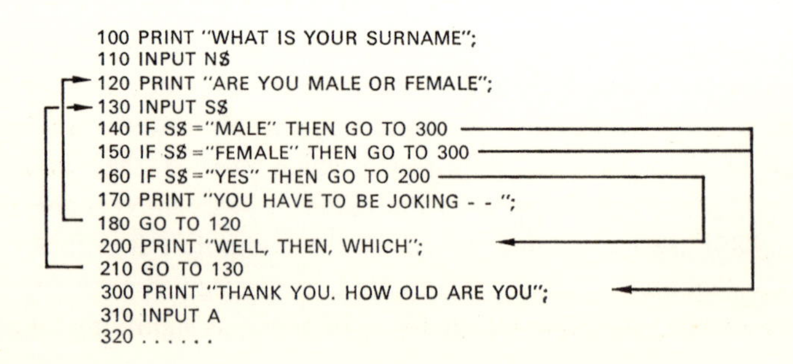

```
100 PRINT "WHAT IS YOUR SURNAME";
110 INPUT N$
120 PRINT "ARE YOU MALE OR FEMALE";
130 INPUT S$
140 IF S$="MALE" THEN GO TO 300
150 IF S$="FEMALE" THEN GO TO 300
160 IF S$="YES" THEN GO TO 200
170 PRINT "YOU HAVE TO BE JOKING - - ";
180 GO TO 120
200 PRINT "WELL, THEN, WHICH";
210 GO TO 130
300 PRINT "THANK YOU. HOW OLD ARE YOU";
310 INPUT A
320 . . . . . .
```

In this case, what happens when the program is RUN depends on what is typed in response to the "?" produced by the instruction INPUT S$; here is a sample dialogue:

RUN

WHAT IS YOUR SURNAME? CHOLOMONDELEY

ARE YOU MALE OR FEMALE? NO

YOU HAVE TO BE JOKING—ARE YOU MALE OR FEMALE? YES

WELL, THEN, WHICH? FEMALE

THANK YOU. HOW OLD ARE YOU? 25

. . . .

At the point reached (there is assumed to be more program, not shown here), the surname and the sex of the person typing the answers are the values of the digit-row variables N$ and S$ respectively, and his (or her) age is the value of the numerical variable A. By means of GO TO and IF ... THEN GO TO instructions, what happens subsequently can be determined by what the values of these and other variables actually are.

Notice that this conversation is only superficially between the person typing the answers and the computer; it is really between him and the *programmer*—the person who wrote the program and who had to foresee and provide in advance for everything which might turn up in the course of the conversation.

III. Programming for C.A.I.

The application to C.A.I. will now be obvious: the teacher, turned programmer, writes a program which converses with a pupil (provision is made whereby a program typed in by one person can be stored indefinitely in the computer, and RUN by other people as often as may be required). All the techniques of *programmed learning*, as otherwise exemplified by programmed texts and teaching machines, can be used in composing computer programs for C.A.I. (It is convenient to distinguish between a teaching *programme* on the one hand, and a C.A.I. *program* which instructs the computer how to present the programme to the pupil.) Here, for example, is a fragment of a C.A.I. program which subjects the learner to practice drill in simple arithmetic:

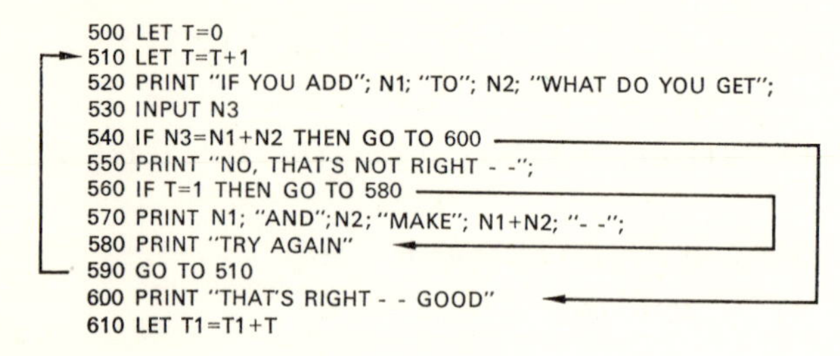

```
500 LET T=0
510 LET T=T+1
520 PRINT "IF YOU ADD"; N1; "TO"; N2; "WHAT DO YOU GET";
530 INPUT N3
540 IF N3=N1+N2 THEN GO TO 600
550 PRINT "NO, THAT'S NOT RIGHT - -";
560 IF T=1 THEN GO TO 580
570 PRINT N1; "AND";N2;"MAKE"; N1+N2; "- -";
580 PRINT "TRY AGAIN"
590 GO TO 510
600 PRINT "THAT'S RIGHT - - GOOD"
610 LET T1=T1+T
```

Instructions obeyed previously to these have given values to the numerical variables N 1 and N 2, which the subject is then asked to add together; the calculating ability of the computer is used to check the answer he types in. The variable T counts the number of times the question is put to him: after an incorrect answer on his first try (T = 1), he is told to try again; after a second and all subsequent incorrect answers, he is given the correct answer and told to try again. By instructions elsewhere in the program, the values of N 1 and N 2 for successive presentations of the addition can be automatically adjusted to the competence of the subject, by making use of the value of the variable T, values of which for each addition are accumulated by the last instruction shown as the value of another variable T 1. Here is a sample dialogue:

IF YOU ADD 53 TO 48 WHAT DO YOU GET? 91
NO, THAT'S NOT RIGHT—TRY AGAIN
IF YOU ADD 53 TO 48 WHAT DO YOU GET? 102
NO, THAT'S NOT RIGHT—53 AND 48 MAKE 101—TRY AGAIN
IF YOU ADD 53 TO 48 WHAT DO YOU GET? 101
THAT'S RIGHT—GOOD

It will be pretty clear by now that a computer, being after all no more than a machine, does exactly and only what the programmer tells it to do, in an extremely slavish and literal manner. With C.A.I., then, it depends very much on the skill and ingenuity of the teacher-programmer whether or not his humanity reaches the learner through the computer. On the other hand, a computer is inherently very flexible: there is no effective limit to the degree of differentiation the programmer can apply to the questions put to the learner and to the answers he gives.

Drill-and-practice by question and answer, as in the example just given, is the lowest-level kind of C.A.I. It is nevertheless not to be despised: it frees the teacher for more creative teaching, and yet gives individual attention to the learner. In addition to providing the practice itself, a program of this kind can (a) monitor the subject's competence and (b) adapt to it so as to give practice where it is most needed; in addition it can (c) keep records of performance and progress for the teacher's benefit, and also (d) accumulate statistics from many learners so that the efficacy of a particular drill strategy can be evaluated. The example given is concerned with simple arithmetic; but drill in many other areas is feasible—one example to which C.A.I. is being applied successfully is computer programming itself.

A C.A.I. program can also *inform*, in the sense that the teacher-programmer can instruct the computer to print out textual matter which the learner is expected to read and study. A typewriter is rather unsuitable for this, as the learner can often read much quicker than it can type and tends to be irritated by its slowness. *Visual-display* output, with which a block of text is displayed on a television-type screen, is really essential for this type of C.A.I.; moreover, *line-drawings* as well as text can be displayed (in principle, the program specifies the X and Y coordinates of the end-points of the lines composing the picture).

The next higher level of C.A.I. is that which engages the student in *problem-solving*. Computers are being used for the solution of problems in many fields: scientific, industrial, managerial, economic, social, and so on. In any such area, if a computer can be of use in solving problems, it can also be programmed to give instruction in problem-solving. An example of this kind of C.A.I. is the computer-controlled "management game", in which the computer simulates management problems for trainee managers and evaluates the consequences of the decisions they make. This example is noteworthy because it may involve several students acting in collaboration or opposition, each at a computer "terminal" of his own but all in conversation with one and the same computer program which interrelates their individual contributions. C.A.I. of this kind is extremely important for another reason: it can induce a proper understanding of the relative roles of computers and people in fields in which computers have a significant contribution to make—an understanding which is not as common as it might be.

IV. C.A.I. in the Future

C.A.I., like a number of other computer applications, is still very much in its infancy. Computers are costly, and advance tends to be quickest in those areas in which their application results in large and immediate savings. C.A.I., as a type of computation, is not a routine matter like say payroll calculation; its efficacy depends at least as much on the teaching skill of the C.A.I. programmer as on the sophistication of the computer hardware at his disposal. There is a vicious circle here: until suitable hardware is made available, teachers are unable to practise its application to C.A.I.; and until they have had the opportunity to do so, the economic justification for supplying the hardware is not established, and an act of faith is demanded of the supply authority. At the present time, apart from a small number of adequately financed experiments, mostly in the United States, C.A.I. development is cost-limited. A recent report from the National Council for Educational Technology has proposed that the initiative be seized by allocating £2 million for a 5-year coordinated development of C.A.I. (which it refers to as "computer-based learning").

Computer hardware, more or less suitable for C.A.I. experimentation, already exists, although it is normally dedicated to other applications. The type of hardware needed is what is called a *multi-access* computer system: the disparity between the rate at which computers compute and the rate at which human users work is so great that a computer serving only a single input/output "terminal" is economically intolerable—the computer would be doing nothing nearly all the time. Some C.A.I. work can be carried out using the spare capacity of available hardware; this is particularly to be encouraged if the hardware concerned operates in an educational environment, such as, for instance, that of a university. Development work under such conditions can, however, only be tentative and preliminary; real advance can only come from properly funded large-scale experimentation such as the N.C.E.T. report suggests.

It is very important that C.A.I. should be regarded as but one aspect of the educational relevance of the computer. The computer is a universal tool, already applied to a great many fields of human activity and potentially applicable in a great many more. The proper relationship between computers and people is close and immediate rather than distant and remote, because only by close contact can their complementary contributions to any application be maximally exploited. It is therefore a matter of urgency that as many people as possible become acquainted with computers by experience of this kind of contact; only in this way will it be generally understood that the effect of computers is to widen and deepen the capabilities of people *as people* rather than to supplant them. This experience is an essential part of everyone's education; it is already available to staff and students at university level, and a start has been made with computers in schools. At present, unfortunately, the type of computer facility so available is far from universally of the direct-access kind described earlier, which is the only kind really suitable for teaching people about computers as well as being essential for C.A.I. As soon as a suitable computer system becomes available in an educational environment, it is then open to the teachers themselves to become familiar with the scope and current limitations of computation and by direct experiment, to explore the possibilities of C.A.I. It is surely essential that C.A.I. should develop with the active participation or at least the informed encouragement of the general body of teachers at all levels of the educational system; it is highly undesirable for C.A.I. to become the preserve of "specialists", and in particular of specialists in computation rather than in teaching. The interest being taken in C.A.I. at Hatfield Polytechnic is exemplary in this respect: it is but an aspect of an interest in the introduction of computation into the school curriculum which originated with teachers in the Hertfordshire schools. Although this interest now centres on the Polytechnic itself, great pains are being taken to enlist the collaboration of the county schools and their staff in its development. It is also desirable that computation in school should not continue to be regarded as a specialisation within a specialisation—as an aspect, that is to say, of the work of the mathematics department. When say the history teacher asks "What relevance has the computer to my subject?", he can be encouraged to look into the possibilities of C.A.I. in his field, and to learn to program so that he can experiment for himself.

Attempts to predict the future of any aspect of computation can be no more than crystal-gazing; after all, the whole history of computation has hardly yet occupied 25 years, and previously inconceivable advances have taken place in that time. Technological advances in respect of the compactness and convenience in use of terminal input/output equipment are taking place all the time; by the end of the century, output in the form of audible speech will almost certainly be available, and possibly the problems inherent in the recognition and conversion to stored digital form of spoken input will have been satisfactorily solved. At the moment, and for some time to come, even with improved terminal equipment, C.A.I. is likely to remain *instruction-dominated*, in the sense that the teacher-programmer is in control, and the pupil-subject has only as much freedom of response as the former has allowed for in the program. True dialogue, in the sense of exploratory discussion between teacher and pupil, must wait upon further development of the art and science of computation in the direction of *machine intelligence*. In principle, this involves programming computers to learn, and then subjecting them to instruction. The computer HAL, featured in the film *2001: A Space Odyssey*, represents in fictional form a reasonable extrapolation of current developments in this direction. It is perhaps not superfluous to point out that all such "machine intelligence" is in fact *human* intelligence: the intelligence of the machine's designers, of its programmers and its instructors, who must assume responsibility for the use to which the machine puts its powers.

Further Reading

1. *Computer-assisted Instruction; a Book of Readings*, edited by RICHARD C. ATKINSON and H. A. WILSON, Academic Press, 1969—essential reading.
2. *Computer-based Learning; a Programme for Action*, National Council for Educational Technology, 1969.
3. HARRY KAY, BERNARD DODD and MAX SIME, *Teaching Machines and Programmed Instruction*, Penguin, 1968—especially Chapter 9.
4. JOHN G. KEMENEY and THOMAS E. KURTZ, *BASIC Programming*, John Wiley, 1967—the best available introduction to the art of computer programming; it uses the B.A.S.I.C. programming language, in which the examples above are expressed, and includes a C.A.I. example.
5. W. L. B. NIXON, *How a Computer Works*, University of London Institute of Computer Science, 1968/9—explains digital computation for beginners.
6. W. R. BRODERICK, *Computers in School*, The Bodley Head, 1968—by a pioneer teacher.

The Uses of Programmed Material in British Schools and Colleges

W. A. Gibby

Summarising our Common Background

As all of us here have some knowledge of "programmed learning", it seems appropriate for us to start by summarising our common background. It might be helpful to do this by posing several questions, one at a time, leaving sufficient time for each one of us to think about a possible answer before a final answer is agreed upon. Strictly speaking, many of our questions require lengthy answers, but we will confine ourselves to basic, simplified answers merely to provide ourselves with a common framework.

When a question is posed, think about an answer and, if you wish, jot down a few single words or a phrase or two, in order to fix the ideas in your mind. Write down just enough for you to be able to give a short answer if called upon to do so. Then after a suitable interval it will be possible for some of you to tell us your answer and we will attempt, as we go along, to build up a brief summary of answers received, together with some points which arise in discussion.

[A note to the reader: The reader may care to try a similar technique for each of the ten questions given on pages 281 to 295. For these questions he might find it helpful if he were to cover up each summary below the question in the text and, once he has written down a few notes of his own, to compare his notes with those of the text.]

QUESTION 1. *What is "learning"?*

Discussion:

(a) Learning is not easy to define,† but human learning is usually said to occur whenever *a person's activity brings about a relatively permanent change in behaviour.*

(b) Most of us would regard learning as a change of behaviour in some desired way. Broadly, references are often made to three kinds of learning:

(i) *Ideational* (or cognitive) *learning:* here we are concerned with the gradual mastery of the field of human knowledge and ideas; this is the kind of learning that is consciously aimed at in schools.

(ii) *Social learning:* this goes on all the time, whenever we are with others.

(iii) *Motor learning:* or the learning of skills.

QUESTION 2. *What is "programmed learning"?*

Discussion:

(a) Ideally, it would be claimed for programmed learning‡ that it is learning which results from material presented to a student in *a series of reasoned statements*, usually relatively short.

(b) Normally, *at each step* of reasoning *a question is posed*, or implied.

(c) Such a step is *a link in a chain of learning*, one complete link being called a *frame*.

(d) A *frame* usually contains:

1. A unit of *information* and, often, some reasoning.
2. A *question*.
3. *Either:* a space for an *answer*,
Or: a set of *multiple-choice answers*.

† Some of us felt that our own ideas on learning had been clarified by working through a programmed introduction to *Learning and Teaching*, by E. Stones.(7)

‡ See the programmed booklet, ref. 4.

(e) A *programme* is *a collection of frames* arranged in an orderly, logical sequence. Each step of this sequence is usually made as short as possible.

(f) At each successive step of a programme, a student *either* supplies *or* selects an answer to a question.

(g) Usually, such a programme aims to provide individual instruction. This method of learning, by presenting a programme, is sometimes referred to as "programmed instruction". The word "instruction" is used advisedly here as many people feel that the kinds of things that can be learned by this method are somewhat limited.†

(h) Material presented in this programmed form is often simply called "programmed material".

QUESTION 3. *On what sort of principles are the various uses of programmed material justified?*

Discussion:

(a) The material to be learnt is usually broken down into *small*, manageable *units*.

(b) A student working through a programme can work on his own and *at his own pace*.

(c) A student makes continuous, *active* and constructive responses to each step of reasoning. He is *informed* immediately *whether each response is correct*.

(d) Programmed material is so written that *success* is claimed to be almost *inevitable*. In more difficult programmes, *remedial sections* aim to correct faulty reasoning or knowledge, the resulting encouragement spurs a pupil on, *motivating him to continue* with the programme.

QUESTION 4. *What is a "Linear Programme"?*

Discussion:

(a) A linear programme is one in which *every student follows the same path* through the programme.

† For a further discussion of this point see ref. 7, part 8 (Aspects of Instruction) and part 9 (The Teacher and the Process of Instruction).

(b) Normally each frame consists of:

A unit of information.
A question.
A space for the student's answer.

(c) The student has to *supply one definite response* or answer. He has to *construct his own answer.*

(d) Immediately after a student constructs his response he *compares it with the correct response* which is given somewhere in the programme.

QUESTION 5. *What is a "Branching Programme"?*

Discussion:

(a) A branching programme† is one in which there is only one *main correct path* through the programme, but where, at each successive step of the programme, it is *possible to provide remedial instruction* for any typical or expected error made in answering the question posed.

(b) Frames which form the main path through the programme can be called "Primary Frames". Those frames which supply remedial instructions for errors made *en route*, through the programme, can be called "Secondary Frames".

(c) Normally each primary frame consists of:

Information/reasoning.
Question.
A set of multiple-choice answers, i.e. several possible answers one only being the correct (or the most suitable) answer.

† At this stage, we are only considering the two basic types of programme. Nowadays, linear and branching techniques are sometimes mixed; occasionally a provision is made (often referred to as "skipping") whereby a student can omit (or "skip") certain sections of a programme because it can be established that he has no need to study that particular part of the programme.

(d) Each secondary frame usually includes:

A discussion of the error made. Remedial treatment. Reference back to the Primary Frame where the error was made so that the student can try again.

(e) Diagramatically, the pattern† of a branching programme could look something like:

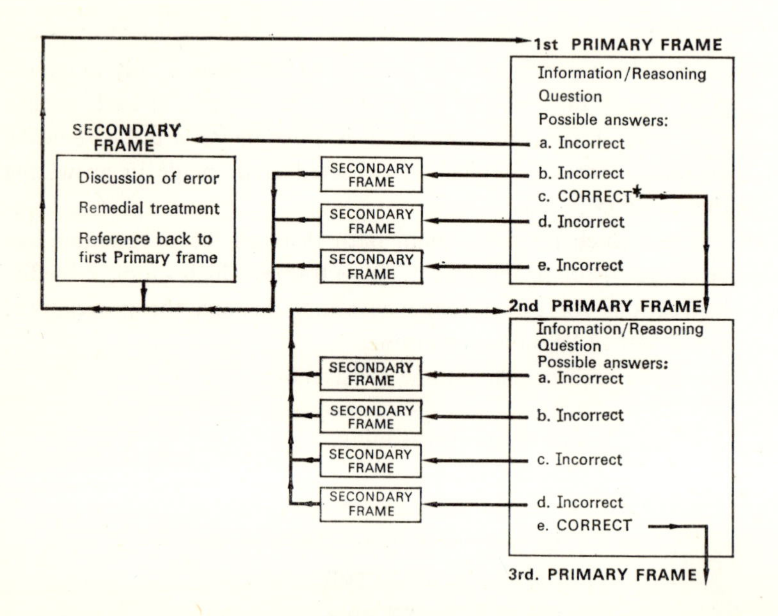

† In each primary frame, five answers (arbitarily) are mentioned here. Usually there are more than three, but the number of alternative answers depends to a certain extent on the kind of errors likely to be made and the proportion of those students likely to make them.

‡ The position of the correct answer is, of course, varied.

(f) In each primary frame a student is asked to select† the answer he feels to be correct or which is the most suitable answer of those given.

(1) If the student selects the *correct* answer: he is guided to the next primary frame, so continuing through the *main* path of the programme.

(2) If the student selects an *incorrect* answer:

(i) he is guided to a secondary frame,

(ii) the secondary frame presents material which attempts to remedy his faulty reasoning or knowledge,

(iii) he is referred back from this secondary frame to the primary frame in which he made the error,

(iv) he works through the primary frame again.

(g) Secondary frames are built in to branching programmes as remedial devices for dealing with common misconceptions and likely errors that tend to arise from the kind of information, reasoning and question of the relevant frame.

QUESTION 6. *What are the claimed advantages and disadvantages of a linear programme?*

Discussion:

Advantages

(a) In order that all students using a programme can work through it with confidence, *each step* will be sufficiently *easy* for almost everyone to give the correct response.

(b) Normally a *student* has to *construct a response* himself, rather than to select‡ an answer.

(c) Linear programmes are easier to write than branching ones.

† Selecting an answer from a group of answers may be very different from supplying an answer. For example, asking a student to supply the solutions to two simultaneous equations may be a quite different exercise from asking him to select the correct pair of answers from several given pairs.

‡ Most of us thought of this as an advantage, but a sizeable minority wished to qualify this attitude. Some thought that there were cases when elimination of any incorrect answers given in a branching programme was itself a useful (but nevertheless different) exercise.

Disadvantages

(a) Often, material becomes *so simple*, at each step, that its very simplicity *irritates* some students as they work through the programme. Accordingly, many students may feel that the minimum of thought is required at each step, with *no real challenge* being offered.

(b) Only a certain *type of material*, of a *limited* kind, appears to lend itself well to the technique used in linear programming. (For example, the learning of either a simple mathematical concept or an easy mathematical skill is, by the very nature of the logical steps involved, the kind of learning that is possible from a linear programme.)

(c) Many students may have to work through steps that are quite *unnecessary* for them as everyone has to follow the one and only single path through the programme. On the other hand, where errors are made by a student there is no built in remedial treatment for these in linear programmes.

QUESTION 7. *What are the claimed advantages and disadvantages of a branching programme?*

Discussion:

Advantages

(a) Likely students' *errors* are anticipated and *dealt with* by the programme itself. To a certain extent, a branching programme records, more typically than a linear one, what might happen if the same material were taught orally. A teacher, trying to deal orally with the material covered by a branching programme, would, in many circumstances, present some information and reasoning, posing questions in a Socratic style, giving remedial treatment to wrong answers as they arose.

(b) *More difficult* and varied *material* can be *presented* with a branching programme than with a linear one. The suggested answers to a particular question need not be as clear-cut as the question–answer must be in a linear programme. Thus a student can be asked to select the most suitable of the answers offered. This allows for wider flexibility of treatment.

Disadvantages

(a) A student *has to select an answer* or response from several given. This might be very different from constructing one's own answer. Most of us felt that supplying one's answer usually required more thought and effort than selecting the correct or most suitable answer from a list of possible answers. However, some of us argued that, in some instances, it was a useful exercise to discover which was the required answer by a reasoned process of eliminating incorrect responses.

(b) *A student's answer* to the question posed in a frame may *not be among those offered* as a possible answer.

(c) *A student may keep on selecting incorrect answers* and therefore be moved backwards and forwards from several secondary frames to the original primary frame. If this happened frequently it could be demoralising.

(d) Often the sheer *volume of reading* becomes *difficult* in itself!†

QUESTION 8. *Through what kinds of devices can programmes be presented?*

Discussion:

(a) Programmes can be presented through some kind of *mechanical device*. Some of these have often been referred to as "teaching machines", this term arising because, initially, such devices were actually thought of as teacher-substitutes. There are several types of mechanical devices,‡ including the following:

† *All programmes* (whatever the type or combination of types) *demand* a certain level of *reading comprehension* and, of course, *verbal ability*. Some writers of programmes, and many users of programmes, tend to take too little account of this.

‡ There are many attempts (beyond our scope for this session) being made to present programmed material through special devices which combine oral presentation—on tape—with visual presentation (on film) of both verbal and non-verbal material.

(i) A fairly complicated *machine* (which looks something like a television set) presents frames on a screen. The programme is on film and passage through the programme is controlled by a series of buttons.

(ii) A *simple* kind of *mechanical device* (usually rectangular in shape) presents frames through a screen or window. The programme is sometimes written on rolls of paper or on sheets of cardboard. Passage through the programme is normally controlled by rollers or levers. Some of the devices provide a small rectangular gap where answers can be written. Subsequent movement of a frame and its answer is so organised that a student can compare his written answer with the suggested answer.

(b) Programmes can also be presented in *book form*. In the case of *linear programmes*, presentation in book form is usually quite *easy*. Normally, several frames appear on each page of the book. Various methods are used for putting the correct answer in an appropriate place which is, at first, out of view when a particular frame is being worked. In the case of *branching programmes, presentation* is more *difficult* as both primary and secondary frames (by their very nature) are usually much longer than the ordinary frames of a linear programme. As, in a branching programme, any individual student only reads *some* of the secondary frames, it does look as though such a programme is untidily presented in book form and it usually appears cumbersome as well.

(c) Whether *programmes* are presented *through a mechanical device or a book*, any such *device* is sometimes referred to as a "*self-instructional device*".†

QUESTION 9. *To what extent does the possibility of "cheating" invalidate the effectiveness of programmes?*

† DeCecco on p. 7 of ref. 1 defines a self-instructional device as "... a mechanical or paper device which presents a set of planned sequential materials to be learned and which the student can complete in the absence of a live instructor and at his own rate of speed".

Discussion:

(a) In the limited experience of those of us who have used programmes with students,† we feel that *cheating* is rarely a serious problem.

(b) It is usually *easier to look too soon at an answer* when the programme is given in *book form*, but whether students do so in school and college situations appears to depend *partly* on the *attitudes of the teachers* concerned with using programmes and *partly* on the *material itself.*

(c) On the whole, students tend to treat *programmed material* as a *learning device* and not as a test or competitive situation, and where a teacher encourages this point of view, it is usually accepted by students that the whole point of a programme is lost if they look at a suggested answer before attempting to supply or select a response themselves.

(d) Some of us have come across cases of *cheating* which seemed to occur because the actual *material* of the programme itself was too *difficult* or too *verbose* or too *boring* for the students concerned.

QUESTION 10. *In what ways does a programme read like a verbatim report of a teacher's actual tuition of an individual student? Are there obvious differences?*

Similarities

(a) A *programme* could *record* the kind of thing that a *teacher* might *say, write, or draw.*

(b) A programme could include *factual material, diagrams,* visual and verbal *illustrations,* a *question-and-answer* technique (Socratic style), together with various *provisions for anticipating* (or dealing with) any *errors* that are likely to be made at each stage.

(c) *Remedial sections* are *possible* in some programmes and these will explain the nature and extent of an error and how it can be corrected.

† Our experiences have meant that, between us, we had used programmes at Primary, Secondary and Further Educational levels.

(d) A *programme* might be *linked with* various types of *apparatus* (e.g. some scientific or mathematical equipment) and a student presented with some practical work, the programme providing a "blueprint" of the various steps required. Thought-provoking questions can then be added.

Differences

(a) *A teacher can adapt his approach* according to the specific nature of a student's response and to meet the need of a particular moment and environment; *a programme is presented in its final form.*†
(b) *A teacher can encourage* a student in a way that it is difficult for a mere programme to do. However, most programmes are presented in such a way that the majority of students experience continuous success. A few programmes do try, somewhat artificially, to encourage students still further with remarks like: "Splendid, you did well to get that right." This tends to be unconvincing because it is produced impersonally, as it were, for everyone who has reached that point, irrespective of his ability and effort; also it will be read by those who did not get the previous answer correct.
(c) All *good teachers* make a flexible, balanced *use of visual and audio methods* of presentation; few programmes do so.
(d) Some *teachers* do tend to *leave too little for a student to do*; all *programmes insist* on continuous active *participation* throughout the programme.

Having established a simple framework of our apparent common background, we can turn to the main question for this session:

What uses are being made of self-instructional devices in British schools and colleges?

† However, all good programmes will have been tried out extensively and at east some apparently necessary adaptions will have been made at successive draftings. As we shall see later, many programmes are more beneficial when some form of teacher guidance is employed.

The listed references (on p. 309) include various research reports (ref. 10 to 15) which record and discuss many projects concerned with the uses of self-instructional devices in British schools and colleges.

There have not been many large-scale research projects. But one of the most recent of these, ref. 12, does deserve our detailed attention. Over a period of 3 years (1964 to 1967), Professor D. M. Lee directed a series of investigations which looked at ways in which programmed learning, through the use of published programmes and by a freer use of the technique itself, could be of service in classrooms. Continuous work was carried out, over the 3-year period mentioned, in thirty-four secondary schools and nine primary schools, in the London Area, with approximately 5500 children, in the age range 9 to 15, taking part.

In trying to answer this question, concerning the uses of self-instructional devices, we are able to divide it into five subsiduary questions. In the main, we will examine in some detail the evidence contained in the work of Professor Lee, but, also, we will refer, briefly, to other evidence.

QUESTION 1. *What are the pupils' reactions to such devices?*

Professor Lee found that the most favourable attitudes to the use of available published programmes were shown by those pupils in classes of higher intelligence. Those in lower intelligence groups tended to show more signs of boredom and a lack of willingness to persist with a programme than pupils in higher intelligence groups—but this may have been because many of the programmes were written with the brighter children in mind.

The lower the intelligence of the pupils, the more they appeared to miss the human contact of the teacher. When children, over the whole age range, were asked whether they preferred to learn:

from a programme alone,

or from a teacher alone,

or from a teacher and a programme combined,

it is interesting to note that Junior school pupils, on the whole, preferred to learn from a teacher alone, but most Secondary school pupils tended to prefer to learn from a teacher and a programme combined. However, there were notable exceptions. The ablest, older Secondary school pupils preferred the programme alone, whilst the least able, at all ages, preferred the teacher alone. But, on the whole, it was clear that the pupil–teacher relationship was disrupted when programmes were used without any kind of teacher guidance. In fact, most children appeared to regard self-instructional devices merely as an *additional learning tool* used by the teacher. Clearly most of them did not wish to regard these devices as "self-instructional" in the precise sense that there was to be no assistance whatsoever from the teacher at any stage of using the device.

QUESTION 2. *How is learning from self-instructional devices related to other forms of learning and teaching?*

Professor Lee supervised various investigations in an attempt to examine the effectiveness of differing kinds of teacher-participation in using programmes with children. This was done in a variety of situations: sometimes a child was given help, or various adaptations of a programme were used, or there was even a complete reorganisation of the material of a programme to suit the needs of a particular group. In some cases, follow-up lessons were given after a child had worked through a programme at home. In yet other cases, supplementary work was added to some programmes. Teachers found this necessary for the quicker workers. These and other methods of teacher-participation generated as good, but usually better, degrees of learning than a child's sole use of a specific programme. Other researchers have also been interested in this issue; particularly Wallis (whose work is summarised on p. 163 of ref. 10) and Noble (pp. 107–18 of ref. 15). In both these references there is evidence to show how it is possible to integrate human and programmed instruction in teaching schemes.

None of Professor Lee's evidence suggested any one particular form of teacher-participation to be superior to any other. A central point was what each individual teacher thought best for a particular group. In addition, it seems important to note the finding that both pupils and teachers felt that the use of programmed learning activities should be confined to not more than one or two periods per day.

QUESTION 3. *How useful are self-instructional devices and what are the attitudes of teachers towards them?*

Professor Lee's investigations showed that significant, if modest, gains in learning were demonstrated for most of the programmes used. It was clear that programmed material, in any form, had something to offer even though programmed learning as *a purely self-instructional device* (in its precise sense, assuming no teacher assistance whatsoever) appears to have rather limited use in schools.

When the relative effectiveness of using a machine and a book in presenting a particular programme was compared, there was no indication that presentation through a machine had any teaching advantage other than, perhaps, a short-lived novelty use of the machine. Using machines was certainly more costly than using books. Some machines tended to break down and it was usually troublesome to have them repaired quickly. Thus, if questions of cost and convenience are of primary importance, the presentation of programmes in book form has an advantage.

There was some evidence to suggest that research into possible means of group teaching with programmes needs more extensive investigation. Several British research workers have shown an interest in this area. In particular, the work of Leith (reported on p. 158 of ref. 10) appears to suggest that the advantage of working with a self-instructional device at one's own rate is sometimes exaggerated. Other relevant recent British research, dealing with paired and group instruction, is discussed on pp. 158–9 of ref. 10. One of the important features of this work is the stress put on group interaction when two or more people work together.

Professor Lee produces some interesting evidence concerning the reaction of teachers to the technique of programming. Whereas they accepted the technique as a useful one in providing an additional learning aid, they felt that far too many published programmes were badly constructed. They argued that almost all programmes (even the best ones), without some teacher-participation, had only a limited use. Teachers tended to appraise a programme in terms of its relevance to a particular class and to a specific teaching scheme. It was felt that there was insufficient emphasis on developing sequences of ideas allowing for creative and stimulating activities. Some British researchers have attempted to answer this criticism. Two particularly useful contributions are reported by Professor Peel and MacDonald–Ross[14].

Many teachers were concerned that programmes allowed no reference back, made answers too readily available and left little scope for a child to summarise what he had learnt. Several of these points and similar ones have been considered by MacDonald–Ross and he suggests possible ways of overcoming some of these difficulties in his article[14].

There is clear evidence to show that many teachers think that they should learn the technique of programme construction. If a group of teachers could write short programmes, then they would probably be useful as an aid to their own teaching schemes. Clark and Leedham (ref. 15, pp. 41–50 and 97–106) give some evidence of ways in which teachers have produced their own programmes and of the ways they have used them with their pupils.

Professor Lee's work is also concerned with the possibility that some subject areas of the school curriculum are more amenable to effective programming than others. She suggests that mathematics and, to a lesser extent, certain aspects of science work lend themselves fairly well to the kind of method of structuring used in programming. In other subjects, the indications are that most programmes are effective more in proportion to the skill and care given to their preparation. But ways in which techniques are being used are improving. In the future, Professor Lee suggests that teachers will find more effective uses of programmes in a wide range of subjects. This will occur if they concentrate on producing their own material by using an imaginative combination of various linear, branching and skipping techniques, and if they attempt to devise programmes which are likely to induce more creative and less routine learning.

The total evidence of this long term study has enabled her to make the following suggestions:

(a) that both published and teacher-made programmes could be used effectively, but that the way in which such programmes should be selected or written and used was a matter for an individual teacher's preference and the demands of his subject;
(b) that each programme should conform to certain agreed standards, e.g. there should be clear statements of the background assumed, age and ability aimed at, additional aids needed, how best to use the programme, the conditions under which the programme was validated (i.e. tried out and proved to be satisfactory in serving the purpose for which it was intended) and so on;
(c) that teacher-made programmes should be used more because they are written to fit a particular teaching scheme;
(d) that programmes of all kinds can be used for most children in the age range 9 to 15 years, but that these were clearly unsuitable for poor readers;
(e) that all programmes should be used on an incidental basis (certainly for not more than two periods per day) and that many varied and flexible methods of teacher-participation were possible and indeed desirable;
(f) that programmes presented in book form are, in general, preferable to those presented through a mechanical device;
(g) that all uses of programmed learning depend on a continuous flow of suitable programmes, so that it seems essential for some teachers to learn how to produce programmes.

It has been necessary to consider the work of Professor Lee in some detail. This large-scale research project on the uses of programmed material in the schools was carried out over a period of 3 years and it probably has more to offer us in terms of practical advice than any other recent British project of its kind. Now it is time to turn from the uses made of programmed material in schools and to consider its uses in colleges.

QUESTION 4. *What uses are being made of self-instructional devices in higher and further education?*

Recently there have been many investigations into the uses being made of programmed material in higher and further education,† but almost all of these studies have been on a small scale. In ref. 15 (pp. 315–20) Hogg describes some work with chemistry students at Aberdeen. In each of three successive years, a series of linear programmes on certain aspects of organic chemistry were issued to degree students. The programmes were intended for private study. The large majority of students showed a favourable attitude to the use made of the programmes and some thought that learning from this programmed material was often more useful than learning in small group tutorials. Many found that the programmes clarified difficult topics, increased comprehension and were simple to follow. But Hogg advises caution. His sample was small and only a limited subject area was covered and the material was of the kind that lends itself to programming. In the light of his comments, he advises cautious acceptance of his findings.

Also, in ref. 15 (pp. 321–8), Unwin and Spencer report on uses that were made of programmed material for revision purposes and for remedial instruction with technological undergraduates. They report positive gains. These research workers point out possible ways in which self-instructional material can, in certain circumstances, save both student and staff time. Although the sample was small, Unwin and Spencer did feel that their results were sufficiently encouraging to suggest that the use of programmes for revision and remedial purposes may have an effective contribution to make in higher education.

The effectiveness of programmed material for the use of students in higher and further education has been more fully reported in

† See refs. 11 (pp. 265–316) and 15 (pp. 315–30).

ref. 11 (pp. 265–316), where we find a series of investigations which were, in the main, on a somewhat larger scale than those already cited by Hogg, Unwin and Spencer. The reports of Stroud (pp. 295–8) and Stones (pp. 307–16) are of particular interest.

Stroud describes an experiment at Coventry College of Technology, where the whole of the first-year mathematics syllabus in various engineering courses has been presented in programmed form. Students worked through the programmes, which were mainly linear ones, and they appeared to enjoy the work. Many students thought that the programmes resulted in more efficient learning than the more normal lecturing technique. The vast majority of students:

(a) wanted to continue using programmes during the remainder of the course,
(b) claimed that they concentrated more with a programme,
(c) approved of the idea of working at their own pace and receiving an immediate check of responses,
(d) preferred to work with programmes individually, rather than in groups.

Stones comments briefly on work carried out with groups of college and university students using programmed material in studying educational psychology. He claims that, in trials involving hundreds of students, ranging from first-year college of education students to post-graduate university students, the following list of findings emerged:

(a) No significant difference was discerned between the programme in book form and the programme in a machine.
(b) No significant difference was discovered between learning by programme and learning by lectures given by the programmer.
(c) Students learn satisfactorily whether they are supervised or not whilst working on the programme.
(d) Students learning by programme work faster than students being lectured.
(e) No significant difference in gain, score or attitude was observed between students working in groups or working individually.

(f) Students' attitudes towards the programme have been uniformly good under a variety of conditions.

Stones further describes, in detail, some interesting experimental work concerning the attitude of university students towards their use of a particular programme of his[7]. Here, the findings confirmed the positive attitude which students showed to other programmes, but the students, in using this particular programme on learning and teaching, preferred a programmed element to a lecture element for this particular topic. Almost all of them preferred to have an individual programme and disliked working in groups. Clearly, the nature of the topic may have influenced their attitude here.

Undoubtedly, some evidence is beginning to emerge to suggest that, for certain topics (particularly where the material is easily structured) and for special purposes (for example, for revision or for remedial instruction), there is a use for programmed material in higher and further education.

In view of the cumulative evidence that some programmes are useful and benefical to learning, it appears desirable that more teachers and lecturers should know something about the production of programmes, how they have been employed in the past, and more importantly, to learn something of the technique of writing programmes. This point leads us to our final question.

QUESTION 5. *How are programmes produced*† *for self-instructional devices?*

This question can be answered by posing four subsiduary ones.

(a) *How many people are involved in producing a programme?*

An important feature of programme writing is that material must be tried out fairly extensively before the programme is put into general use. The pupils taking part in such trials contribute vital information to those responsible for producing the programme. The adult members of a team writing programmes for use in schools are best

† See refs. 2 (chap. 4), 3 (chaps. 7 and 11), 6 (chap. 9), 8 and 9.

drawn from teachers themselves. Three teachers are more productive than a teacher working alone. One teacher will almost certainly be responsible for writing the programme, but others are needed to help at the planning and testing stages. In any case, group discussion at all stages of production are essential. If a programme is to be published, or used fairly extensively in schools, careful validation is necessary. Apart from the children taking part in the various trials, a writing team would comprise: at least three teacher-advisers who would be concerned mainly with syllabus cover, methods of teaching the material and the testing; a programme writer, who would act on the advice of the teacher-advisers and produce a draft form of the programme. Normally, he is a person who is specially trained to write programmes. Other members of the team would include a psychologist, who would advise on such matters as sampling in the trials, on testing and on certain methods of presentation, programme format and techniques to use. Finally, there must be a team leader or coordinator responsible for overall planning and the arrangements for continuing discussion at all stages of development of the programme. So, a team of at least six adults and a fairly large number of children would normally be involved in the production of a programme usable by a wide population.

(b) *What consideration should be given to the age, sex, background and competence of pupils in preparing a programme?*

It is vital to state the ability range, the reading comprehension assumed, the age, the sex and the general background of the intended population. If a teacher is writing for one school, it is clearly easier to take account of all the variables. A team, writing for a much larger population, may find the task considerably more difficult.

(c) *How are the objectives of a programme defined?*

Once a team has been selected and the age, sex and background of the intended population are known, the members of the team spell out, in specific detail, precisely what they want a child to be able to *do* by the time he has worked through the programme. In other words, they list the knowledge they want him to acquire, the kind of situations where he is expected to apply his learning, the extent of the understanding they hope the child will achieve, the skills and abilities needed to perform any required operations and the insight needed to solve problems and so on. In psychological terms, the team lists the objectives of the proposed programme in terms of pupil behavioural change.

(d) *What are the subsequent stages involved in programme writing?*

There are possibly five subsequent stages after the objectives of a proposed programme have been defined.

(i) The *first* stage consists of a detailed scrutiny of the syllabus content of the programme in terms of the concepts to be taught. The team considers probable methods of presentation of the material, the type of programme technique to use for various stages and the extent—if any—to which the material can, or should be, linked with apparatus or specific aids such as, for example, an abacus.

(ii) The *second* stage is the preparation of the first draft of a test which children would take after they had worked through the programme. This draft of the test is carefully related to the original objectives and syllabus coverage of the proposed programme. This test is called the post-test and is redrafted after trials with children. In its final form, it will attempt to establish what a child has learnt by working through the programme. It is necessary to produce a pre-test (one given before the programme is worked) parallel in content, form and difficulty to the final draft of the post-test. The child works a pre-test merely to see if he needs to do the ensuing programme. If he does it, then the difference in scores on the two tests (the pre-test and the post-test) represents some index of gain in learning resulting from working the programme.

(iii) In the *third* of the five stages, the programmer writes the initial draft of the programme itself. He has full details of the results of earlier stages and incorporates the team's accumulated advice into the writing. Often he consults team members again during the writing and, when the draft is ready, the whole team co-operate to edit the material.

(iv) The *fourth* of the five stages consists of the first try-out. There is a pilot trial of the programme with a small group of, say, ten children. These children are drawn from the kind of population which will work the finalised programme. Normally such a trial is carried out with one pupil at a time sitting with a team member. The child is encouraged to comment at all stages where he experiences difficulties and to discuss these with the team member. These comments are recorded on tape and discussed at subsequent meetings of the team. This leads to redrafting of the programme. At a later stage, a redrafted form of the programme is tried out with a fairly large group of children. The children work individually but can be supervised together in the same room. Again, the supervisors record any difficulties and these provide further evidence for redrafting.

(v) The *last* of the five stages consists of trying out the programme with a wider population, checking to see that the vast majority of those working the programme have recorded an agreed gain in learning. This continuous process of drafting and redrafting, and the checking of the effectiveness of the material, is often referred to as a validation procedure. It is important to stress that at every stage of this procedure all individual members of the team constantly refer all accumulated evidence back to the team as a whole for discussion and consequent action. The best kind of programmes are usually those in which full details of the validation procedure are publicly available.

Because of the importance now attached to the fact that teachers and lecturers should write their own programmes, courses and workshops† are being set up to enable anyone who wishes to write programmes to do so.

† Full details of current courses and workshops can be obtained from the Association for Programmed Learning (see ref. 17).

It is necessary to comment on the relative place which a teacher's use of programmed material has in relation to his overall responsibilities for his pupils' learning. Any teacher strives to give the children a rich and varied experience. In organising any course or series of activities, the teacher first decides what he wants to teach and why he wants to teach it. Only then does he select methods to achieve his objectives. But he must be clear about what he wishes his pupils to do and, more important, what they are capable of achieving. Only when a teacher has arrived at these decisions does he look for useful aids and devices to assist him in his teaching. In order that his pupils should learn effectively, he will find, in certain circumstances, that some of his methods of presentation will require augmentation with some specific visual and/or audio aids. He needs to select carefully. He will decide to use a balanced pattern of techniques, aids and organisation, all aimed at promoting the most effective learning situations. But the decisions that form this pattern are essentially for the teacher himself. Within this framework, programmed learning might be one of many aids that he may wish to use. If he uses such material he will probably offer some guidance to the children on how to use a particular programme.

In many of the investigations mentioned in this chapter, there emerges sufficient evidence to encourage teachers and lecturers to use published programmes and, in addition, to write and use their own.

Knowledge is expanding so explosively that many pupils and students are often faced with the responsibility of not only acquiring much more knowledge than was required by their parents in their younger days, but also of developing an ever-increasing number of related new skills and abilities. To cope with this burden, teachers are beginning to use more flexible and effective aids in their teaching. Perhaps self-instructional devices have a useful contribution to make to this ever-changing general pattern of classroom and lecture-room organisation. But, most importantly in the present context, a teacher who decides to use programmed material should be selective in his choice of programmes, imaginative in the way he uses them, anxious to produce his own and willing to experiment. Such a teacher is capable of working out for himself the relative place and relevance of programmed material.

References

(a) *General*

1. DeCecco, J. P. (ed.), *Educational Technology. Readings in Programmed In struction*, Holt, Rinehart & Winston, 1964.
2. Kay, H., *et al.*, *Teaching Machines and Programmed Instruction*, Penguin, 1968.
3. Leedham, J. and Unwin, D., *Programmed Learning in the Schools*, Longmans, 1967.
4. Milton, O and West, L. J., *Programmed Instruction. What it is and how it works* (A programmed booklet), Harcourt, Brace & World Inc., 1961.
5. Skinner, B. F., *The Technology of Teaching*, Appleton-Century-Crofts, 1968.
6. Stones, E., *An Introduction to Educational Psychology*, Chap. 9, Methuen, 1966.
7. Stones, E., *Learning and Teaching. A Programmed Introduction*, Wiley, 1968.

(b) *Programmed writing*

8. Pipe, P., *Practical Programming*, Holt, Rinehart & Winston, 1966.
9. Rowntree, D., *Basically Branching. A Handbook for Programmers*, MacDonald, 1966.

(c) *Research reports*

10. Butcher, H. J. (ed.), *Educational Research in Britain*, U.L.P., 1968 (pp. 152–66: Kay on "Programmed Instruction". Note particularly the references to the works of Leith and Wallis.)
11. Dunn, W. R. and Holroyd, C. (ed.), *Aspects of Educational Technology*, Vol. 2, Methuen, 1969 (particularly articles by Stroud and Stones).
12. Lee, D. M., *Programmed Learning in Schools.* (Unpublished report, Institute of Education, London.)
13. Leedham, J. and Unwin, D., *Programmed Learning in the Schools*, Chap. 14, Longmans, 1967.
14. Tobin, M. J. (ed.), *Problems and Methods in Programmed Learning*, National Centre for Programmed Learning, University of Birmingham, 1967 (particularly articles by Peel and MacDonald-Ross).
15. Unwin, D. and Leedham, J. (ed.), *Aspects of Educational Technology*, Vol. 1, Methuen, 1967 (particularly articles by Clark, Leedham, Noble, Hogg, Unwin and Spencer).

(d) *Details of published programmes*

The most up-to-date information on the availability of published programmes can be obtained from:

16. Dillon's University Bookshop, 1 Malet Street, London, W.C.1.
17. The Association for Programmed Learning, 27 Torrington Square, London, W.C.1.

The Midlands Mathematical Experiment

C. Hope

M.M.E. really started when the Harold Malley/Harold Cartwright Schools were established in 1961 and the Headmaster then appointed, Mr. R. H. Collins, was asked to attempt to organise a school which was different. Mr. Collins approached me for a new syllabus in mathematics based on my experiences with the O.E.C.D. working parties in 1959(1).

In response to this request I constructed a scheme of work in new topics to cover the secondary school course to O-level(2). This scheme excited some interest amongst some secondary-school headmasters in the Birmingham area and I was invited to speak at one of their meetings. This address is published in the first M.M.E. report(3). The outcome of these preliminary activities was a conference of mathematics teachers held at the Harold Cartwright school in Solihull early in 1962. The discussions at this conference showed how little the teachers present really knew about the new topics. In spite of this lack of knowledge, the conference decided that an experimental scheme to O-level was necessary. In September 1962 six schools embarked on the scheme of work involving some 600 pupils in the first forms. It fell to me to construct draft materials in duplicated form which after being used in the schools were published as draft texts(4). Early in 1963 an O-level syllabus was submitted and accepted by the Joint Matriculation Board(5).

It is interesting to reflect on the climate of mathematical education in the two or three years after 1959. In 1959 the A.T.C.D.E. Mathematics Section organized the first conference on Modern Mathematics at which group theory, matrix algebra and numerical analysis were the three major themes. Generally it was held that only the way-out groups of teachers were interested in topics such as these. There were a number of teachers, influenced by the experimental programmes in America and in Europe, who were encouraged to experiment with new topics but the large majority of teachers did not know what a group or a set was and had little knowledge of any of the topics which are now commonplace in the school curricula. The colleges of education (or training colleges as they were then called) were fostering methods of teaching new topics and were introducing them gradually into their syllabuses. Against this background, any experiment in curriculum reform was bound to involve not merely the production of new text materials but it meant educating teachers so that they gained understanding, expertise and confidence in what they were teaching and it also meant evolving methods of teaching the new mathematics so that the mathematics, the teacher and the pupils would come together successfully in the classrooms.

One of my responsibilities in M.M.E. at that time was to choose some new topics for the syllabus. One of the aims governing my choice was to reflect contemporary interests in mathematics. One must remember that contemporary interests includes an interest in arithmetic as well as an interest in sets. It includes an interest in such things as percentage and ratio as well as in such things as vector algebra and matrices. There is a tendency for some radical thinkers to think that modern mathematics excludes all the old, which is not so. Contemporary interest in mathematics must include all the interests used by industry and commerce, by professional mathematicians and by society generally. In the traditional syllabuses one might summarise the contemporary interests as the algebra of real numbers and the calculus. (We are, I believe, the only educational system that has tried to teach calculus to pupils younger than 16 years of age.) In some parts of the world calculus is regarded as a "modern" topic for schools. "Modern" is a relative term.

In our M.M.E. syllabus, it was decided to include the algebra of the real numbers and the calculus. For the new topics, we decided to look for those which seemed to have widest general use. The idea of a vector space permeates a large part of contemporary usages in both pure and applied mathematics. We therefore sought to include a selection of mathematics which gave some appreciation of vector spaces, linear transformations and matrices, vector algebra. What distinguishes M.M.E. principally from other projects in this country is the use of directed line segments and the vector algebra of directed line segments in proving geometrical results. This formed the pure-mathematics side of the syllabus. The applied aspect set out to reflect contemporary uses of mathematics as it occurs in commerce, science, industry and administration. Such topics, novel in the early 1960s, were Boolean algebra as applied to switching problems and probability and statistics. The overall aim was to achieve some sort of education for everyman to O-level. At the outset one had no overriding thoughts of O-level but to safeguard the future of the pupils, an O-level syllabus had to be constructed and a great deal of the syllabus had to be crystallised before one really knew, as a result of direct classroom experience, what was appropriate to boys and girls following a course leading to O-level. In 1969, as a result of experience, the syllabus is undergoing its first revision to produce a course suitable for the average pupil.

The foregoing might be represented as the mathematical aims of the experiment but alongside these we took as our principal educational aim to evolve exploratory approaches. We wanted to start off as many of our topics as possible with an open situation from which children could start off in different directions arising out of their perceptions and interest. We wanted to encourage as much discussion as possible, to challenge the pupils so that lessons in mathematics became an adventure in ideas in which each pupil could exercise some influence on the direction the lesson was taking. A great deal of the success or otherwise depends on the teacher concerned. Some teachers are quite formal and exercise a firm control over the children's activities so that the outcomes are predetermined and inevitable. Some teachers are quite informal and are putting into practice the spirit of the experimental text. Such teachers are always finding something new and many of our ideas of presentation and stimulation have come from the schools in which they teach. M.M.E. is trying to find an educational instrument to encourage new attitudes and an appreciation of usages in contemporary mathematics and contemporary usages of mathematics for everyman 11 to 16. M.M.E. does not try to find a blueprint for the future, a rather futile endeavour, but seeks to find teaching methods for new topics, to investigate whether these new methods are appreciated by students, if students themselves can see the new topics as problem fields and are able to invent problems in the field within which they are working. We see the experiment as a continuing one providing methodology, teaching materials, experience and documentation for new topics which will ultimately be introduced into the established syllabuses of the various examining boards. The examples which follow serve to illustrate how, with some of the new topics, M.M.E. attempts to achieve its aims and objects.

Our first example is the introductory sequences leading to the geometry of directed line segments. The children start off with navigational problems (see Fig. 1) in which they plot a course for a ship or plane to travel from one place to another. Of course, there are many possible ways of going from Harport to Barport: 3 miles

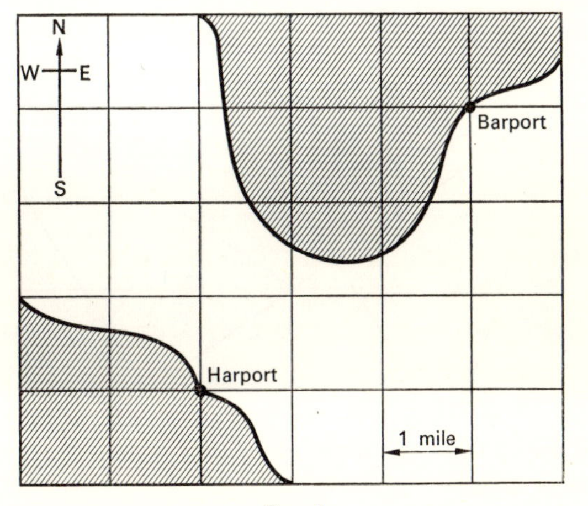

FIG. 1

east, 3 miles north or 1 mile north, 3 miles east, 2 miles north. The children are encouraged to make their own maps. We found that a surprisingly large number of children in the Midlands had little conception of ship navigation but had travelled in aeroplanes. To these children navigating a plane between London and Frankfurt was a realistic problem and we obtained maps and timetables from the various airlines on which problems could be based. Later the children are encouraged to specify courses by distances and bearings so that the notion of ships or planes in different geographical locations taking the same course, so many miles in a given direction, is established. In the course of this work, they learn all about acute, obtuse and reflex angles, about how to construct triangles and, given favourable weather, the pupils' activities may be linked with outdoor work. The next stage is to build up the idea of a directed

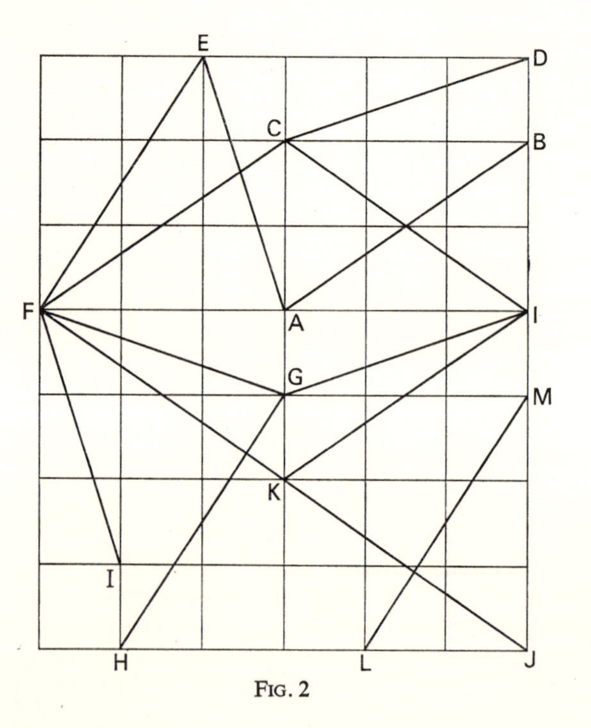

FIG. 2

line segment, its representation by an ordered pair and vectors as equivalence classes of directed line segments. The pupils are given a chart similar to that in Fig. 2 and challenged to discuss what they perceive. A pupil may state that *FE* is equal and parallel to *HG*. Encouraged to think of the line segments as representing courses, the journey from *F* to *E* is described as (2 E., 3 N.) and the notation *FE* = (2 E., 3 N.) is introduced. The children take readily to the equivalences,

$$\overrightarrow{FE} = \overrightarrow{HG} = \overrightarrow{LM} = (2\text{ E.}, 3\text{ N.}).$$

They see that $\overrightarrow{EF} = (2\text{ W.}, 3\text{ S.})$, $\overrightarrow{IK} = (3\text{ W.}, 2\text{ S.})$, $\overrightarrow{GF} = (3\text{ W.}, 1\text{ N.})$ and so on. The diagrams actually given are much bigger and contain more line segments so that a good deal of activity goes into classifying them into sets, characterised by the same ordered pair representation. For example,

$$\overrightarrow{FC} = \overrightarrow{AB} = \overrightarrow{KI} = (3\text{ E.}, 2\text{ N.}),$$

$$\overrightarrow{GI} = \overrightarrow{CD} = (3\text{ E.}, 1\text{ N.}).$$

To go from *H* to *G* followed by *G* to *I* is written

$$\overrightarrow{HG} + \overrightarrow{GI}.$$

This route takes us from *H* to *I* but there are other routes, for example: the direct route $\overrightarrow{HI}$: $\overrightarrow{HL} + \overrightarrow{LM} + \overrightarrow{MI}$. There is some investigation of the alternative routes and the children learn to write

$$\overrightarrow{HG} + \overrightarrow{GI} = \overrightarrow{HI} = \overrightarrow{HL} + \overrightarrow{LM} + \overrightarrow{MI}.$$

When these directed line segments are translated into ordered pairs we have

$$(2\text{ E.}, 3\text{ N}) + (2\text{ E.}, 1\text{ N.}) = (5\text{ E.}, 4\text{ N.}).$$

The two courses are "added" together to get the resultant course, a procedure which may be verified with the other routes and their alternatives. When we consider

$$\overrightarrow{GF} + \overrightarrow{FE} = \overrightarrow{EG}$$

we have

$$(3\text{ W.}, 1\text{ N.}) + (2\text{ E.}, 3\text{ N.}) = (1\text{ W.}, 4\text{ N.}).$$

Similarly

$$\overrightarrow{IF} + \overrightarrow{FC} = \overrightarrow{IC}$$

gives

$$(1\text{ W.}, 3\text{ N.}) + (3\text{ E.}, 2\text{ N.}) = (2\text{ E.}, 5\text{ N.})$$

and the idea of westings as negative and eastings as positive arises naturally. Indeed in some schools where this work on eastings and northings has preceded lessons on directed numbers and the integers the children have "discovered" the need for and the additive properties of the integers, a fortunate occurrence which the teachers have not been slow to exploit.

$$\overrightarrow{HG} + \overrightarrow{GH} = \overrightarrow{HH}$$

leads to the null vector (0, 0).

By these sorts of procedures, entirely experimentally and intuitively the teachers and pupils build up a working knowledge of the group of ordered pairs of numbers and the vector space ideas. Figure 2 might well have been drawn on a parallelogram lattice and brighter children see that the vector space properties are not dependent on the square lattice.

In an open situation such as this there are bonuses. One, the directed numbers, we have already met. Occasionally some youngster will come up with the observation that $\overrightarrow{FK}$ is perpendicular to $\overrightarrow{HG}$.

Now

$$\overrightarrow{FK} = (3, -2) \text{ and } \overrightarrow{HG} = (2, 3) \text{ and } 3 \times 2 + (-2) \times 3 = 0.$$

Similarly $\overrightarrow{FJ}$ is perpendicular to

$$\overrightarrow{HG}, \overrightarrow{FJ} = (6, -4) \quad \text{and} \quad 6 \times 2 + (-4) \times 3 = 0.$$

A few more examples and the pupils establish experimentally the general rule. If $\overrightarrow{XY} = (x, y)$ and $\overrightarrow{PQ} = (p, q)$ then $\overrightarrow{XY}$ is perpendicular to $\overrightarrow{PQ}$ if $xp + yq = 0$. The pupils learn to write this $\overrightarrow{XY} \, . \, \overrightarrow{PQ} = 0$ and we have a preliminary glimpse of the scalar product.

Having established an ordered pair basis for directed line segments the next stage is to use a single symbol to describe an equivalence class of directed line segments and to build up the vector algebra of vector geometry. There is a conceptual gap between the geometrical figure drawn on squared paper and the same figure on plain paper when vector methods are to be used to determine its properties. The brighter pupils bridge the gap quite readily but the remainder need a good deal of experience with plain and squared paper representations of a problem considered together, even going to the length of putting a tracing of the figure over a squared lattice in an extended process of weaning the pupils from the squared paper which they find such a firm basis for their vector ideas.

Figure 3 gives us an example of a fairly open situation involving the traditional topic of fractions. We have the fractions down to eighths. In practice, a large chart which gives fractions down to twentieths is given but Fig. 3 should give some idea of the possibilities. The pupils are allowed to play around in many ways, colouring in the chart in various ways, comparing results and discussing what can be seen. If a blank chart is held horizontally at eye level the curves and arches shown in Fig. 2 become apparent and sequences of fractions may be picked out, for example, the sequence $\frac{1}{3}, \frac{2}{5}, \frac{3}{7}, \frac{4}{9}, \ldots$ and we join these with a curve. The symmetry of the diagram suggests the curve generated by the sequence $\frac{2}{3}, \frac{3}{5}, \frac{4}{7}, \frac{5}{9}, \ldots$ In between the two sequences lies $\frac{1}{2}$ intuitively suggested by the two curves:

$$\frac{1}{3}, \frac{2}{5}, \frac{3}{7}, \ldots \qquad \ldots, \frac{4}{7}, \frac{3}{5}, \frac{2}{3}.$$

We are on the threshold of the real numbers. There is a gap between the two sequences. The chart allows us to choose many such pairs of sequences and we may be led to ask, if we "shake" all the long horizontal rectangles down so that the bottom one contains lines giving the subdivisions into all possible fractions will there be any gaps? This is the problem originally raised by the Greek discovery of the irrationality of 2 and a satisfactory solution was obtained in the nineteenth century. Teaching an appreciation of the real numbers, whether the real numbers are complete, is one of the major problems in mathematical education which has not yet been resolved in a satisfactory manner. The fraction chart at least allows one to suggest that at least there is a problem for teachers to consider. Another discovery which comes from the diagram is the graph of $y = 1/x$ ... If we continue the half line upwards we make a rectangle whose width is AA'. Similarly continuing the third line, the quarter line, etc., we obtain a set of rectangles. Each of the rectangles has an area of 1 unit and the graph generated by the vertices $A, B, C, D, \ldots$ is the graph of height against length for a set of rectangles each of area 1. Other curves on the diagram suggest similar interpretations.

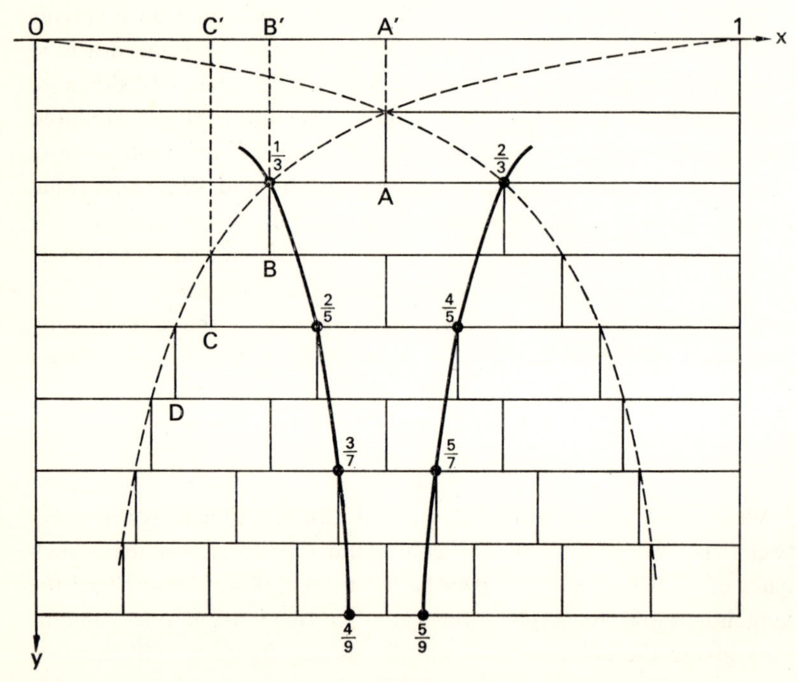

FIG. 3

The fraction chart thus may lead to an everyday understanding of fractions (an aspect omitted in the discussion above), their addition and subtraction, or it may lead to a discussion of limits of sequences or it may lead to a discussion of areas of rectangles. Children may, in such a situation as this, be allowed to take some part in choosing a direction for the course of lessons to be followed.

These two examples illustrate the exploratory techniques in action. It is well to remember that in the late 1950s teachers were not convinced that these techniques were good techniques to use. M.M.E. at least has provided classroom experience which validates the use of child-centred exploration as a teaching technique. The children construct the problems and in so doing glimpse the solution. Ten years ago we had to convince people. Nowadays, of course, exploratory techniques are accepted practice.

We now turn to some problems which illustrate the techniques of some of the new topics:

Problem 1. ABCD, PQRS are two parallelograms. *G, H, L, M* are the mid-points of *AP, BQ, CR, DS*, prove that *GHLM* is a parallelogram.

Figure 4 gives a diagram of the problem. Using vector geometry we may proceed algorithmically: Choose an origin *O*. Let

$$\overrightarrow{OA} = \underline{a}, \quad \overrightarrow{OB} = \underline{b}, \quad \overrightarrow{OC} = \underline{c}, \quad \overrightarrow{OD} = \underline{d},$$

$$\overrightarrow{OP} = \underline{p}, \quad \overrightarrow{OQ} = \underline{q}, \quad \overrightarrow{OR} = \underline{r}, \quad \overrightarrow{OS} = \underline{s}.$$

Then

$$\overrightarrow{AP} = \overrightarrow{AO} + \overrightarrow{OP} = \overrightarrow{OP} - \overrightarrow{OA} = \underline{p} - \underline{a}$$

and

$$\overrightarrow{AG} = \tfrac{1}{2}\overrightarrow{AP} = \tfrac{1}{2}\underline{p} - \tfrac{1}{2}\underline{a}.$$

Finally

$$\overrightarrow{OG} = \overrightarrow{OA} + \overrightarrow{AG} = \underline{a} + \tfrac{1}{2}\underline{p} - \tfrac{1}{2}\underline{a} = \tfrac{1}{2}\underline{p} + \tfrac{1}{2}\underline{a} = \tfrac{1}{2}(\underline{p} + \underline{a}).$$

Similarly

$$\overrightarrow{OH} = \tfrac{1}{2}(\underline{q} + \underline{b}), \quad \overrightarrow{OL} = \tfrac{1}{2}(\underline{r} + \underline{c}), \quad \overrightarrow{OM} = \tfrac{1}{2}(\underline{s} + \underline{d}).$$

Hence

$$\overrightarrow{GH} = \overrightarrow{OH} - \overrightarrow{OG} = \tfrac{1}{2}(\underset{\sim}{q} + \underset{\sim}{b}) - \tfrac{1}{2}(\underset{\sim}{p} + \underset{\sim}{a}) = \tfrac{1}{2}(\underset{\sim}{q} - \underset{\sim}{p}) - \tfrac{1}{2}(\underset{\sim}{a} - \underset{\sim}{b})$$

$$= \tfrac{1}{2}\overrightarrow{PQ} - \tfrac{1}{2}\overrightarrow{BA},$$

$$\overrightarrow{ML} = \overrightarrow{OL} - \overrightarrow{OM} = \tfrac{1}{2}(\underset{\sim}{r} + \underset{\sim}{c}) - \tfrac{1}{2}(\underset{\sim}{s} + \underset{\sim}{d}) = \tfrac{1}{2}(\underset{\sim}{r} - \underset{\sim}{s}) - \tfrac{1}{2}(\underset{\sim}{d} - \underset{\sim}{c})$$

$$= \tfrac{1}{2}\overrightarrow{SR} - \tfrac{1}{2}\overrightarrow{CD}.$$

Using the given information that *ABCD*, *PQRS* are parallelograms, we have

$$\overrightarrow{BA} = \overrightarrow{CD} \quad \text{and} \quad \overrightarrow{PQ} = \overrightarrow{SR}.$$

Hence

$$\overrightarrow{GH} = \overrightarrow{ML}$$

and consequently *GH* is parallel to *MC*, and *GH*, *ML* are equal in length. *GHLM* is a parallelogram.

The attraction of vector geometry is the algorithmic quality of its methods and vector geometry was originally chosen for M.M.E.

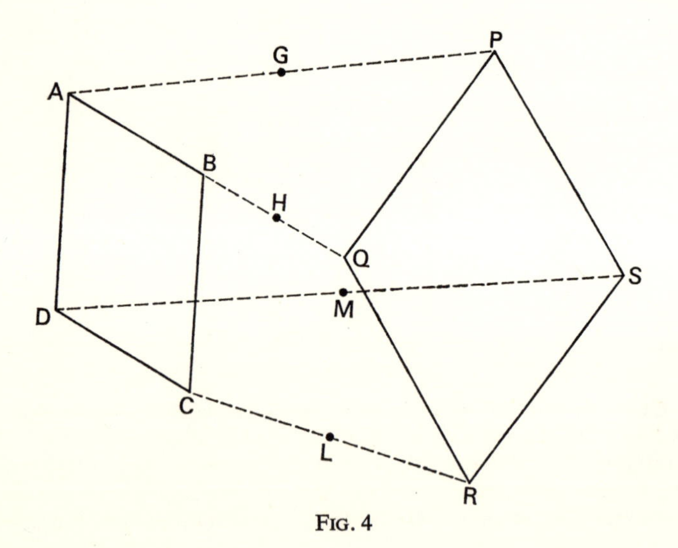

FIG. 4

for this reason; its methods were likely to be accessible to the majority of pupils. Other geometries are generally "mystical", i.e. a problem is extremely difficult until the flash of insight which gives the solution. The problem is then easy. Whilst this is a quality of mathematics which everyone should appreciate, we did not want to make it too substantial a part of the diet. Vector geometry is teachable to most people.

The techniques used in Problem 1 enable one to deal with incidences and ratios. When one wishes to discuss metrical properties, such as lengths of lines, sizes of angles, then the scalar product has to be introduced.

The reader will be familiar with the definition of the scalar or inner product of two directed line segments

$$\overrightarrow{OA}, \overrightarrow{OB},$$

$$\overrightarrow{OA} \,.\, \overrightarrow{OB} = OA \,.\, OB \cos\theta.$$

Where OA, OB are the lengths of $\overrightarrow{OA}$ and $\overrightarrow{OB}$ respectively and θ is the angle between them from $\overrightarrow{OA}$ to $\overrightarrow{OB}$.

Problem 2. The altitudes of a triangle are concurrent. Let OA, OB be perpendicular to the sides BC, AC respectively of ABC. We wish to prove that OC is perpendicular to AB.

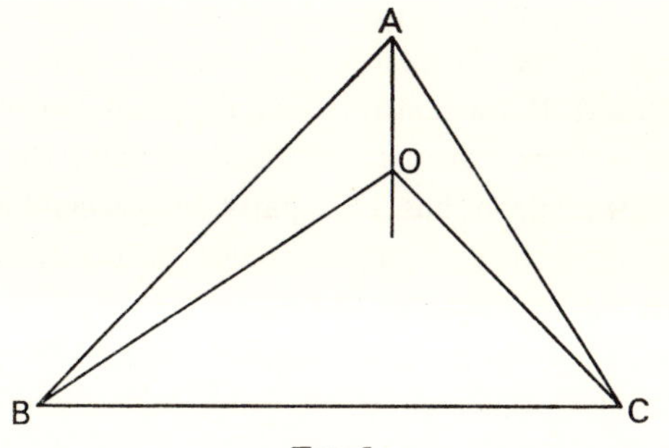

FIG. 5

We put $\overrightarrow{OA} = \underset{\sim}{a}$, $\overrightarrow{OB} = \underset{\sim}{b}$, $\overrightarrow{OC} = \underset{\sim}{c}$ then $\overrightarrow{BC} = \underset{\sim}{c} - \underset{\sim}{b}$, $\overrightarrow{CA} = \underset{\sim}{a} - \underset{\sim}{c}$, $\overrightarrow{AB} = \underset{\sim}{b} - \underset{\sim}{a}$ and we have since

$$\overrightarrow{OA} \text{ is perpendicular to } \overrightarrow{BC},\ \underset{\sim}{a}\,.\,(\underset{\sim}{c} - \underset{\sim}{b}) = 0,$$

$$\overrightarrow{OB} \text{ is perpendicular to } \overrightarrow{CA},\ \underset{\sim}{b}\,.\,(\underset{\sim}{a} - \underset{\sim}{c}) = 0.$$

Hence

$$\underset{\sim}{a}\,.\,\underset{\sim}{c} = \underset{\sim}{a}\,.\,\underset{\sim}{b} = \underset{\sim}{b}\,.\,\underset{\sim}{a} = \underset{\sim}{b}\,.\,\underset{\sim}{c}$$

and

$$(\underset{\sim}{b} - \underset{\sim}{a})\,.\,\underset{\sim}{c} = 0. \qquad (1)$$

From this we conclude, $\underset{\sim}{b} = \underset{\sim}{a}$ or $\underset{\sim}{c} = 0$ or $\underset{\sim}{b} - \underset{\sim}{a}$ is perpendicular to $\underset{\sim}{c}$. If $\underset{\sim}{c}$ is zero then BC and AC are altitudes of the triangles and the result follows.

If $\underset{\sim}{b} = \underset{\sim}{a}$ then $AB = O$, i.e. CA, CB coincide and the triangle ceases to exist.

The remaining possibility implies that OC is perpendicular to AB. The equation (1) gives us a counter example to the solution of quadratic equations with real numbers. There is in this case a third possibility.

In both problems considered above we started with a geometrical situation. We then produced an algebraic model of it and manipulated the model to get a new form. This new form is interpreted in terms of the geometrical situation to give us a result which we require. The models we have constructed not only describe plane figures but are equally descriptive of three-dimensional figures. Parallelogram *ABCD* (Problem 1) may be placed anywhere in space and the result would still be valid. If we remove *O* in Problem 2 from the plane of *ABC* we obtain a tetrahedron *OABC* and we have the result, in this context, if a tetrahedron has two pairs of perpendicular opposite edges then the third pair of edges is also perpendicular. Generally there has been an educational denial of the third dimension. The infant school builds solid models, the junior school flattens them onto paper and the secondary school ignores generally the existence of solid figures after the first form apart from certain volume problems.

In M.M.E. we started with this plane outlook but in the past 3 or 4 years we have recovered the pupils' birthright and are making more and more use of three-dimensional models and problems.

Sets have never been a fetish in M.M.E. We have introduced notation and used it sparingly. Sorting into sets, Venn diagrams (all fairly trivial stuff really) is done in the first year but the Boolean algebra is left to the third forms at the stage when pupils are beginning to assert their individuality, to rebel against routines. We start with switching algebra. It brings a renewal of interest and involvement to both boys and girls, a touch of reality with mathematics actually doing something with real things. Using a flashlamp bulb, a battery and two switches, the pupils establish the following truth tables:

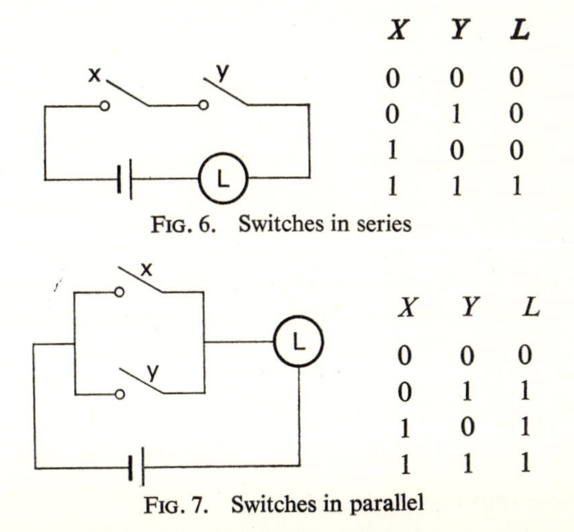

X	Y	L
0	0	0
0	1	0
1	0	0
1	1	1

FIG. 6. Switches in series

X	Y	L
0	0	0
0	1	1
1	0	1
1	1	1

FIG. 7. Switches in parallel

1 represents a switch, X or Y, being closed and so allowing current to flow. 0 represents the switch being open and no current flows. The reader will notice that binary two-figure numbers give all the possibilities for two switches. We use 1 to denote the lamp is lit and 0 to show the lamp is not lit. A switch is said to have the value 0 or 1 according to whether it is open or closed. For a series circuit we may represent the truth table by $X \cdot Y$ where the result is obtained by taking the product of the values of the switches. Similarly a parallel

circuit may be represented by $X + Y$ provided we agree that $1 + 1 = 1$. This gives us an algebra which we may use to design switching circuits and in particular computers. To design a computer we need switches such as x, y. Switches which are either all on or all off are represented by the same symbol x, say. A switch which is always in a state opposite to that of a given switch x we shall represent by x^1. Hence when $x = 1$, $x^1 = 0$ and when $x = 0$, $x^1 = 1$. If we want to design a computer to add 1 and 1 together in binary arithmetic we need two switches x, y. The following brief outline serves to illustrate the principles of switching algebra. First we draw up a truth table to show all possible combinations of two different switches chosen from x, x^1, y, y^1. This table is called a design table.

x	y	x^1	y^1	xy	x^1y	xy^1	x^1y^1
0	0	1	1	0	0	0	1
0	1	1	0	0	1	0	0
1	0	0	1	0	0	1	0
1	1	0	0	1	0	0	0

The reader will note that each of the last four columns consists of three zeros and a one with the one on a different row in each column. Now turn to our computer. It has to perform the following calculations:

0	0	1	1	switch x
0	1	0	1	switch y
00	01	01	10	L_2L_1 = two lamps.

We turn these into the truth table:

x	y	L_1	L_2
0	0	0	0
0	1	1	0
1	0	1	0
1	1	0	1

To obtain a circuit equivalent to L_1, we go back to the design table and select from the last four columns, x^1y and xy^1. These give the truth table for L_1:

x^1y	xy^1	L_1
0	0	0
1	0	1
0	1	1
0	0	0

By adding x^1y and xy^1 we obtain the one values in the right places. Hence

$$L_1 = x^1y + xy^1.$$

Similarly

$$L_2 = xy.$$

The reader will have realised that once more we have taken the physical situation represented by the computer and made an algebraic model which we have manipulated to get the last two equations. Our task now is to interpret the model in terms of the physical situation, i.e. to draw up the circuit.

$L_1 = xy^1 + xy^1$ is given in Fig. 8,

$L_2 = xy$ is given in Fig. 9

and the final circuit for the computer in Fig. 10. Six switches are necessary but by using double-pole double-throw switches we may reduce the number to two.

Two schools, whose pupils were enthused by this work, are now building their own computers.

Another algebra was that of matrices.

In passing it may be noted that Richardson wrote an article for the *Mathematical Gazette* about 1940 in which he called for the introduction of groupoids, groups, etc., and that 3 years later at a joint conference of the Science Masters' Association and the Mathematical Association part of the discussion centred on linear algebra and matrices, but some remarks, adverse in tone if not in content, terminated that particular discussion. Matrices is now a well-established topic in mathematics syllabuses. Matrices form a ring with divisors of zero and so the equation $AB = 0$ may not imply either $A = 0$ or $B = 0$ where 0 is the zero matrix. The fact that there are factors of zero different from zero is useful and finds application in some business applications (see *Finite Mathematics and Business Applications* by Kemeney *et al.*) because such matrices are easy to multiply and an nth power must be zero for some n. Matrices with the lower triangle all zeros are used in costing applications.

We have tended to stress the algebra of matrices and the matrices associated with linear transformations in geometrical applications but we are proceeding now to investigate these wider business applications.

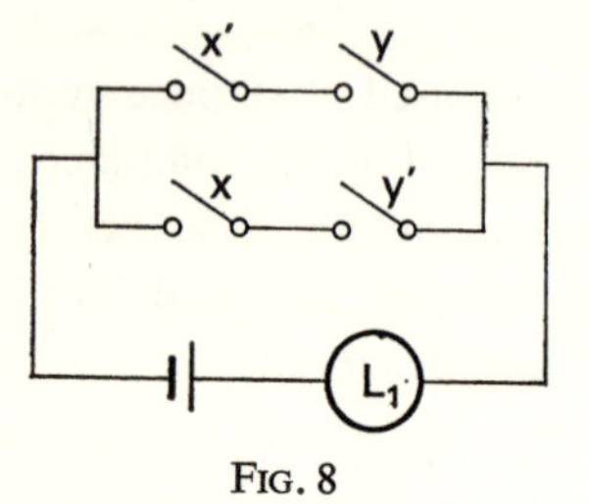

FIG. 8

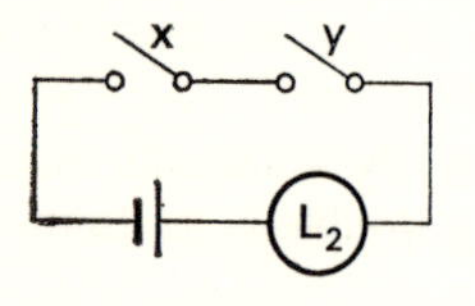

FIG. 9

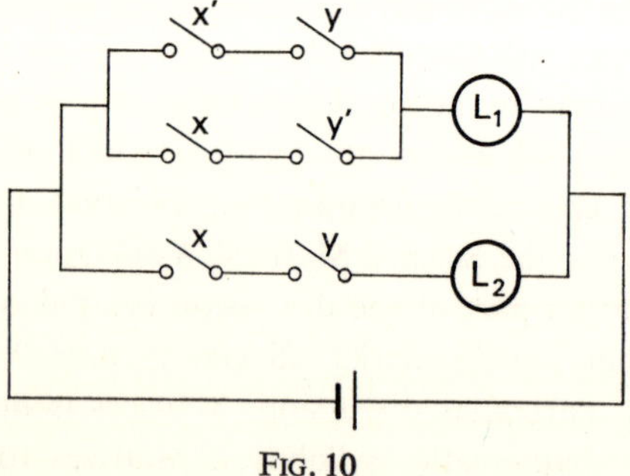

FIG. 10

Probability was again a new venture for O-level courses. Some work done with primary children suggested that it was a possible topic. The children recorded the sum of the scores of two dice by colouring in a square to make a block graph of the results. When a large number have been recorded the familiar Napoleon's hat results. A dent in the hat usually indicated that one of the addition facts was not known accurately. A similar experiment used the product of the scores. This went down well and the children's discussions indicated that they could find an intuitive reason for the shape. Probability goes down well although better results are obtained in the classroom than in the examination room. We try to get the pupils to sort out their assumptions and this we find difficult for most people go straight to a standard model without bothering to consider the conditions of the problem. The answer to the teaching difficulties appears to be more and more discussion about a variety of problems rather than practising various examples of a stereotype problem. Pupils then gain that insight and confidence which enables them to assign the right probabilities to the various events, to choose the right model and the appropriate manipulation to gain a solution. From simple problems we go as far as germination experiments where seeds are planted in threes and one seeks the probability of a certain percentage actually flowering using a binomial model. It is vitally important to discuss the results, to examine the arguments which decide whether a certain result is significant or not.

The above brief outline should give the spirit and flavour of the O-level course. In the early days because of the schools and the fact that we were in the Midlands M.M.E. tended to get a large number of pupils who would ultimately take C.S.E. The teaching materials were received enthusiastically by the pupils. Many of the pupils took the Mode 1 paper of the West Midlands C.S.E. which has so many options that pupils could be prepared for the examination. About 500 pupils are taking the Mode II (M.M.E.) C.S.E. of the West Midlands Examination Board and the results are quite satisfactory.

O-level had to be followed by A-level[7] and if I have concentrated unduly on the O-level it is because it is the sort of mathematics more easily taken during the lecture and discussed at the end. The A-level is a sort of crystallisation of the O-level topics. The first question on the A-level paper of July 1968 set by J.M.B. was to show that a matrix mapping was a homomorphism, to find the kernel of the homomorphism and from this to find the general solution of a system of linear equations. This example gives the reader an idea of the area in which we are experimenting. The example in fact is, we believe, on the basis of our experience so far, a better Further Mathematics question than an Ordinary Mathematics question. We have tried to validate the techniques involved in isomorphisms and homomorphisms. There is a need for the experience of teaching these topics to find out if sixth-form students can make something of them and whether they are worth-while topics for the sixth form. The M.M.E. A-level is principally concerned with this sort of experimentation which will provide information on which subsequent reforms may be based. We carry forward the work on Boolean Algebras. We develop vector spaces, linear transformations and matrices more abstractly and use vector geometry in three-dimensional problems and introduce the vector product and its applications. We have not found the answer to teaching the real number system and analysis and our calculus course has tended to be rather traditional. We are working on this problem of analysis and hope to have some results in 2 or 3 years. We have tried to make students aware of what they are doing when they are using calculus methods. In October 1967 I interviewed over forty candidates for admission to Worcester College of Education and again another forty in October 1968. During the course of the interview the candidates were asked various simple questions on mathematics and in the main their grasp of techniques was good but no candidate was able to give a satisfactory reply to the problem

$$\text{“Find } \int_{-1}^{1} \frac{dx}{x^2} \text{”.}$$

The answers were 0, 1, -1 and which depended on arithmetical slips. These were good candidates. Twenty of them went on to university in 1968.

The A-level is still in the experimental stage. An experimental course was conducted with two schools, a girls' and a boys', over 1966–8 but the students had not a great deal of mathematical ability. In 1967–9 three schools were involved. It would be a good plan to choose the students following the course so that from the onset there was a degree of certainty that some mathematical ability was present, but this has not proved possible and some students who would be better studying some other course follow the M.M.E. course because the custom of the school demands it.

The materials are not yet available to the general public. They are very much in the developmental stage. The abstract forms of the bookwork are developed and some teaching techniques have been successfully tried. The problem situations occupy our attention at the moment and there is a full-time development officer now engaged on the work.

On the whole we have found the M.M.E. worth while. There is a large number of schools of all types taking part in the work. The teachers and pupils have been enthusiastic about the new topics. The children are full of life, they answer back freely, ready to discuss a topic raised by the visitor and they are proud of the work recorded in their notebooks. It is a matter for great regret that the examination techniques employed by G.C.E. and C.S.E. boards leave a great deal to be desired in the way of feedback to the schools. We must somehow involve school teachers more and more in the evaluation of pupils' achievement.

We are almost ready to start again with new topics for the seventies and eighties. When one starts something like the Midlands Mathematical Experiment, one is not starting something one can finish. Daily one gets ideas from people, suggestions from literature, new topics come to light and much as one would like to incorporate these new ideas, the present syllabus is full enough already. One has to begin to think in terms of another syllabus. M.M.E. is evolving, changing and developing. A number of people have been discovered trough M.M.E., people who have something to contribute to teaching techniques, to exploration of new topics and to in-service courses.

References

1. *Synopses for Modern Secondary School Mathematics* (O.E.C.D.), Paris, 1961. *New Thinking in School Mathematics* (O.E.C.D.), Paris, 1961.
2. *M.M.E. Report 1961–62*, Harrap.
3. *M.M.E. Report 1961–62*, Harrap.
4. *The Midlands Mathematical Experiment 'O'-level*, Book I, Book II, Book III, Harrap, 1963, 1964, 1965.
5. *Syllabuses for O-level and C.S.E.*, M.M.E. Ltd.
6. *Chart M.M.E./4*, M.M.E. Ltd., 1966.
7. For A-level syllabus, etc., see *M.M.E. Report 1963–65*, Harrap.

Scottish Mathematics Group

W. Brodie

The views expressed must in some cases be those of the writer and may not be universally held. Many references have been taken from Reports issued by the Scottish Education Department, from *A Review of the First Five Years* published by the Scottish Mathematics Group in August 1968 and from a chapter on mathematics written by A. G. Robertson in *Scottish Education Looks Ahead* published in 1969 by W. & R. Chambers of Edinburgh.

For many teachers in Scotland the change in the teaching of mathematics began with the issue by the Scottish Education Department's Committee on Mathematics of a report on *Recent Changes in Honours Courses in Mathematics*, in which was stated:

> Few people can be unaware that there is a shortage of teachers of mathematics. This shortage, which is already very serious, threatens to become much more so in the years that lie ahead and to cause ever-increasing difficulties in our schools and colleges unless some means can be found of improving very substantially the supply of mathematicians.
>
> Many different groups of people in all parts of the educational system are interested in the problems facing the schools in the teaching of mathematics and, on the invitation of the Scottish Education Department, representatives of some of them attended a meeting early last year. As a result of this meeting a Committee, composed of representatives of the universities, central institutions, colleges of education, directors of education, teachers' associations and the Department, was set up and was given the task of reviewing the position and of considering possible measures, both long- and short-term, to relieve the shortage of mathematics teachers. Among these problems is the nature of the course in mathematics at present offered in schools. If mathematics in schools can be made more interesting by the introduction of some modern ideas, this of itself might encourage more boys and girls to continue their study of the subject to a higher level.

The breadth of this committee had an important bearing on what resulted because it meant that in principle what was to take place received favourable consideration and help from all those originally represented on the committee.

In June 1963 two residential courses taking forty persons at each were held, one in St. Andrews and one in Aberdeen. At these, teachers, lecturers and H.M.I.s went back to university. In an intensive course they learned something of Set Theory and other branches of mathematics.

From the January 1963 booklet and these courses many began to realise that, in less than two decades from the end of the war, set theory or set language had not just crept in but had rushed into almost every branch of mathematics. Even if they could not apply Boolean algebra they were beginning to understand it a little and to observe terms like "if and only if", mapping, group, ring, field, vector space and of course a Venn diagram. It is interesting to think that teachers of considerable experience were just hearing of something which is now taught to primary children. Teachers were seeing how honours courses were changing and were beginning to appreciate that the electronic digital computer was throwing out some old methods and introducing newer methods in numerical analysis.

On the advice of the Committee on Mathematics the Scottish Education Department set up the Mathematics Syllabus Committee to prepare alternative syllabuses on Mathematics for the Ordinary and Higher Grade examination of the Scottish Certificate of Education. This Committee consisted of principal teachers of mathematics in fifteen schools, principal lecturers in two colleges of education, and four members of H.M. Inspectorate.

At this point it may be in order to digress to explain the system for Secondary education in Scotland. It is quite different from that of England. At the age of between $11\frac{1}{2}$ and $12\frac{1}{2}$ years most pupils transfer from the Primary to the Secondary school. Until recently, according to the apparent ability of the pupil, he was sent for a 4- or a 5-year course or a non-certificate course. Those considered suitable for a 4- or 5-year certificate course totalled something in the region of 40–45 per cent of the secondary school population. The 4-year course has at its end the S.C.E. "O"-grade examinations, and the 5-year course has at its end the S.C.E. "H"-grade examinations. These examinations will be first attempted at the ages of 16 and 17 years. The "O"-grade is of a standard approximating to the G.C.E. "O"-level and the "H"-grade is between "O"- and "A"-level. The 4-year course for "O"-grade comparison with the English 5-year course for "O"-level is very important. Further mention of this will be made when some of the problems for a comprehensive course are examined.

The draft Ordinary Grade syllabus was discussed informally in December 1963 with university, college and further education representatives and the resulting syllabus was published in a First Interim Report in April 1964. In the following few months an experimental text, with teachers' notes, was written by the Scottish Mathematics Syllabus Committee and printed for use in the pilot schools. Two schools were preparing, without the aid of books, for the course one year ahead of all others. School trials started in September 1964 for fifteen "pilot" schools and forty-five "outer ring" schools. This sample of 7000 pupils comprised rather more than 20 per cent of the first-year certificate population in session 1964–5.

The letter to the experimental schools stated "a recommendation will be made to the S.C.E.E.B. that an alternative Ordinary grade examination be set in 1968, and an alternative Higher grade examination in 1969. ... While standard textbooks will be necessary for some parts of the work, teachers' guides and pupils' example books will be issued to all those taking part". Note here the suggestion that standard textbooks would be necessary for some parts of the work. This indicates that at first it was thought much of the work could be taught from traditional texts. In practice this did not take place and all the necessary material appeared in the experimental books.

The Interim Report previously mentioned stated that the redrafted syllabus had been drawn up "bearing in mind that the school mathematics course should be interesting, relevant and enjoyable for all the pupils and at the same time should form a sound foundation for those pupils who will continue the study of mathematics at a later stage".

It was decided that the teachers and lecturers on the Syllabus Committee should form the Scottish Mathematics Group, revise the text and have it published commercially. This was done and a Blackie/ Chambers consortium published *Modern Mathematics for Schools*. In December 1964 it was decided to publish in half-yearly editions. This enabled the first book to be available for the beginning of Session 1965–6 and permitted sufficient experience to be gained with the experimental texts. The relationship between the experimental texts and books of *Modern Mathematics for Schools* is shown in Fig. 1. A great deal of discussion ensued in the Group on the possibility of publishing teachers' guides. These had been issued along

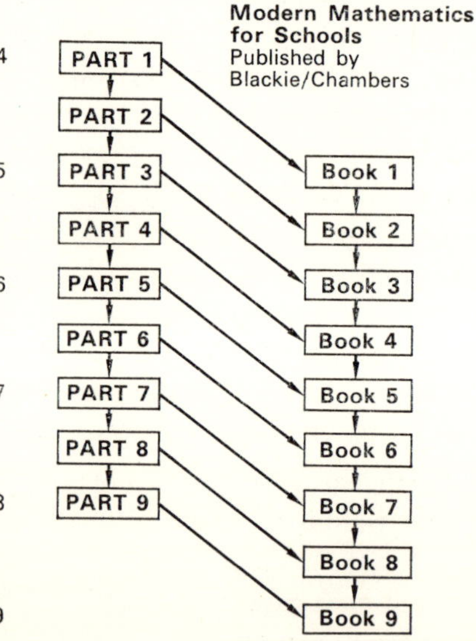

FIG. 1. The programme for testing, revising and publishing the nine main texts originally planned

with the experimental books but it was believed that teachers were not making use of them. This was evident from the fact that teachers frequently asked questions which had already been answered in the teachers' book. It was decided that necessary and sufficient guidance should be given where appropriate and these guides put into the pupils' books. When the books ultimately reached the schools the teachers' notes were printed on differently coloured paper and the pages stuck into the pupils' books at appropriate places. At this time it was felt that a guide had to be given to the teachers as well as the pupils. Many of the methods used were new to many of the teachers. Efforts were being made to make the work more pupil-centred and teachers had to realise how the new methods might be handled and also to know the theory behind the work. Initially it was possible for a teacher to be learning as much as the pupils. It is surprising how quickly the change has come about. The teachers are to be congratulated for rising to the challenge. They made great personal sacrifices to prepare themselves. Not only did they study the books ahead of the pupils but they attended courses given by colleges of education in the evenings and in the summer term. Local courses were frequently arranged. Local Education Authorities, the S.E.D., the Mathematical Associations in Edinburgh, Glasgow, Dundee, Aberdeen, and the Borders all co-operated to assist the teachers. It was an enormous project involving staff and local authorities. The latter assisting greatly with material assistance, textbooks, graph paper, new teaching aids, slide rules and even calculating machines.

Prior to this time it had been very difficult to persuade an education authority to supply calculating machines or slide rules. These had not been used in schools to any great extent and so authorities were suspicious of any applications for them. The official blessing given to this alternative syllabus helped to bring calculating machines and slide rules into the schools.

Teachers could not fully appreciate the course until they had worked through the book with a class. Frequently pupils could see the answers as quickly as their instructors because their minds were not cluttered up with preconceived ideas. The teachers in the pilot and outer ring schools tended to keep very rigidly to the experimental books because they were assessing these. This was contrary to their usual practice of using many textbooks largely for examples only and working through an examination syllabus in whatever order they felt best. Even when the work had passed beyond the experimental schools it was still new for the teachers who consequently felt obliged to keep very closely to the text. This has now been largely overcome with experience.

The Committee also decided to keep the Algebra, Geometry and Arithmetic in separate sections although close attention was paid to the order of development so that the branches could fit together. This gave the teacher much more freedom than he might have found in the rigid order of most integrated texts.

The *Modern Mathematics for Schools* books were written after receiving comments on assessment forms from the experimental schools. The schools were also visited by members of H.M. Inspectorate and the principal teachers of mathematics of the schools met once a year to discuss their progress and difficulties. Frequent reports made it possible for each school to have some idea of where it was in relation to other schools so far as speed was concerned. At one time it was reported that pupils in all the schools were at varying stages which extended over 10 pages in Algebra, 15 in Geometry and 25 in Arithmetic in both first and second years. For third-year cases the range was nearer 40 pages in each. Each school had its own problems and it would have been unreasonable to expect each to be at the same place. The knowledge that not all were ahead of you was a great comfort to many teachers.

The publication of *Modern Mathematics for Schools* was simply to enable schools in general to start the new syllabus. It was expected that other texts based on the alternative syllabus would be published in due course by other firms.

The Syllabus Committee studied the trends in the development and teaching of mathematics in a number of other countries as well as in the Scottish universities and technical colleges. Thereafter it critically examined, assessed and redrafted the school syllabuses against this broad background, bearing in mind that they should be interesting, relevant and enjoyable for all pupils—while at the same time the syllabuses should form a sound foundation for those who would continue the study of mathematics at a later stage.

The general aims of the new syllabuses are to provide a course that will interest pupils in mathematics and will train them to use the language in which popular, technical and professional texts are now being written. These syllabuses give pupils an early introduction to the concept of a set, the structure of a number system, the use of vectors, and the idea of a group. The aim is to relate mathematics to the solution of up-to-date problems by means of linear programming, the use of matrices, iterative processes and calculating machines—in short to include the kind of mathematics that reflects modern developments and leads to useful links with later work at school, college or university; and in the long term to increase the number of persons qualified in mathematics.

About one-third of the content of the algebra course is new, but much of the remaining two-thirds is approached from a new angle or is placed in a new context.

The language of sets, inequations as well as equations, the meaning of a variable and the idea of function are emphasised. The concept of mapping is introduced and used where most appropriate.

The algebra has less emphasis on manipulation and more understanding of mappings. The modern idea of a function $f: x \dashrightarrow f(x)$ as a set of ordered pairs or as the operation taking a variable in a number system to another variable in the number system is more important. A great deal of graphical representation is needed to assist in ideas of variation and later of calculus.

Systems of numbers N, W, Z, Q, R which are infinite sets are studied as also are modular number sets (finite) and systems of isometries in geometry. Hence we can show a little of group structure and isomorphism. This can be used in a study of logarithms.

As in algebra, about one-third of the material in arithmetic is non-traditional. This new work includes such matters as the laws of operation; the number system; number bases other than ten; an iterative method for square roots; elementary probability and a greater emphasis on statistics.

Rather less emphasis is given to computation and more to approximation and the appropriate limits of accuracy of measurement. The use of slide rules and calculating machines is encouraged where appropriate. Nevertheless the need to provide sound experience of social arithmetic and of mensuration is fully recognised.

With the introduction of the slide rule and consequent three-figure accuracy, it was considered desirable to standardize the mathematical tables to be used on a three-figure pattern. This has streamlined calculations in arithmetic and trigonometry, and has enabled more emphasis to be put on ideas and methods, and less on drill with tables. In any event, most of the data used in schools are accurate to three figures at the most—so that calculations based on them will be at the same level of accuracy. The inculcation of this idea is one of the aims in arithmetic.

The geometry course as set out in the writing breaks with tradition in a number of ways; in its starting-points, in its development and in its objectives.

A sophisticated mathematical text would begin with its axioms and its minimum definitions and then would proceed, with the appropriate degree of rigour, to deduce from these a sequence of theorems working in the main from the general to the particular.

The *Modern Mathematics for Schools* books, however, are not written for mathematicians, but for boys and girls who are doing mathematics by learning to think mathematically about their experience of the real world. This real world provides the background against which it is natural to build up the ideas and properties of the geometrical transformations of reflection, translation, rotation and dilatation.

These transformations are examples of the concept of "mapping" which is of great importance in the mathematics course as a whole. There is indeed constant interplay between algebraic and geometrical ideas; an obvious instance of this is the work with coordinates which is begun early in the first year.

Euclidean geometry is not so rigorous as is sometimes believed because it depends very much on visual assumptions. To put it on an axiomatic basis renders it far too difficult for school work. The older approach to geometry demanded a precision of thought and language too sophisticated for most pupils. Facts which are to be put into a logical sequence must be clearly grasped by some other means first of all.

The method used in *Modern Mathematics for Schools* in geometry is one based on what we would call two axioms, although the pupils never hear the word axiom. These are: (1) the rectangle fits into its own outline in four different ways, no two of which have a given vertex in the same place; (2) congruent rectangles can be used to cover completely a plane surface. From these the properties of various shapes can be discovered. The important point is that these can be built up by the pupils, who, by guided discovery, appreciate the various properties.

In all this writing the Group have had two objectives so far as the first 2 years of each course is concerned.

One problem which faced the Group at the outset was to make provision for these pupils, mainly girls, who gave up the study of mathematics after 2 years. In order to provide this group with a satisfactory mathematical education it was decided to have in Books 1 to 4 a self-contained course. At the same time these books had to be such that the other pupils could proceed smoothly to Books 5, 6 and 7 for the Ordinary Grade course. In the event the course has proved so interesting that more and more pupils are continuing with mathematics beyond the second year. In some schools the number of girls choosing to do this has proved an embarrassment. It is too early to say whether a greater number of pupils will study mathematics beyond Ordinary Level, but already there are indications that this will be the case.

To cater for those pupils who do not continue mathematics beyond the second year, however, there is a separate volume containing the arithmetic sections from Books 5, 6 and 7, i.e. to Ordinary Grade/O-level: *Modern Arithmetic for Schools*. It was published in 1968.

In geometry, coordinates are introduced very early in the course and this has been very successful. Naturally this flows into the more advanced work and combines with the ideas of sets. The idea of relations and mappings is given in Book 3 and of course continues throughout.

At the end of 3 years the teachers in experimental schools reported on the alternative syllabus. They were asked to comment on the syllabus, the experimental writing and on the pupil reaction. The general reaction was favourable although the geometry did give rise to some complaints. It might be interesting to comment that in speaking to teachers in England the replies often were that the geometry was good but the algebra and arithmetic were rather old hat.

In the report given after 3 years the teachers in the experimental schools commented on how they found not only the syllabus but also on the writing and on the pupils' reaction. The response was as follows where the symbols A to D were used as follows: A—satisfactory; B—mixed, generally satisfactory; C—mixed, generally unsatisfactory; D—unsatisfactory.

	Syllabus				*Writing*				*Pupil reaction*			
	A	B	C	D	A	B	C	D	A	B	C	D
Algebra	50	7	1	0	50	7	1	0	33	24	1	0
Geometry	15	32	10	1	6	28	18	6	5	28	20	5
Arithmetic	46	10	2	0	30	24	3	1	31	24	3	0
Trigonometry	46	10	0	0	24	27	5	0	20	30	6	0

The views on the geometry may have been due to the attempt to emphasise method and under-emphasise facts. Discussion may have arisen in the classroom but often the lack of distinct facts to be memorised as a result was felt to be a fault. It may also have been that the fundamental change in method required the training of the teacher as well.

Past experience has shown that only a mathematically able child can appreciate a strictly axiomatic development of geometry. The teaching of geometry in this axiomatic way would tend to be successful with only a small minority. All modern trends are away from the teaching of minority groups and aim to give a child some experiences of particular cases before moving to generalisations. There is no point in rushing to form concepts. *Modern Mathematics for Schools* aims to produce a good background and a good attitude to the subject. During the first 2 years there is an attempt to give the pupils opportunities to acquire geometrical knowledge, ideas and tools. Then in the third year the idea is to introduce more difficult work. Sometimes it would appear that the third year required some facts to be known but it took too long for the pupils, trained in methods and not facts, to reassemble in their minds the necessary facts. "I do and I understand" does not always mean "I understand and so I remember". It may be necessary to introduce more "drill" than was originally anticipated.

The Scottish system of 4 years for "O"-grade and 5 years for "H"-grade have a particular effect on the syllabuses. At the end of the second year pupils make the decision to continue studying a limited number of subjects—usually five to eight. During the fourth year, it is necessary to pick out those going on to attempt the "H"-grade in fifth so that some preparation can be made during the fourth year. These 4 and 5 years are minimum times and for some are difficult minima. So the first 2 years must be more than just tasting years. The teacher who takes too long in attempting to lay a very firm foundation will leave too much work to be done in the last 2 years. In spite of this, many schools have experimented with topics from the new syllabus in classes of less able pupils. Some of the new material can form bridges between the various ability levels.

As comprehensive schooling becomes more common, the problem of dealing with the wider ability range than that for which the book was written becomes more pressing.

In the present books every chapter was debated and discussed. The final writing invariably involved compromises between the various extremes and shades of opinion. To give an example, it took about $1\frac{1}{2}$ years before the agreed version of some of the work in the books was produced.

The books have proved very popular in many countries. Although originally written by Scotsmen for a Scottish syllabus, they are not mainly Scottish in interest. In fact it is believed that more books are in use in the Midlands and the south of England than in Scotland and the north of England. Blackie/Chambers publish separate editions in Australia and South Africa (in English and Afrikaans) while the original edition is in use in other Commonwealth countries. Separate editions have already appeared, or are in active preparation, for Malaysia, Hong Kong, Singapore and the Caribbean. Rights have been sold to Holland, Norway, Spain and Sweden.

Although the first four books were intended for 2 years' work in Scotland, it was found possible to cover a good amount of the C.S.E. work with Books 1 to 4 and a fifth book called Book 5A.

This is one of the big advantages England has over Scotland so far as comprehensive education is concerned. A C.S.E. course can be spread over 5 years and can be covered approximately with the aid of five books. The S.C.E. course for "O"-grade is normally covered in 4 years yet requires seven books. Changes may require to be made in the future to allow for comprehensive courses.

The contents of the first seven books are set out in Fig. 2 (pp. 344–5) together with that of Book 5A.

This shows how some of the work from Books 5, 6 and 7 has been incorporated in Book 5A. The C.S.E. course and that of the S.C.E. O-grade are not strictly comparable in content nor in depth of study.

The books prepared for the S.C.E. H-grade work are Books 8 and 9 whose contents are now set out in Fig. 3 (pp. 346–7).

This gives some idea of the comparison between S.C.E. H-grade and G.C.E. A-level courses. It is interesting to note that two methods of approach have been given to the work on the exponential and logarithmic functions. This enables the teacher to approach these from whatever angle he prefers.

A list of contents can never completely indicate the type of approach. Yet to pick out any one part does not do justice to the fact that a different method of presentation is used through almost the entire set of books. A few samples will now be taken in an effort to indicate to some extent where the book has introduced changes from texts used 20 years ago. It is not intended to show the actual layout of any page of the text but to give some feeling of what it is hoped the pupils will begin to appreciate. It must be emphasised that these may not be the most significant changes but are picked almost at random.

The chapters in algebra, geometry and arithmetic constantly intertwine. It would not be possible to go completely through all the chapters forming one of these subject sections without looking at the other subjects. Some fact would be needed from a chapter in one of the other sections.

Figure 4 shows the method of building up the various shapes. From the rectangle the right-angled triangle is readily obtained and from it the isosceles triangle and so on. An isosceles triangle is not defined as "a triangle with two equal sides". Instead the student learns that the figure obtained by putting two congruent right-angled triangles back to back with their right angles adjacent is called an isosceles triangle. Immediately, all the properties of such a figure can be observed. Some classes may use magnetic right-angled triangles. Some may have plastic shapes. Others may have cellulose acetate sheets on which they draw one of the congruent right-angled triangles. But others may simply use tracing paper or cut-out shapes from ordinary paper. Whichever method is used, the pupil actively participates in the construction of the triangle and learns the properties of the shape he has formed. Later it may be that the teacher will wish to define an isosceles triangle more formally, perhaps as a triangle with an axis of bilateral symmetry. At no place in the books is it defined in this way, but the pupil will ultimately realise that it is a triangle with this property of symmetry. Although the transformation of reflection does not occur until Book 3 the students have already met the particular cases of the isosceles triangle, the rhombus and the kite. This is another example of moving from the particular to the general.

In their first year, the pupils are likely to be calculating the areas of polygons. The area of the shaded square in Fig. 5 can be readily

	Algebra	Geometry	Arithmetic
1	1. An introduction to sets 2. Mathematical sentences—(i) Equations 3. Multiplication using the commutative, associative and distributive laws 4. Replacements and formulae 5. Mathematical sentences—(ii) Inequations	1. Cube and cuboid 2. Rectangle and square 3. Coordinates 4. Right-angled triangle 5. The isosceles and equilateral triangles	1. Length, area and volume 2. The system of whole numbers 3. Fractions, ratios and percentages
2	6. The distributive law 7. Powers and indices 8. Negative numbers 9. Mathematical sentences: methods of solving equations and inequations	6. Rhombus and kite 7. Parallelograms, triangles and parallel lines 8. Angles—rotation	4. Decimals and the metric system 5. Binary numbers 6. Introduction to statistics 7. Introduction to probability
3	10. Relations and mappings 11. Operations on the integers 12. Number systems 13. Equations and inequations in one variable	9. Locus 10. Calculation of distance 11. Translation 12. Reflection	8. Square roots 9. Proportion 10. Social arithmetic—(i) 11. Number patterns and sequences
4	14. Further sets 15. Linear equations and inequations in two variables 16. Systems of linear equations and inequations in two variables 17. Formulae	13. Specification of a triangle 14. Similar figures 15. The Circle—Rotational and bilateral symmetry Topic to explore—Topology (i)	12. A calculating aid—the slide rule 13. Applications of percentages 14. Length, area and volume associated with the circle 15. Statistics—(ii)
	Revision Section	*Revision Section*	*Revision Section*

FIG. 2

	Algebra	Geometry	Arithmetic	Trigonometry
5	18. Reasoning and deduction 19. The language of variation 20. Further addition and multiplication 21. Functions—the quadratic function and its graph	16. Revision, summary and some deductions 17. Introduction to vectors 18. Theorems and converses	16. Logarithms and calculating machines 17. Areas and volumes 18. Estimation of error	1. The cosine, sine and tangent functions
6	22. Factors and fractions 23. Surds 24. Quadratic equations and inequations	19. Dilatation 20. The Circle—(ii) Tangent and angle properties	19. Social arithmetic—(ii) 20. Counting systems 21. Statistics—(iii)	2. Triangle formulae
7	25. Indices 26. Introduction to linear programming 27. General revision exercises Bks. 1-7	21. Composition of transformations 22. Coordinates, vectors and transformations 23. General revision exercises Bks. 1-7	22. Flow diagrams and computers 23. General revision exercises Bks. 1-7	3. Trigonometry in three dimensions 4. General revision exercises Bks. 1-7

BOOK 7 COMPLETES THE O-LEVEL COURSE

5A (ALTERNATIVE BOOK 5 FOR CSE CLASSES) includes

Algebra: The Language of Variation · Further Addition and Multiplication: Functions: The Quadratic Function and its Graph · Factors · Quadratic Equations

Geometry: Revision and Summary · Introduction to Vectors · The Circle—Tangent and Angle Properties

Arithmetic: Logarithms and Calculating Machines · Areas and Volumes · Estimation of Error · Social Arithmetic · Statistics

Trigonometry: The Cosine, Sine, and Tangent Functions · Three-Dimensional Trigonometry

---

FIG. 2 (*cont.*)

Many teachers will be particularly interested in the development of the series beyond O-level. The following summaries are *not* the complete lists of contents.

8 ALGEBRA

Sequences and Series: the limit of a sequence · series and their sums · geometric sequences and series · the sum to infinity of a geometric series · arithmetic sequences and series
Matrices: matrix notation · addition and subtraction of matrices · multiplication by a real number · multiplication of an $(m \times p)$ matrix by a $(p \times 1)$ $(p \times n)$ matrix · the inverse of a square matrix of order 2 · systems of linear equations
Functions: Composition of functions and inverse functions
Polynomials, the Remainder Theorem and Applications

GEOMETRY

Gradient and Equations of a Straight Line: equation in forms $y = mx$, $y = mx + c$, $Ax + By + c = 0$ and $(y - b) = m(x - a)$ · perpendicular lines · intersection of two straight lines
Composition of Transformations: reflections in parallel axes · isometries · reflections and rotations · transformations in terms of coordinates and as matrices · mathematical structure · groups

TRIGONOMETRY

The Addition Formulae: the trigonometric functions of certain related angles · measurement of angles · the addition formulae · formulae involving 2α

CALCULUS

The Differential Calculus: introduction · rate of change · derivative of f at $x = a$ · the derived function · a differentiable function increasing or decreasing · stationary values · stationary points and elementary graph drawing · maximum and minimum values of a differentiable function on a closed interval · practical problems · a note on increments and differential notation
The Integral Calculus: introduction · area as the limit of a sum · the integral notation · anti-differentiation · particular anti-derivatives · areas, using integration · volumes of revolution · the trapezium rule

FIG. 3

9 ALGEBRA

Quadratic Functions and Equations: introduction · the nature of the roots of a rational quadratic equation · the discriminant, and tangents to curves · factorising a quadratic expression · sketching the graph of a quadratic function · expressions symmetrical in the roots of an equation · irrational numbers of the form $p \pm \sqrt{q}$

Systems of Equations: introduction · revision of systems of linear equations in two variables · systems of linear equations in three variables · systems of equations of which one at least is quadratic

The Exponential and Logarithmic Functions: positive integral indices · rational indices · the exponential function · the logarithmic function · the laws of logarithms · the derivation from experimental data of a law of the form $y = ax^n$ · change of base of logarithms · exponential growth and decay · an alternative approach based on the isomorphism between multiplicative and addition sets

Deductive Reasoning: negation · quantified sentences and their negation · the contrapositive · the idea of a mathematical proof — direct, indirect · 'and' and 'or'

GEOMETRY

The Co-ordinate Geometry of the Circle: the equation $x^2 + y^2 = r^2$ · the equation $(x - a)^2 + (y - b)^2 = r^2$ · the equation $x^2 + y^2 + 2gx + fy + c = 0$ · the intersection of a line and a circle · tangents to a circle

Vectors — 2: revision · vectors in three dimensions · addition and subtraction · multiplication by a number · position vectors · components of a vector in three dimensions · a vector as a number triple · the section formulae · the distance formula · the scalar product · angle · projection

TRIGONOMETRY

Products and Sums of Cosines and Sines: product formulae · sums and differences · equations

The Functions $\begin{matrix} a \cos x + b \sin x, \\ a \cos x^\circ + b \sin x^\circ \end{matrix}$ and some Applications: introduction · the form $k \cos (x - \alpha)$ · maximum and minimum values · solution of equations of the form $a \cos x^\circ + b \sin x^\circ = c$ ·

CALCULUS

Further Differentiation and Integration: trigonometric functions · chain rule for differentiation · some special integrals

FIG. 3 (*cont.*)

calculated by subtracting the areas of the four congruent triangles from the area of the larger square. The result of (100 – 2(16)) units^2 or 68 units^2 can give rise to the question of the length of the side of the shaded square. Thus the mind is being drawn to the calculation of the length of the hypotenuse of a right-angled triangle of sides 8 units and 2 units. The young person is meeting in this

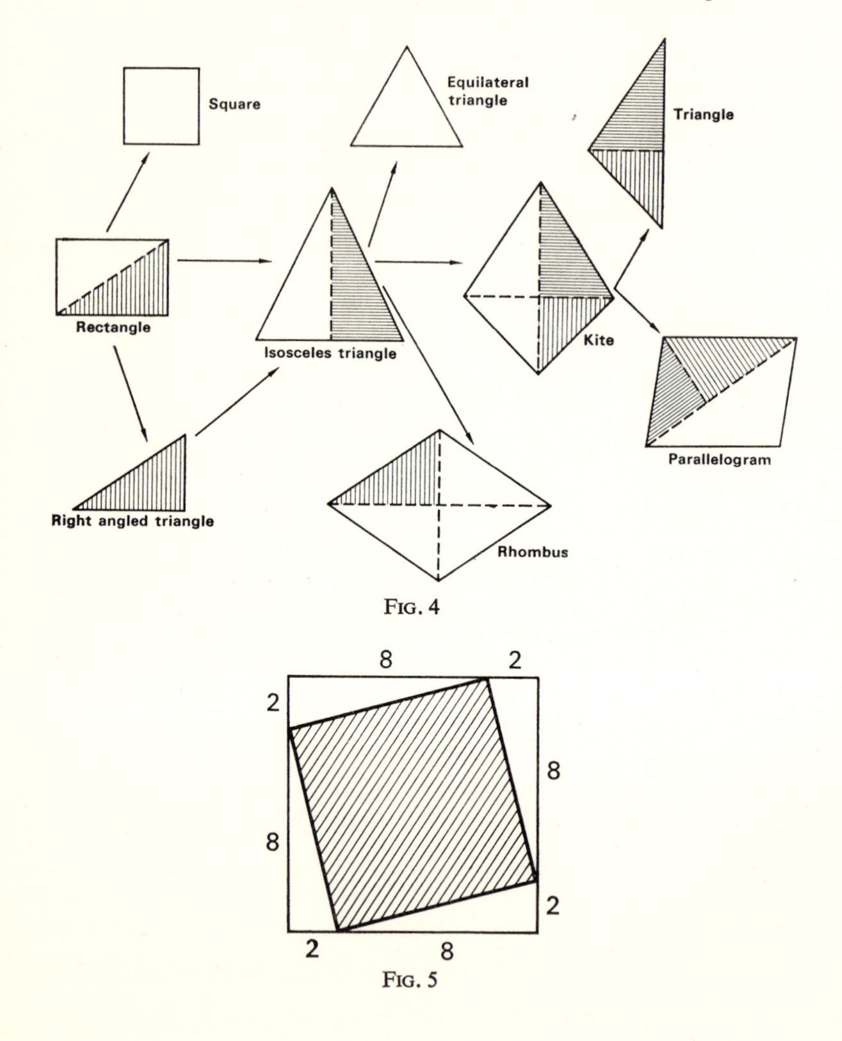

FIG. 4

FIG. 5

special case the idea used later in the Theorem of Pythagoras and also the need to find a square root which is not discussed thoroughly until later. At this stage a guess of 8·1 or 8·2 units is quite satisfactory and gives motivation for the calculation of square roots. The teacher can usefully ask the class what is the length of the other side of a rectangle of area 68 units^2 when one side is 8·1 units. Much later the seeds sown at this stage can assist in the method of finding the square root of a number by an iterative method.

In arithmetic in dealing with square roots it had been customary to use the idea that $(a+b+c)^2 = a^2 + (2a+b)\,b + (2a+2b+c)\,c$. Thus a problem in square roots might have been set down thus:

	4	1	8
4	17	54	36
	16	00	00
81	1	54	36
		81	00
828		73	36
		66	24
		7	12

This was done and the square root of 175436 was found to be something between 418 and 419. The pupil had progressively discovered a value of 400 for a, 10 for b and 8 for c, but seldom was there any understanding of why this gave the correct answer. Now an iterative method is used. The reasoning is much more easily followed and with improved comprehension there is more likelihood of the method of iteration being remembered and used in other situations. If $a, b, c \in R$, $a \times b = c$ and $a < b$ then $a < \sqrt{c} < b$. For the square root of 175436 he can start with $a = 400$ and $c = 175436$. This enables him to calculate $b = 438{\cdot}59$ and so $400 < \sqrt{175436} < 438{\cdot}59$. Using a new value of $a = \frac{1}{2}(400 + 438{\cdot}59)$, i.e. 419·3, he can obtain $418{\cdot}402 < \sqrt{175436} < 419{\cdot}3$. Continuing this method with an a value of $\frac{1}{2}(418{\cdot}402 + 419{\cdot}3) = 418{\cdot}851$ he will discover 418·8506

$< \sqrt{175436} < 418{\cdot}851$ which indicates to him that to six significant figures the answer to $\sqrt{175436}$ is 418·851.

This ability and need to recognise what is a reasonable answer is very important. At a reasonably early stage they are taught to look for approximate answers to give an idea of the magnitude of the answer. A little later the number of significant figures in an answer is considered from the point of view of accuracy and approximations. If 76 and 54 are counts then $76/54 = 1{\cdot}4073$. If, on the other hand, the 76 and 54 are measurements then the number of units in each case lies between two limiting values. For example the 76 may indicate some value between 75·5 and 76·5 while the 54 is somewhere between 54·5 and 53·5. This means the value of $76 \div 54$ in this case will lie between 75·5/54·5 and 76·5/53·5, i.e. between 1·386 and 1·43, using a slide rule. Whereas it may be permissible at times to give 1·4073 as the answer to $76 \div 54$, at other times the only reasonable answer is 1·4, correct to 2 significant figures. This is as accurate an answer as we can truthfully give.

In algebra it was customary in the past to solve equations. Now inequations are also considered and the solutions to these depend on the universal set being used. If $3x = 7$ and $x \in N$ then this equation has a solution set which is empty. But if $3x = 7$ and $x \in R$ then the solution set is $\{7/3\}$. If $3x < 7$ and $x \in W$ then the solution set is $\{0, 1, 2\}$, but if $x \in Z$ then the solution set is $\{\cdots -2, -1, 0, 1, 2\}$ or $\{x \in Z; x < 2\}$.

The meanings of mapping and function are brought out and these are used in many places, not least of all in calculus. In the chapter in which the quadratic function is discussed there is the suggestion that "nested" multiplication is useful (e.g. for a numerical replacement for x in $x^2 + 3x - 5$ the value of the expression would be found from $(x + 3)\,x - 5)$ and the relation between the sketches of $f: x \to x^2$ and $g: x \to 4 - (x - 3)^2$ is suggested as worth investigation.

As the books progress the more modern terms introduced in the earlier books are seen to be more and more useful and meaningful. At the same time the interrelationship of the various parts of the books becomes more obvious. The composition of functions occurs in algebra, geometry, trigonometry and calculus.

At all points there has been an attempt to be adequately rigorous. When dealing with the equation of a straight line through the origin with gradient m it was often considered in the past that it was sufficient to take ANY point $P(x, y)$ on the line and show that the gradient of OP was y/x and hence $y/x = m$. The usual method was then to state that since P was ANY point on the line, this equation held for ALL points on the line and so $y = mx$ held for ALL points on the line. The use of the word ANY is not appropriate since an equation might be true only for one or two points and these could be chosen when taking "any" point. In this particular case the equation $y/x = m$ is not meaningful for the point (0, 0). So the equation must be obtained by stating that when the point P is on the line and distinct from O, then the equation is

$$\{O\} \cup \{P : m_{OP} = m\}$$
$$= \{O(0,0)\} \cup \left\{P(x, y) : \frac{y}{x} = m, x \neq 0\right\}$$
$$= \{(x, y); y = mx\}.$$

It is difficult to give a proper idea of the nature of the books. They must be examined—even more than an examination is needed, they must be used—before a real appreciation can be made.

As has been indicated, changes have taken place in the teaching methods over the last 10 years. Not only in Mathematics teaching have ideas changed but also in what education really means when the whole population is being considered. Whereas it might at one time have been considered sufficient to have a multilateral school with different types of courses we now have comprehensive schools with common courses. These books appear to have satisfied a long-felt want for a change in the methods of teaching mathematics for certificate-pupils. They were written for the top 35 per cent of the school population and have been used with a much larger percentage. The introduction of common courses in comprehensive schools has raised difficulties. Does a book written for the top 35 per cent meet the needs of 85 per cent of the pupils? At the time of writing the Scottish Mathematics Group is considering the possibility of altering the books so that they may meet the needs of a much larger group of the school population. This is one of the difficulties of living in an age when education is so dynamic. Our books cannot remain static but must be ever ready to make changes.

School Mathematics Project; The Initial Stages

A. P. PENFOLD

THE first artificial satellite, Sputnik I, was launched on October 4, 1957, by the Russians, and concern was already being expressed both in the U.S.A. and in the U.K. about the preparations that we were both making for this new era of mathematical applications, and possibly for new approaches to the teaching of mathematics in schools. The Oxford Mathematical Conference[1] in April 1957 "was designed to broaden school teaching by providing contact (with those who develop science) specifically in the form of representatives from schools, universities, technical colleges, industry and research organisations meeting and talking together for a sufficiently prolonged period to understand each other's work". The Liverpool Mathematical Conference in April 1959 had as its theme "Mathematics in Action" and the Southampton Mathematical Conference in April 1961 concentrated "On Teaching Mathematics".

The School Mathematics Study Group (S.M.S.G.) in the U.S.A. was already in being during this period and in the foreword to the preliminary edition of *Mathematics for High School, Intermediate Mathematics, Part I*[2], published in 1959, it was stated that "One of the prerequisites for the improvement of the teaching of mathematics in our schools is an improved curriculum—one which takes account of the increasing use of mathematics in science and technology and in other areas of knowledge and at the same time one which reflects recent advances in mathematics itself". Similarly, the foreword of each of the published texts of the School Mathematics Project[3] since 1964 echo this theme with "This project was founded on the

1964

belief, held by a group of practising school teachers, that there are serious shortcomings in traditional school mathematics syllabuses, and that there is a need for experiment in schools with the aim of bringing these syllabuses into line with modern ideas and applications".

The Chairman of the Southampton Mathematical Conference was Professor Bryan Thwaites of Southampton University, who had attended the Oxford Conference in 1957 as a schoolmaster representative from Winchester College and who, in his inaugural address as Professor of Applied Mathematics, had strongly stressed the serious shortage of teachers of mathematics. There is much in the published deliberations of the Southampton Mathematical Conference which must have influenced Professor Thwaites in his subsequent initiation of the School Mathematics Project. The Report of the Sub-Committees on Mathematics in the General School Course probably best illustrates this influence by the three conclusions and nine suggested modifications to existing syllabuses (pp. 30–32)[(4)]. The conclusions were:

(a) The unification of the course should be encouraged. "Alternative A" type syllabuses with separate external examination papers in arithmetic, algebra and geometry prejudice this and should be discontinued. However, it would be deplorable if, as a result of such a change, the geometrical content became relatively unimportant.

(b) Since Set Theory provides a common language in which the concepts of many branches of mathematics can be expressed, such language should be utilised, whenever appropriate, from the earliest years.

(c) Although at this level too much rigour cannot, in fact should not, be expected, clarity of thought and precise use of the mathematical language so far developed is vital. This must not, however, be allowed to stultify mathematical experience. Correct and desirable concepts will often be formed in the more informal and intuitive parts of the work which come early, and it may well be that only later, with greater maturity, will the more formal verbal expression be demanded.

The suggested modifications to the syllabuses were:

(a) Greater emphasis should be placed on an appreciation of the structure of algebra—stressing the commutative, distributive and other similar properties—rather than on the acquisition of techniques. A greater understanding of the structure will inevitably enable this acquisition of techniques to be made.

(b) Inequalities should be considered at the same time as equalities. There is no inherent reason for not doing so. In fact, positive advantages are to be gained.

(c) A wider conception of function should be instilled, and many examples of functions and relations which are not obviously mathematical should be introduced.

(d) Much practice should be given in the construction of an equation or a set of equations as the model for a problem, without the insistence that the equations be solved. It is also of great value to give practice in the reverse process—of suggesting what could be a physical problem for which a given equation or set of equations is the model.

(e) Since variation permeates so much of mathematics, especially in association with the concept of function, it is better dealt with in the several settings in which it naturally arises, rather than as an isolated topic.

(f) Only very occasionally, and never in examinations, should pupils be expected to answer questions which involve "heavy manipulation".

(g) If extraction of square root by an arithmetical method is taught at all, it should be done as an example of an iterative process. The method of making an intelligent guess followed by successively closer approximations is basically similar to many of the numerical methods of solving more complex problems.

(h) Some idea should be given of the importance of statistics, of the limitations of statistical method, and of some of the inferences which may and may not be drawn from a given set of data.

(i) There should be some mention of the historical development of mathematics. This might be supplemented by, or even take the form of, "thumb-nail sketches" of the work of the more important mathematicians, introduced whenever a suitable and natural opportunity arose. In any case some of the history and the elementary ideas of the calculus should be included.

Nearly 10 years later, it is difficult to dispute any of these conclusions or suggested modifications, but at the time there must have been many teachers of mathematics who would have had doubts about or ignorance of their implications. But with possible foresightedness, Professor Thwaites set about the task of implementation. With the active co-operation of the senior mathematicians at Charterhouse, Marlborough, Sherborne and Winchester, writing of material suitable for the last two years of an O-level course was begun. It is doubtful if the School Mathematics Project would have made such an auspicious start but for the combined efforts of T. D. Morris, D. A. Quadling, H. M. Cundy and T. A. Jones in producing a remarkable amount of material for use in schools from September 1962. In the meantime, Professor Thwaites was concerned that the Project would not eventually have the necessary impact on the teaching of mathematics throughout the country if it were restricted to these four public schools.

During the early part of 1962 Professor Thwaites persuaded the senior mathematicians of four other schools, Battersea Grammar, Exeter, Holloway Comprehensive and Winchester High School for Girls to accept the Project material, so that the original Project was started in eight schools, and although Holloway Comprehensive was quickly replaced by Abingdon School during the first year, these eight schools remained as the central organisation of the Project. Conditions for carrying out the teaching of the experimental material must have varied considerably, even within these eight schools, and my own experience at Battersea Grammar School was probably as difficult as it could have been.

Of four full-time teachers of mathematics, two were engineers by training, at least one of whom was horrified at the prospect of radically changing his approach and the content of his mathematics teaching. However, the headmaster, W. J. Langford, a mathematician of some repute, was prepared to spend part of his time assisting in the launching of the material. But even this generous co-operation was not sufficient to staff all classes, and in any case, I was not sure that I should be able to staff them for a protracted period. It is important to realise that all of us were unfamiliar with much that was demanded of us in our teaching of mathematics, and the pressures during that early period not only to familiarise ourselves with the material but to attempt to teach it at the same time were very great.

The decision which I made in fairness to my staff, the pupils and to the project was to submit only two classes out of four to S.M.P. material, whilst the other two classes continued with a traditional syllabus. The classes were normally setted, but I decided to make the second and third sets of comparable ability, so that we might have some indication at the end of 2 years of performance at O-level in different syllabuses of two comparable groups. At the same time, the top set who were already at the end of their third year almost ready to take the traditional O-level examination would be entered for the traditional O-level examinations at the same time as the S.M.P. examination without any deliberate preparation during that 2 years for the traditional course. Desirable though these comparisons might have been, events militated against any fair comparison.

From the eight project schools, the expected number of entries for the first O-level S.M.P. examination was of the order of 600 in the summer of 1964, and because of such small numbers, predictions of the expected numbers of passes from past experience of traditional courses were given to the examiners to assist them in deciding a pass mark. For example, at Battersea Grammar School, we did not expect more than forty-five out of sixty-five candidates to pass. The actual entry for S.M.P. O-level was 919 which included from some of the project schools, large groups who would not normally have taken the examination until a year later as well as entries from other schools who were working to the syllabus but not specifically using the S.M.P. texts. Our passes were fifty-seven instead of forty-five out of the sixty-five entries, which was appreciably better than we expected. The explanation of such an improvement over expectation is difficult to make, but we suspected that the two teachers involved had been obliged to devote more energy to the project material than they would have been devoting to the traditional course, and their enthusiasm and energy had rubbed off on the candidates. It is doubtful whether such energy could be sustained indefinitely.

We, therefore, did not consider that the better result than expected was necessarily entirely due to the improvement in the mathematical content of the course. Any comparison between the performances of members of the top set on the two examinations was not possible, due to circumstances beyond our control. The class were told *after* they had taken the S.M.P. examination that they had been entered for the London O-level examination, and that their attendance for examination was voluntary. At the time, they expressed readiness to attend unanimously, but since it was the custom during the examination period for the candidates to be on leave, we were frustrated by the coincidence of a vital Test Match at Lords on the first day of the London examination, and only ten of the thirty-two candidates sat the examination. The rank ordering of these ten candidates on both examinations were comparable and all ten passed both examinations, so that we had scanty evidence for comparison.

I think that it is important to present a picture of the conditions under which the experimental groups carried out their work for the first 2 years on S.M.P. material. At the end of the summer term in 1962, 211 packages each containing seventy foolscap sheets of typed material arrived at the school for the two classes, together with seventy folders. These had to be converted into seventy folders each containing 211 sheets for distribution to the pupils for the first 6 months of the project. This activity alone was demanding on the available man-power and a post-examination A-level group co-operated willingly in the task, at the same time skimming through the material which they were assembling. The first reactions of this group were of commiseration with the younger group of guinea-pigs, since so much of the material was unfamiliar even to them.

When work began on the folders in September 1962 a problem arose immediately of having to carry around the school and to and from home such a bulky volume. Each teacher resolved the problem in his own way, the method adopted by the headmaster taking the top set, being to allow each boy to keep the main folder at home with a supplementary folder for transporting the current chapter for classwork.

The sequence of topics in the first draft folder had chapter headings as follows:

A. The Slide Rule B. Organised Knowledge C. Sets
D. Reflection, Rotation, Translation E. Graphs and Diagrams
F. Shearing G. Inequalities H. Mensuration

In Chapter A the pupils were encouraged to construct two sliding paper scales firstly for addition and then for multiplication, the approximate nature of the devices being emphasised with discussion of significant figures and limits of accuracy in any calculations. Chapter B gave a variety of mathematical problems and discussed how their solutions might be organised, and then concentrated on the advantages in Euclidean Geometry of an organised approach particularly in dealing with congruence of triangles and in use of Pythagoras' theorem. Chapter C was the longest chapter in the folder, and was comprehensive in its use of set language and set notation in a wide range of situations. Chapter D attempted to take a fresh look at the bulk of Euclidean plane geometry from the standpoint of reflection, rotation and translation and introduced the novel approach to standard ruler and compass constructions using only a band (i.e. a pair of parallel edges). This approach was very popular with the pupils and teachers alike.

Chapter E was the first introduction to the collection of statistical data and the graphical representation of such data, and the remaining Chapters F, G and H gave a fresh approach to area and volume through shearing as well as the mensuration of familiar plane and solid figures, the chapter on inequalities concentrating on the permissible combinations of simple inequalities.

Throughout the teaching of this material the text laid emphasis on class discussion and, although we had always encouraged this attitude to mathematics prior to the project, we found that the effect on the classes was to produce an even more critical consideration of their work, together with a side effect of criticism in other subject areas. At all stages, the classes were encouraged to comment favourably or adversely on the work and these comments were combined with the teachers' reactions in regular feedback to the S.M.P. office at Southampton. The same procedure was adopted throughout the whole of the first year's experimental text, the results of which were a drastic reorganisation of the text together with a change in the style of presentation for the second wave of classes. The cumbersome foolscap folders were replaced by four slim, stapled booklets produced by an offset litho process, and were a much more manageable proposition. The sequence of topics in these booklets, although drastically rearranged from the foolscap version, were still subject to regular criticism, but it is interesting that the second revision which appeared as the published Book T was not substantially different from the booklet version, either in content or in order of presentation. This must have been one of the earliest attempts to produce a text after nearly 2 years of experiment and criticism by a large body of users.

To those of us who were involved in the initial stages of the Project, our reactions are probably best summarised in the *Director's* Report[7] for the year 1962–3.

> We feel that the increased excitement and enthusiasm is greater than can be accounted for solely by novelty: thus if one of our aims is to make mathematics a more attractive subject at school, then it is our belief that some genuine progress has already been made. Indeed, some teachers have been most agreeably surprised at the changed atmosphere in the classroom as a consequence of the concentration on discovery and ideas rather than technical skill.

Although the four original public schools had greater staff/pupil ratios in mathematics than the rest of us and hence greater flexibility for releasing staff or lightening their teaching load in order to write material, it became a matter of urgency to appoint relief teachers to the Project so that the writers could have even lighter timetables, without losing all contact with the classroom. In the first 2 years, six such relief teachers were appointed, most of whom subsequently became full-time teaching staff. No such appointment was possible at Battersea Grammar School as a voluntary-controlled school, so that although we were anxious to take a more active part in the writing of material, we had to be content with joining revision committees which proved valuable in both directions. There was also much to be gained by the regular meetings of representatives of each of the eight participating schools not only to report on the current position but to plan in concert the development of the Project.

It is perhaps easy at the present time not to realise that S.M.P. is really three projects. The first, started in 1962, was the 2-year course to O-level, represented by Books T and T 4[5]. The second, started 1 year later, was the 5-year course to O-level, represented by Books 1 to 5[6], and the third, started in 1968, was the C.S.E. course represented by Books A to H, of which the first four books have so far been published. The A-level course is intended as a continuation of either the 2-year or 5-year course, with the insertion of two additional mathematics books if the O-level course is completed one year early. A major policy decision was made during 1963 and published in 1964 in which the sixth-form course would be as a single subject mathematics course with the possibility of a handful of candidates offering further mathematics.

The intention of this course of action was the majority belief that mathematics was indeed one subject, and in any case the traditional view of Applied Mathematics (meaning Mechanics) was fast becoming a limited view of the subject. It was also considered by the majority of the Project members that by restricting Mathematics to one subject at A-level, then potential mathematicians would be assured of a broader basis for their A-level course. I have never accepted the thesis that studying one subject necessarily restricts a person's education; moreover, a subject well taught must stray outside the artificial boundaries which surround it. Although I accepted the situation that Applied Mathematics must be broadened, I was not satisfied that telescoping the time devoted to Mathematics as a whole could be justified for the majority of candidates, especially if so many of them would be hard-pressed to find a satisfactory group of subjects to offer at A-level outside Pure Mathematics, Applied Mathematics and Physics.

In a questionnaire to all the mathematics departments of universities in the United Kingdom, the Director was anxious to discover whether a single subject mathematics qualification in S.M.P. mathematics would be acceptable for admission to an honours mathematics course, and although the replies were varied, there were a substantial number who were sympathetic to the idea, provided that the performance of any one applicant was of a sufficiently high standard. In the event, although such approval had been given, some of the early applicants did meet difficulties when making applications. It is hoped that at the present time many of these difficulties have been overcome. Our own experience at Battersea Grammar School was of a different nature. We had been accustomed under a double mathematics course to gain of the order of half a dozen admissions to mathematics honours courses, and although the number of students studying the first S.M.P. A-level course rose from twenty-five to thirty-five, the number who went on to read honours mathematics declined. This was mainly due to the students having been obliged to choose a third subject such as Geography for their A-level course, and discovering that either Physics or this third subject was the subject which they wished to study further at university. It is to be hoped that although one of the objectives of the project was to produce more mathematicians, this unfortunate side effect has now been overcome.

In a commentary on the A-level syllabus in the *Director's Report*(8) for 1963–4, it was stated that

> As with the O-level syllabus one of the main changes from convention lies in the increased emphasis on algebraic structure. ... Side by side with the algebra goes the analysis and this part of the syllabus is governed by the evident need to cover the standard techniques of calculus. It is also governed

> by the fact that the S.M.P. has stopped short of introducing a systematic treatment of calculus in the O-level course; thus pupils will be picking up the language of calculus while already at a more mature stage in algebra. But ... we are anxious to avoid a treatment of calculus by rules-of-thumb of the kind which any fourth-former can pick up in an hour or two.

In teaching the beginnings of calculus in the sixth form with a background of O-level S.M.P., I endorse the sentiments of this comment. The pupils were already familiar with the concept of a mapping and the need to have clearly defined the domain and range of the mapping. In this way, the very sensible approach to the idea that a derived function was a scale factor from the domain to the range of the original function was easy to demonstrate visually, as well as being sound in the discovery of limiting value of the scale factor.

I have dwelt at some length on the background of the early stages of S.M.P., and given little indication of the effects upon the teaching of O-level mathematics. I think that it is true to say that as far as the teacher was concerned, we discovered that what had previously been considered as diversions from a set course were now quite naturally incorporated as part of the course, since the ideas involved were fundamental to the change of approach. Moreover, diversion even from the text was encouraged, of which the following examples are an illustration.

In dealing with the problem of a measured length being an approximation to the abstraction of the true length, and at the same time to inculcate the intelligent reading of scales, I marked on the blackboard a length AB and randomly marked a point C, approximately two-thirds of the distance from A to B. I then challenged the first form in front of me to estimate to one decimal place the fraction that AC was of the length AB. As expected, two values were ascribed, namely 0·6 and 0·7 with a heavier weighting on 0·7. I then asked the class to produce a second estimate to two decimal places, and a finer frequency distribution resulted. When challenged how the concensus of opinion of the class might be represented, they soon suggested a mean value in each case and discussed the apparent small discrepancy in the two results. I then suggested that we might get a more reliable estimate by successively bisecting AB and the subsequent section containing C. If at any stage, the point C was in the right-hand section of a bisection, we ascribed the symbol 1, and if it was in the left-hand section we ascribed the symbol 0. Of course, if the length AC was two-thirds of the length AB, then the resulting representation of the fraction would have been 0·101010 ..., and we were quite naturally handling a binary system of counting, which had already been met in another context. It will have been noted that at each stage, the bounds of the ratio of AC to AB have been successively restricted between 0·5 and 1·0, 0·5 and 0·75, 0·625 and 0·75, 0·625 and 0·6875, and so on. In this one simple experiment, a number of ideas emerged, not least of which is that measuring length is essentially counting of a number of standard lengths and that group measurement of the same length leads to a distribution of results. But there were of course many other useful concepts which are bound up in the whole exercise. With no knowledge of a binary scale nor of distribution of results, the whole experiment would have been much more restricted in its aims.

When progressing from the measurement of length to the measurement of area, the project material in Book I emphasises that the process is essentially a counting one, and stresses the number of patterns of the same size and shape which will fill an irregular shape. The idea of a regular unit of area is delayed until the counting process has been established, and the extension to volume becomes a natural one.

The emphasis in the first draft of experimental text on two-dimensional geometry in terms of transformation geometry has been somewhat muted in the published material, but, nevertheless, it is still true that geometry is still seen initially as the study of translation, reflection and rotation, and their eventual representation in matrix form. Perhaps one of the most useful ideas to emerge is the matrix form of a simple rotation and the combination of two rotations to give immediately the addition formulae in trigonometry. In this branch of elementary mathematics, the traditional order of presenting the trigonometrical ratios of tangent, sine and cosine is reversed, since they are considered from the outset not as ratios but as component multipliers, and stress is made on empirical derivation of these multipliers to two decimal place accuracy. The insistence at O-level on no more than three-figure accuracy has probably taken out of computation the tedium for which it is sometimes accused.

In this respect, I regret the limitations which seem to be accepted on accuracy when using a slide rule. I have always believed that it is as easy to develop a rule-of-thumb approach to the use of the slide rule without real understanding as it is to learn simple derived functions in calculus without real understanding. To illustrate, I have persuaded classes to find the recurring figures in the decimal equivalents of 1/7 and 1/23 by using the slide rule and the residues or remainders at each stage of the computation. For example, when expressing 1/7 in decimal form, using the slide rule, the mark 1 on the sliding scale is placed coincident with the mark 7 on the fixed scale. The slide rule is now set for two possible uses, namely division or multiplication by 7, but remainders can also be read from the slide rule at each stage by counting as follows:

10 on fixed scale has 1 as nearest single digit on sliding scale to the left, which is opposite 7 on fixed scale, i.e. 3 units below 10 on fixed scale, which means that 10 divided by 7 is 1 with a remainder of 3.

30 on fixed scale has 4 as nearest single digit on sliding scale to the left, which is opposite 28 on fixed scale, i.e. 2 units below 30 on fixed scale, which means that 30 divided by 7 is 4 with a remainder of 2.

If this process is continued in this sequence, then the result is:

Fixed scale	10	30	20	60	40	50
Slide scale	1	4	2	8	5	7
Fixed scale	7	28	14	56	35	49
Remainder	3	2	6	4	5	1

and the slide scale digits give the sequence required, i.e. $1/7 = 0{\cdot}142857\ldots$ In the same way, the reciprocal of 23 can be computed with no more than the counting of remainders.

Fixed scale	10	100	80	110	180	190	60	140	20	200	160
Slide scale	0	4	3	4	7	8	2	6	0	8	6
Fixed scale	0	92	69	92	161	184	46	138	0	184	138
Remainder	10	8	11	18	19	6	14	2	20	16	22

Fixed Scale	220	130	150	120	50	40	170	90	210	30	70
Slide scale	9	5	6	5	2	1	7	3	9	1	3
Fixed scale	207	115	138	115	46	23	161	69	207	23	69
Remainder	13	15	12	5	4	17	9	21	3	7	1

Thus $1/23 = 0{\cdot}0434782608695652173913 \ldots$ has been deduced entirely from the slide rule. Although I would not advocate that the slide rule should be used for all calculations requiring more than three-figure accuracy, I am convinced that such computation well-handled could play as important a part in general education mathematically as any widespread attempt to bring computer science down to the level of the majority.

At about the same time as the experimental group at Battersea Grammar School were discovering the intricacies of graphical solution of elementary linear programming, we were presented with a practical problem which helped considerably in the appreciation of the method. The local running-track was temporarily out of action and we had to design a running-track on the school field. The greatest length available was 170 yards and the greatest width was 140 yards. It was generally agreed that the shape of the track should be traditional, that is, two semicircles joined by two parallel straight lines, and that the two variables of the problem were the radius, r, of the semicircles and the length, d, of the straights. After much discussion, the class decided that both variables could not sensibly be less than 35 yards; in the one case, so that athletes could comfortably negotiate the bends and in the other case, so that each athlete in a distance race was not at a disadvantage if the final straight was too short. It was also considered desirable if the length of the track were to be 440 yards.

The conditions were therefore (i) $d + 2r \leqq 170$, (ii) $2r \leqq 140$, (iii) $r \geqq 35$, (iv) $d \geqq 35$, (v) $d + \pi r = 220$, the first four of which determine a region of ordered pairs (r, d) to be satisfied. The fifth condition, however, could not be satisfied, and perhaps it was salutary that the problem as stated had no solution, since all previous and subsequent problems which the class had to tackle had solutions. In this practical problem, we were obliged to change one of the conditions, the length of the track to 330 yards. A set of solutions now existed and when a scale drawing for each of several ordered pairs (r, d) was made, much discussion took place of the most pleasing shape of the track, out of which quite unexpectedly came the golden section. The rectangle formed by the straights and the diameters of the semicircles had dimensions $2r$ by d, and the rectangle enclosing the whole track had dimensions $(d + 2r)$ by $2r$. If the ratios of the dimensions of these two rectangles are equated, then $2r/d = (1 + \sqrt{5})/2$. For a track of 330 yards, this means a radius of 37·7 yards and a straight of 46·6 yards, both within the restrictions imposed. The enthusiasm with which the class set about the task of actually marking the track on the school field made all the preliminary investigation worth while.

Although the examples chosen are not strictly from the context of S.M.P. material, they are nevertheless within the spirit of the O-level course, and moreover they demonstrate a dynamic approach to the teaching of mathematics which can do nothing but good. Class reaction as has already been stated has been favourable in these circumstances, and although it has been of some concern to the Project that so many schools have adopted the S.M.P. texts before the organisation was satisfied that the scheme was effective, it could be interpreted that independent observers were already satisfied that the O-level syllabus offered by S.M.P. was more appropriate to present-day needs than the traditional syllabus.

In attempting to assess this impact, information can be found in the *Director's Report*[9] for 1967/68 of the number of candidates for O-level S.M.P. Mathematics from 1964 to 1968, which rose from nearly 1000 in 1964 to nearly 11,000 in 1968, and has reached 20,000 in

1970. Although the rate of growth has not been as great in the last few years, it is still faster than comparable situations when examination boards introduced integrated mathematics syllabuses alongside those containing separate arithmetic, algebra and geometry papers soon after the end of the Second World War.

For the 1969 examination, the O-level syllabus had already received substantial revision, mainly as a result of the material which had subsequently appeared in Books 1 to 5, such as networks and their associated matrices. It had always been a conviction of the Director from the beginning that the spirit of S.M.P. should be for continuous revision of the texts to keep pace with not only developments in mathematics but developments in the teaching of the subject. Perhaps herein lies the undoubted success of the Project, which, in the original drive for change by the Director, Professor Thwaites, backed up by his strong team of schoolmaster writers under the able coordination by Dr. A. G. Howson of Southampton University, was able to make a big impact initially and was able to sustain and modify this drive as circumstances demanded. S.M.P. is now an accepted form and style of mathematics which will have a lasting effect on the teaching of the subject in this country.

References

1. *Oxford Mathematical Conference, Abbreviated Proceedings*, p. 5, Technology, 1957.
2. School Mathematics Study Group, *Mathematics for High School Intermediate Mathematics* (*Part 1*), Foreword, S.M.S.G., 1959.
3. *School Mathematics Project Book T*, Foreword, Cambridge University Press, 1964.
4. BRYAN THWAITES, *On Teaching Mathematics*, pp. 30–32, Pergamon Press, Oxford, 1961.
5. *School Mathematics Project Book T 4*, Cambridge University Press, 1965.
6. *School Mathematics Project Books 1-5*, Cambridge University Press, 1966–9.
7. The School Mathematics Project, *Director's Report* 1962–3, University Printing House, Cambridge, 1964.
8. The School Mathematics Project, *Director's Report* 1963–4, P. & G. Wells Ltd., Winchester, 1964.
9. The School Mathematics Project, *Director's Report* 1967–8, P. & G. Wells Ltd., Winchester, 1968.

Nuffield Foundation Mathematics Teaching Project

A. G. VOSPER

THE Nuffield Foundation Mathematics Teaching Project was inaugurated in September 1964. At that time a number of projects designed to reform Secondary School mathematics courses had already been in existence for 3 or 4 years, and great changes in the methods of teaching mathematics in Junior schools had been in progress during the same period. These changes consisted essentially in extending, first to younger junior children, and later to older junior children, the activity-based methods that had long been common in the Infant schools. A very influential figure in this development was Miss E. E. Biggs, H.M.I., who since 1959 has been largely engaged in running courses for teachers on the new methods of teaching mathematics in Primary schools.

In 1964, after consultation with the Ministry of Education, as it then was, the Trustees of the Nuffield Foundation offered to finance a project, designed to consolidate the progress that had been made, for the age-range 8–13, and invited Dr. Geoffrey Matthews, then Head of the Mathematics Department at St. Dunstan's College, to become its organiser. The upper age limit of 13 was chosen to ensure an adequate link-up between the new project and the existing secondary projects, and to cover the transition from Junior to Secondary school; although today there are many Middle schools, for the age-range 9–13, in 1964 nearly all children changed from a Junior to a Secondary school at the age of 11. Dr. Matthews accepted the post on condition that the lower age limit for the project was reduced to 5, as he believed that an integral part of a child's mathematical development took place during his first 3 years at school. The Trustees accepted these revised terms of reference, and the project was planned to last from 1964 to 1970.

In 1969 Dr. Matthews became Shell Professor of Mathematics Education at the Centre for Science Education, Chelsea College of Science and Technology, occupying the first chair of Mathematics Education to be established in this country, and from that time he has worked half-time for the Nuffield Foundation Mathematics Teaching Project, whose lease of life has been extended to 1971 to take account of the changed circumstances of its Organiser. In 1970 it remained the only nationally organised mathematics project concerned with work in Primary schools.

At its inception, Dr. Matthews made the following statement of the aims of the Project:

> The object of the Nuffield Mathematics Teaching Project is to produce a contemporary course for children from 5 to 13. This will be designed to help them connect together many aspects of the world around them, to introduce them gradually to the processes of abstract thinking, and to foster in them a critical, logical, but also creative, turn of mind.
>
> A synthesis will be made of what is worth preserving in the traditional work with various new ideas, some of which are already being tried out. These cover presentation as well as content, and emphasis will be placed on the learning process: sleight-of-hand will not pass as a substitute for genuine understanding. A concrete approach will be made to abstract concepts, and the children will make their own discoveries whenever possible. The work of the project will be set against the present background of new thinking concerning mathematics itself. But there will be no novelty for its own sake: any topic or aspect of the subject introduced must be more intelligible and purposeful (and so easier to teach) than what it replaces. Perhaps the most important message of "modern" mathematics at this level is its ubiquity, the fact that doing sums is only a fraction of the programme envisaged.
>
> The Project will follow the general lines of those already started by the Nuffield Foundation in other subjects. Teams of teachers will be appointed to assist the Organiser, who will also be able to seek guidance from a Consultative Committee. An integrated range of teaching resources will be prepared, including ample teachers' guides, pupils' materials, and visual and other aids. These materials will be subjected to widespread testing with pupils of various ages and levels of ability, revised in the light of experience, and retested, and finally the resulting range of resources made generally available to all who care to make use of them.

The Project has developed in some ways not referred to in the above statement, but in 1970 the statement may still fairly be taken as indicating what the Project has so far tried to do, and what it is proposing to do in the time left to it. The statement does not refer to Teachers' Centres, described later, which have played a a vitally important part in the Project's development. However, their establishment was planned from the beginning in conjunction with the Project; although the Nuffield Foundation has no financial responsibility for them, they have been so closely associated with the Project that they are rightly regarded as an integral part of it.

The first team, which started work in September 1964, consisted of Dr. Matthews and six other persons, a College of Education lecturer, a Local Authority Inspector, a Secondary-school teacher, a Junior-school headmaster, another Junior-school teacher and an Infant-school teacher. The names of the members of this first team, and of the teams that have followed it, are given in Appendix B. The first team embarked upon writing a number of Teachers' Guides for the infant and lower junior age-ranges, and a Guide called "I Do and I Understand" describing the general philosophy of the Project and dealing with such matters as the place of the child's activity, discovery methods, the use of the environment, and classroom organisation. It was decided that initially at any rate no pupils' materials should be produced, though these have been produced more recently for older children. The first team felt that textbooks for children's use were not appropriate for young children and that Teachers' Guides should be written to help teachers to devise for themselves work suitable for their own classes. The team were agreed that the Guides should emphasise the importance of the child's understanding, the importance of his learning at his own rate, and consequently the advantages of children working in small groups rather than as a whole class, though of course it was recognised that there were occasions when class work, especially class discussion, was appropriate. It was appreciated that many teachers would need more help than a series of books could provide, and it was here that the Teachers' Centres, whose establishment had formed part of the original plans discussed by the Trustees of the Nuffield Foundation with the Ministry of Education, would be invaluable as places in which teachers could attend courses and meet to discuss difficulties. These Centres were set up first in the fourteen pilot areas chosen to provide trials of the Teachers' Guides as they were written. In June 1964, the Department of Education and Science, as it had then become, wrote to all local education authorities in England and Wales asking them if they would like to take part in the Project. The proposal made to them was that they should be given any trial material produced by the team members, on condition that they

1. selected suitable schools in their area where the material could be tried out;
2. sent feed-back comments on the material to the Project headquarters;
3. provided a place and staff for a Teachers' Centre.

More than half the L.E.A.s in England and Wales wished to join in, but as this was far too large a number for the team to keep in effective contact with, thirteen were selected to start work on the trial materials in September 1965, and one further area was selected in Scotland. The pilot areas were chosen to cover as many different types of area as possible in educational and social background. Pressure from the areas that had not been selected was such that in 1966 another seventy-five areas were allowed to join as "second-phase areas", with the difference that these areas were asked to buy the trial materials as they were produced at roughly cost-price. Since then, many other areas, using the published, not the trial, versions of the Guides, have joined in as "Continuation Areas". A list of the pilot areas is given in Appendix A, and a list of the second-phase areas, other than those in Scotland, is given in the *Schools Council Field Report* No. 1, referred to in Appendix G.

In 1964–5 the Project team members were engaged in writing Teachers' Guides, basing them on work already being carried out in many Primary schools, and on work they, and teachers known to them, were doing in a few schools that came to be known as pre-pilot schools. Before reaching the form of a trial edition, distributed to pilot, and later second-phase areas, Teachers' Guides in typescript form were used in a number of these pre-pilot schools, and were criticised and discussed at team meetings, and by members of the Consultative Committee. This committee had been set up under the chairmanship of Professor W. H. Cockcroft, of the University of Hull, in 1964 to give general guidance to the Project, and it has continued to meet regularly to advise on matters of policy. Its membership is given in Appendix C.

During the first year, team members also helped to run courses for teachers, administrators and local inspectors sponsored by the Schools Council. The Schools Council was set up in October 1964, and it assumed a general responsibility for the arrangements for conferences and in-service courses in connection with the Project. After September 1965 team members continued to help to run courses, and of course to write, and in addition they regularly visited the pilot areas. Some contact was also established with second-phase areas from 1966 onwards, and with continuation areas from 1967 onwards, especially by means of regional and national conferences, but the closest contact was maintained with the pilot areas, and these continued to supply the most detailed comments on the trial editions of the Guides. Each pilot area comprised at least nine Primary schools and at least one Secondary school. The trial editions of the Guides were thus tried out in well over 100 schools. As a result of comments from the pilot areas, some Guides were extensively rewritten before being published, whilst some were published in a form very little different from that of the trial edition. Because it was felt that considerable changes might be called for, trial editions of the Guides were made available only to pilot and second-phase areas and colleges of education. It was felt important that colleges of education should be kept abreast of developments, and, too, that students during periods of teaching practice in schools using the trial editions should themselves have access to them. They have indeed been useful in colleges as source books of ideas for students teaching in schools not associated with the Project. Pressure on the

Project to make all the Guides more widely available has often made it difficult to strike the right balance between publishing a Guide prematurely before sufficient feed-back has been obtained, and holding it back longer than necessary in order to make comparatively trivial amendments.

The first two Guides to become generally available were published in May 1967. They were *I Do and I Understand*, mentioned earlier, and *Pictorial Representation*, a Guide dealing with graphical work throughout the Primary school. These have both sold very widely (over 80,000 copies each by May 1970), and remain two of the most successful that the Project has produced. Seventeen further Guides had been published by June 1970, together with a book for parents entitled *Your Child and Mathematics*, by Professor Cockcroft, the Chairman of the Consultative Committee. Publication of all the books of the Project is by a consortium consisting of W. & R. Chambers, Edinburgh, and John Murray, London; the contents of the Guides already published are summarised in Appendix D.

The Guides are in three main series, *Computation and Structure*, *Shape and Size*, and *Graphs Leading to Algebra*, and there are a number of "weaving Guides" on topics not included in the main series. Associated with the three main series are three distinguishing symbols, a circle, a triangle and a square respectively. It was thought that these would be convenient distinguishing marks if the titles of the various Guides did not immediately make clear to which series they belonged. However, except for the first Guide in each series, *Mathematics Begins*, *Beginnings*, and *Pictorial Representation* respectively, all Guides in each of the series have the same title as the series itself, and the distinguishing marks are not really necessary. Five Guides in the *Computation and Structure* series are to be produced, and four have already been published; four are to be produced in the *Shape and Size* series, and three of these have already been published; and three in the *Graphs Leading to Algebra* series, and of these two have already been published. In accordance with the general aim of the Project that children should work at their own rate as far as possible, no indication is given with a Guide as to the precise age-range with which it should be used. It is hoped that teachers using the Guides will decide this for themselves. However, it would perhaps not be misleading to say that of the Guides in the three main series, *Pictorial Representation* $\boxed{1}$ contains work for all levels from infant to upper junior, *Beginnings* ▽1 and *Mathematics Begins* ① will be found most useful in the Infant school, *Shape and Size* ▽2 and ▽3 and *Computation and Structure* ② and ③ will be used most in lower and middle Junior school classes, whilst the remaining books contain work mostly suitable at the upper Junior and lower Secondary levels. It is expected that teachers will omit topics and sections that they do not think suitable for their classes, and also introduce other material not treated in the Guides. The Project does not wish the Guides to be treated as in any way authoritative, or to constitute a new orthodoxy. They are essentially books of suggestions which have been tried out widely before publication and found useful, but which the individual teacher may well be able to improve upon.

The contents of the earler Guides are by now fairly well known, but a few comments about their format may be useful. As has been emphasised, they are primarily for teachers, and are not normally suitable for giving directly to children. Most of the Guides, except the very early ones relating to children who cannot yet read, include suggestions for assignment cards. The language in these is intended to be suitable for children's use, but of course it is hoped that teachers will make up their own cards, perhaps on the lines suggested in the Guides, rather than use directly any of the specimen examples; in any case, the number of these is too small to suffice without considerable supplementing. Although the Project has produced problem cards, discussed later, for children in the lower forms of Secondary schools, it has firmly decided against producing sets of assignment cards for young children.

Most of the Guides are illustrated with examples of children's work, most of it produced during the pre-pilot trial stage. Such illustrations were something of an innovation when they first appeared, but they have been widely welcomed as part of the process of helping teachers to know what their colleagues, and their colleagues' pupils, are doing, a process helped forward most greatly of course by the opportunities for discussion provided by the Teachers' Centres. The children's work was not selected to show the best that might be done, but rather the quality of what might normally be expected, and a fairly wide range of attainment has been illustrated. It is probable that it is somewhat above average in general tidiness and attractiveness, but an attempt has been made not to allow the desirability of these features to over-ride the need to keep the illustrations really characteristic of what most children might do.

The earlier Guides were able to draw upon a great deal of experience in favour of the activity-based methods they advocated, and they included comparatively few radical innovations. Together with an emphasis upon the importance of the child's own activity, and discussion of his work, went a caution against premature insistence upon written recording, and especially the use of mathematical symbolism. Written recording should, of course, be undertaken from time to time when children are ready for it, but as far as possible in the children's own words. Children are encouraged to illustrate relationships, including numerical relationships, in a variety of graphical ways, including particularly the use of an arrow joining signs for the related objects, together with a descriptive phrase characterising the relationship. In the case of numerical relationships, it is recommended that the introduction of the symbols of elementary arithmetic, $+$ and $=$, for example, should be postponed until the child has acquired a clear idea of the concepts involved. Thus children are led gradually to the use of the normal symbolism $3+1=4$, that used to be introduced quite early in the Infant school, through the use of more informal notations such as

$$3 \xrightarrow{\text{add one}} 4 \quad \text{and} \quad (3, 1) \xrightarrow{\text{add}} 4.$$

In this connection, the importance of the concept of an "ordered pair" of objects (in the previous sentence, the ordered pair of numbers 3, 1) was drawn out, and the concept has appeared as one of particular significance in a lot of work of other kinds in later Guides, notably that on the integers in *Computation and Structure* ④ and on coordinates in *Graphs Leading to Algebra* [2]. Great emphasis is laid in the early Guides on relationships of many kinds, such as "bigger than", "longer than" and "heavier than", and comparatively less emphasis than in the past on the relationship of numerical equality. This change of emphasis is followed up in later Guides with more work on inequalities of various kinds than is customary in most traditional mathematics courses.

Some of these innovations, and others, described below, in later Guides, some of which are of a fairly radical character, did cause difficulties to teachers, and it was here that the Teachers' Centres proved invaluable, as places where teachers could meet to discuss their difficulties, study the Guides, and perhaps attend a course given by a team member or other teacher experienced in the new methods of teaching mathematics. These activities, of course, still continue in Teachers' Centres, and many others are now common. Teachers meet to pool ideas on writing assignment cards, to make use of reprographic equipment, in some places to make apparatus; in some they have access to a computer. (Experiments with a desk-top computer have been carried out in a number of Junior schools as part of the Project and there may well be considerable developments in this direction.)

Teachers often meet to discuss matters not connected with the Nuffield Project. Many Centres were from the first associated also with the Nuffield Foundation Junior Science Project, and many more are now concerned with subjects other than mathematics. The Centres do not, of course, all flourish equally. Those that are most lively are those that have been most encouraged by the L.E.A. by the provision of good premises and a full-time salaried warden. In some cases, teachers can only come to a Centre in their own time; naturally a Centre flourishes more when teachers are released from school to attend, and some L.E.A.s are prepared to release teachers for a half-day a week for an extended period such as a term or longer; in such cases, teachers often come twice a week, once in their own time and once in school time, and this has proved a very good arrangement. Before the setting up of Teachers' Centres, it was comparatively rare for teachers in the same school to discuss their common teaching problems, or even to discover that they had common teaching problems, and even rarer for teachers from different schools to do this. Now the opportunities for discussion have generated an enthusiasm for teaching that has affected other teachers in an area, and made them want to join the Project. This has helped the Project to spread, and when the Project formally ends in 1971, the Teachers' Centres even more than the Teachers' Guides will help to keep its ideas alive and, still more, to develop.

The ways in which the Project works have remained broadly the same throughout the period of its existence, but there have been a number of developments. In 1964 a wide range of structural apparatus was available commercially for the use of young children, and with all the possibilities for mathematical work with materials that occur naturally in the child's environment, such things as household containers, pieces of ribbon, buttons and cotton reels, the first team felt there was no need to devise new pieces of apparatus for use in the Infant school. However, the Project did develop in 1965 a piece of apparatus called the Multi-board, consisting of a large nail-board with a variety of ancillary materials, such as number squares, number strips, Napier's rods and wooden blocks, useful for work in number, coordinates and algebra; a descriptive leaflet is supplied with the apparatus, explaining how it may be used, and several references to it are made in the Guide *Graphs Leading to Algebra* [2]. The Project also produced, in 1967, the Pandora Box, useful for work in classification and sorting, and referred to in the forthcoming Guide to be entitled "Logic and Sets", which, however is not likely to be published before 1971. The box itself, whose name comes from the words "and" and "or", is commercially available, together with a descriptive leaflet, and details of its suppliers, the suppliers of the Multiboard, and prices, are given in Appendix F. Some other pieces of apparatus, some simple circuit boards and possibly a very simple computing device, and perhaps a piece of apparatus for conducting simple experiments in mechanics, are also being developed, but these are not yet available. None of this apparatus is in any way an essential part of the Project, and participation in the Project does not necessarily entail the purchase of any special apparatus.

The Project has also sponsored the production of a number of films. The now widely known *I Do and I Understand* was made in a Blackpool Junior school in 1964 for the Project by one of the first team members shortly before he joined the Project. The title of the film, the same as that of one of the first Teachers' Guides, is taken from a Chinese proverb: "I hear and I forget, I see and I remember, I do and I understand". Like all proverbs, this is no doubt an over-simplification, but in many ways it epitomises an important part of the Project's philosophy. In 1967, the B.B.C. broadcast a series of five television programmes, entitled "Children and Mathematics"; of the first three of these programmes, one was presented by Dr. Matthews, and the two others by two of the first team members; the fourth was concerned with Check-ups, discussed below, and the fifth was filmed in a Teachers' Centre. These programmes have been made into films, and details of their availability, together with that of other films, are also given in Appendix F. A successor to *I Do and I Understand*, entitled *Into Secondary School*, was made in 1968, showing work in the spirit of the Nuffield Project being carried out in two Secondary schools, Grass Royal School in Yeovil, and Elliott Comprehensive School in Putney. At present the Organiser of the Project is working on an infant film; though most of the material at present being written is for upper Junior or lower Secondary pupils, contact with and interest in work at infant level has not been lost, and besides the projected infant film, the series of proposed Checking up Guides, next to be discussed, has ensured that the Infant schools have remained very much a concern of the Project.

With children working in small groups, largely at their own rate, it becomes impossible to check up on their progress by conventional tests. The need for some kind of check up is, however, strongly felt by most teachers, and parents, and early on it was foreseen that the Project would have to do something to meet this need. In 1966

Jean Piaget, at the Institut des Sciences de l'Éducation at Geneva, expressed interest in the Project and general approval of its aims, and offered help to the Project from his team of research workers. For many years Piaget and his team had been investigating concept-development in children, especially in mathematics, and they had devised many techniques for studying this in individual children. Most of these techniques were quite elaborate and required a skilled psychologist to employ them. It was hoped, however, that they could be simplified sufficiently for teachers to use them in the classroom with individual children, spending no more than a minute or two with each individual child. Three volumes on checking up, obtained in this way, are projected, but their production has proved much more difficult than was expected. The trial edition of *Checking up I* (called Check-ups I) contained simple procedures for use in the Infant school for checking up on whether a child had acquired such concepts as that of order, or the invariance of number, and included with these procedures precise instructions on questions the teacher was to ask and on how the child's answers were to be interpreted. The Guide aroused more controversy than any other, with the possible exception of *Computation and Structure* ④, discussed later, and it was radically rewritten before being published. Many teachers felt that the procedures, though simpler than those used at Geneva, were still too elaborate for use in the classroom, that the questions could easily be misunderstood by the child, and that the child's answers could easily be misinterpreted by a teacher not skilled in this kind of investigation. Some teachers felt that the Check-ups could easily be misused as a basis for a course, and still others felt that Check-ups at this level were not needed at all. As now published, most of the Check-ups are much looser in form, with less detailed instructions, and more general guidance to the teacher on how to judge a child's progress from his ordinary work. How far Check-ups are essential at infant level is perhaps debatable, but there seems little doubt of the need for them with older children. *Checking up II* has just been written, and how far this and the as yet unwritten *Checking up III* will meet the need remains to be seen. The continued collaboration between the Nuffield Project and the Institut des Sciences de l'Éducation is, however, very welcome, and whatever the difficulties, it offers the greatest hope for a solution of the problem of checking up on the progress of a child when he is working freely at his own rate that has yet appeared.

Initially, it had been intended to continue the three main series of Guides, and the ancillary weaving Guides, in the same format as the early ones, to cover the full age-range 5–13. This came to seem unsuitable because of the difficulties associated with introducing the work of the Project into Secondary schools. At least one Secondary school was associated with the Project in each of the pilot areas, but as nothing was written that was of direct interest to Secondary schools for the first 3 or 4 years of the Project's existence, it was naturally difficult to retain their interest, and in particular to ensure their continued participation in the work of the Teachers' Centres. With some few exceptions, therefore, Secondary schools in the main had very little connection with the Project by about 1968, and to produce Guides entitled *Computation and Structure* ⑥, or *Shape and Size* ⑤ for use in Secondary schools seemed unrealistic, because it was not likely that secondary teachers would wish to embark on using a series at volume 6, nor that they would wish to make themselves fully familiar with the 5 volumes that preceded it. Further, the Project, because of the mathematical content of some of the later Guides, seemed to link up well with the new syllabuses proposed by such projects as S.M.P. and M.M.E., and it was not at all clear that in terms of mathematical content the Nuffield Project had any very distinctive contribution to make at secondary level. However, activity and discovery methods recommended by the Project had proved valuable for many children right up to the top of the Junior school, and it seemed likely that the experience the Project had gained in this field could be usefully applied to work in at least the lower forms of Secondary schools. It was therefore decided in the first place to produce a number of series of problem cards for first- and second-year secondary children, the first pupils' materials that the Project had produced. These have come to be known by the colours in which they are, or are to be, printed: a fairly simple Green Set, and a more difficult Red Set, already published, and a Purple Set, intermediate in difficulty, which has been written but is not yet in the form of a trial edition. The problems are all fairly open-ended, giving the pupil situations suitable for investigation in a variety of ways, and so far as possible, at different levels of sophistication. Each set of problems is accompanied by a Teachers' Book consisting of a full commentary on each of the problems, with suggestions for developing the problem further. It was not envisaged that work on these problems would be more than a small part of the mathematics done from 11–13 in Secondary schools, but it was hoped the problems would foster a spirit of individual inquiry, and encourage pupils to work at their own rate and level, continuing into the Secondary school work in this way for pupils who had already met it in their Primary schools, and introducing it for those who had not.

Most of the existing secondary projects, and the B.B.C. Television series "Maths Today", which appeared while the Red Problems were being written and influenced the content of a number of them, of course encourage the same kind of thing, but they naturally lay considerable emphasis on the logical development of the subject at this level, and Nuffield team members felt that problems of the kind described could be a useful adjunct to lower secondary work, even in schools working with one of the new projects. The Green and Red sets of problems were tried out extensively in a number of London Secondary schools, as well as in the few Secondary schools in the pilot areas, and modified considerably before publication.

Besides the sets of problem cards, the Project is producing a number of small booklets for use in Secondary schools, to be called units or modules. Still with the intention of supplementing, rather than offering an alternative to, the texts of the various secondary projects, these booklets will treat particular topics in which it is felt that the usual work done in Secondary schools could usefully be supplemented by more practical activities, or in which most pupils experience difficulties, perhaps as a consequence of an insufficiently concrete approach. These booklets will appear in a number of different forms, but most will contain enough material for two or three weeks' work, and will include a series of cards with a collection of related problems for issue to children, instructional sheets for children which will be given out as needed, and a commentary for the teacher, with suggestions for further development and perhaps a book-list. Occasionally some simple materials or equipment may be included. Booklets of this kind have already been written on Number Patterns, Decimals, Simultaneous Equations, Speed and Gradient, Area and Coordinates, and others are under consideration. The content of some of these, of course, considerably overlaps that contained in some of the Teachers' Guides, but most will take the subject matter somewhat further than it was taken in the earlier treatment. Some of the Guides already published, particularly *Graphs Leading to Algebra* [2] and *Computation and Structure* ④, and some still unpublished, contain a good deal of material which it is hoped will be useful in Secondary schools for those who care to use it, but none of them, except the general Guide called *Into Secondary School*, was written specifically for use in Secondary schools. It is certainly hoped that all the later Guides will be suitable for use in Middle schools.

Something will now be said about some of the more controversial mathematical suggestions in one or two of the later Guides, though it should be clear that in the context of the aims of the Project as a whole these are not of major importance. Some attention has been focused upon them, however, and some discussion would not be out of place. One of the most radical proposals concerns the treatment of subtraction and division, and the related suggestions that positive and negative integers might be introduced and used before rational numbers or fractions, and that both integers and rational numbers might be introduced, and defined, as equivalence classes of ordered pairs of natural numbers. In the early Guides, as we have seen, introduction of the usual symbols for addition and multiplication is deferred much longer than is usual, and those for subtraction and division are not introduced at all, for the operations themselves are avoided until a late stage. Mathematical and pedagogical reasons for deferring the introduction of subtraction and division are not hard to find. The set of natural numbers, the first set of numbers with which children become familiar, and the only one for which they have much mathematical use during their first few years at school, is not closed with respect to subtraction nor with respect to division; to obtain sets which are closed with respect to subtraction and division, larger sets of numbers are required, the positive and negative integers and rational numbers respectively. Furthermore, subtraction and division are neither commutative nor associative; this causes difficulties to children who early become accustomed to the commutative and associative properties of addition and multiplication, and when subtracting two-digit numbers, for instance, frequently attempt to subtract the wrong way round. There are, of course, on the other hand reasons why subtraction and division should be met early, notably the early occurrence in the child's experience of "taking away" and "sharing"; however, Project members felt that when these practical problems could not be dealt with by essentially simpler procedures, such as complementary addition, at least trial should be made of the effects of deferring their mathematical treatment until later than is at present usual, until indeed sets of numbers with respect to which subtraction and division are closed can be introduced.

Because addition is essentially easier than multiplication, and so subtraction than division, it was further decided to suggest in the Guides that positive and negative integers, and subtraction, might be introduced before fractions, or rational numbers, and division; with this end in view, a large part of *Computation and Structure* ④ is devoted to the integers, and most of *Computation and Structure* ⑤, not yet published, is devoted to fractions. Finally, because children have such difficulties later on with the rule of signs, and with the manipulation of fractions, it was decided to try to introduce these new numbers in as mathematically sound a way as possible, in the hope that a fuller understanding of their nature would assist in later work. There are thus five main innovations proposed in this field:

1. deferment of subtraction until positive and negative integers have been introduced;

2. deferment of division until fractions have been introduced;
3. treatment of positive and negative integers before fractions;
4. introduction, and definition, of integers as sets of ordered pairs of natural numbers;
5. introduction, and definition, of fractions as sets of ordered pairs of natural numbers.

As *Computation and Structure* ⑤ is not yet in a trial edition, no feed-back from the pilot areas has yet been received on this Guide. Pilot areas gave a cautious welcome to the more controversial features of *Computation and Structure* ④, and it was decided to publish it, with considerable amendments made as a result of the feed-back that was received. The Project has, of course, no wish to be dogmatic in the matters discussed above. Innovation (3) above is naturally suggested by devoting a large part of *Computation and Structure* ④ to the integers, and most of *Computation and Structure* ⑤ to fractions; but although it is natural to take the Guides in any one series in the order indicated by their numbers, there is no need to do this rigorously, for they are made as independent of each other as is practicable, and in particular many teachers may well prefer to do a lot of the work suggested in *Computation and Structure* ⑤ before some of that in *Computation and Structure* ④. Innovations (4) and (5) above can readily be ignored by those who wish, for traditional treatments of integers and fractions, avoiding the use of sets of ordered pairs, are also included in the relevant Teachers' Guides; however, in the treatments involving ordered pairs, every attempt is made to build up the relevant sets naturally, from a variety of experiences, and not to introduce them in a way that would be appropriate only at a more advanced level. Innovations (1) and (2), too, may of course also be disregarded by teachers who wish to introduce subtraction and division earlier than the Guides recommend; some suggestions about subtraction are included in *Computation and Structure* ③.

Space prohibits mention of most of the other mathematical innovations, such as the discussion of modular, or "clock", arithmetic in *Computation and Structure* ③ and ④, the introduction of vectors in *Shape and Size* ∇4, and the brief treatment of groups in the Appendix to *Shape and Size* ∇3. A brief account of the provisional contents of the Guides as yet unpublished is included in Appendix E.

However, something must be said about the terminology and symbolism used in *Graphs Leading to Algebra* [2], and the sequence of topics in this book and its successor in the same series. *Graphs Leading to Algebra* [2] was much influenced by the American Madison Project; it was felt that the introduction of the concepts of open sentence, place-holder, and truth-set make for simplicity and clarity, though they are mostly unfamiliar in this country at Junior-school level. The use of a box instead of x or y for a place-holder is more customary, but this, together with the new terminology, and the fact that all the graphs in the book consist of disconnected sets of points, since the only numbers used are natural numbers and integers, gives *Graphs Leading to Algebra* [2] a very unfamiliar appearance. However it was found that surprisingly many important ideas relating to coordinates, graphs, and algebra could be illustrated using no numbers other than natural numbers and integers, and this, taken in conjunction with the decision to treat integers before fractions in the *Computation and Structure* series, led the Project to defer to *Graphs Leading to Algebra* [3] any graphical or algebraic work involving fractions and real numbers. One consequence of this is that comparatively few practical experiments leading to graphs are included in *Graphs Leading to Algebra* [2], for such experiments invariably require measurement, rather than counting, and so need rational numbers. As a result, many teachers will undoubtedly prefer to deal with the work contained in the first few sections of *Graph Leading to Algebra* [3] before that in the last few sections of its predecessor; however, as with the inversion of order discussed in connection with the *Computation and Structure* series, there is nothing to prevent this, and it is certainly not the intention of the Project to discourage it.

It would be misleading to conclude with a discussion of technical innovations, particularly ones likely to be very controversial. Such innovations naturally bulk more largely in the suggestions made to teachers of older children than they do in those made to teachers of infants, but they are not important for their own sakes. All the controversial proposals of the Teachers' Guides may in time be rejected; unless they are obviously foolish or harmful, and it is hoped that the trials have been sufficiently extensive for this at least not to be the case, they will have served their purpose if they have helped to stimulate teachers into thinking about their subject, and the best ways of teaching it, more deeply than before. The ideas in the Guides must be constantly rethought in the light of experience. The Guides themselves will certainly be superseded by other books when the Project ends, and it is rather through the continuing work of the Teachers' Centres that the Nuffield Mathematics Project will have its most lasting and, it is hoped, beneficial influence on the teaching of mathematics, and indeed on teaching generally.

APPENDIX A. PILOT AREAS

Birmingham	Moseley/King's Heath area
Bristol	Withywood and Brislington areas
Cambridgeshire	Cambridge city and neighbouring rural area
Cardiff	Llanrumney area
Doncaster	An urban area of the borough
Edinburgh	Rosebank
Hampshire	Winchester area
Inner London	Ladbroke area, North Kensington
Kent	Folkestone area
Middlesbrough	Area on eastern outskirts of town
Newham	Former West Ham area, London E. 15
Northumberland	Whitley Bay area
Somerset	Yeovil area
Staffordshire	Kidsgrove area

APPENDIX B. TEAM MEMBERS

1964–5
J. W. G. Boucher
G. B. Corston
H. Fletcher
Miss B. A. Jackson
D. E. Mansfield
Miss B. M. Mogford

1965–6
J. W. G. Boucher
G. B. Corston
H. Fletcher
Miss B. A. Jackson
D. E. Mansfield
Miss B. M. Mogford

1965–6 *Miss R. K. Tobias
*A. G. Vosper

1966–7 D. R. Brighton
Miss I. Campbell
*H. Fletcher
D. E. Mansfield
J. H. D. Parker
Miss R. K. Tobias
*A. G. Vosper

1967–8 E. A. Albany
D. R. Brighton
Miss I. Campbell
*Miss R. K. Tobias
*A. G. Vosper

1968–9 *E. A. Albany
D. R. Brighton
*A. G. Vosper

1969–70 *E. A. Albany
D. Jones
J. H. D. Parker
*A. G. Vosper

* Part-time for 12 months or full-time for less than 12 months.

APPENDIX C. CONSULTATIVE COMMITTEE

J. W. G. Boucher
Professor W. H. Cockcroft, Chairman
R. C. Lyness, H.M.I.
Miss B. M. Mogford (1964–6)
H. S. Mullaly (from 1966)
R. Openshaw
N. Payne (from 1967)
D. R. F. Roseveare
J. Shanks (from 1966)
A. G. Sillito (died 1966)
P. F. Surman
Dr. D. R. Taunt
Mrs. D. E. Whittaker (from 1967)
F. Woolaghan
Professor J. Wrigley

APPENDIX D. SUMMARY OF THE CONTENTS OF THE BOOKS PUBLISHED BY JUNE 1970

Introductory Guide:

I Do and I Understand. General philosophy of the Project; the place of activity, discovery, use of the environment; class organisation, equipment, records; appendices by two teachers on how they started, and on the role of colleges of education. Published May 1967.

Computation and Structure Series:

① *Mathematics Begins*. A Guide parallel to ▽1 *Beginnings* (see below) but with greater emphasis on number. Relations, sorting, matching, sets, ordering, inclusion. Operations and the use of symbols. Published September 1967.

(2) *Computation and Structure*. Natural (i.e. counting) numbers; length, weight, capacity; counting, addition; place-value; time, money. Book-list. Published September 1967.

(3) *Computation and Structure*. Addition, a note on subtraction, multiplication; simple sharing; factors and primes; introduction to fractions. Published May 1968.

(4) *Computation and Structure*. Extension of place-value; decimal notation. Modular arithmetic. Introduction to positive and negative integers as sets of ordered pairs of natural numbers; applications. Large numbers and indices. Published May 1969.

Shape and Size Series:

▽1 *Beginnings*. The concepts of volume, capacity, length, area, symmetry, time and number emerging from play, everyday experiences and other activities. Published September 1967.

▽2 *Shape and Size*. Volume and capacity, fitting three-dimensional shapes together; from three dimensions to two; symmetry and area, fitting two-dimensional shapes together; angles, perpendiculars, parallels; classification of two-dimensional shapes. Published September 1967.

▽3 *Shape and Size*. Two-dimensional frameworks, rigid and non-rigid; parallelism and angles; patterns on circles. Polygons, tesselations, bilateral and rotational symmetry. Polyhedra. Transformations. Appendix on groups. Published May 1968.

Graphs Leading to Algebra Series:

[1] *Pictorial Representation*. The mathematics involved in pictorial representation; stages in the development of graphs, especially bar and line charts; the care needed to avoid drawing misleading graphs. Published May 1967.

[2] *Graphs Leading to Algebra*. Introducing coordinates using natural numbers only; open sentences, their truth-sets and graphs; graphs of inequalities; extension to work using positive and negative integers (no fractions). Published May 1969.

Other Guides:

Desk Calculators. A brief guide to ways in which calculators can be used in teaching children number patterns, place-value, multiplication in terms of repeated addition and division in terms of repeated subtraction. Published September 1967.

How to Build a Pond. A facsimile reproduction of a class project. Published September 1967.

The Story So Far. A brief summary of, and index to, ①, ②, ③, ▽1, ▽2, ▽3, [1], Chapter 1 of [2] and *Desk Calculators*. Published May 1969.

Environmental Geometry. First stages in the study of shape and size in the environment. Published May 1969.

Probability and Statistics. The play stage; games leading to ideas in probability; sampling; recording, and the use of pictorial representation; averages; measuring probability. Published May 1969.

Problems (*Green Set*). Fifty-two open-ended problems devised for children beginning at the Secondary school. The problems are available on cards for issue to individual children, and the accompanying Teachers' Book discusses each problem and possible developments in detail. Published May 1969.

Problems (*Red Set*). A set of thirty-five problems similar to, but rather harder than, those in the Green Set, with a similar accompanying Teachers' Book. Published June 1970.

Into Secondary School. A companion to *I Do and I Understand* concerned with work in the first 2 years in the Secondary school. Published June 1970.

Checking up I. Suggestions, based closely on Piaget's work, and written in conjunction with the Institut des Sciences de l'Éducation at Geneva on how teachers in Infant schools can check up on children's progress. Published June 1970.

Your Child and Mathematics, by Professor W. M. Cockcroft. Not a Teachers' Guide; written for parents by the Chairman of the Project's Consultative Committee. Published May 1968.

APPENDIX E. SUMMARY OF THE PROVISIONAL CONTENTS OF PROJECTED GUIDES AND OTHER BOOKS UNPUBLISHED BY JUNE 1970

Some of the following are in trial editions, some completed in manuscript, some incomplete.

▽4 *Shape and Size*. Introduction to vectors; extension of work in ▽2 and ▽3 on symmetry and transformations.

⑤ *Computation and Structure*. Addition of decimal numbers. Introduction to positive rational numbers as sets of ordered pairs of natural numbers. Rational numbers represented by points on the number line. Ordering, and the four arithmetical operations on rational numbers.

[3] *Graphs Leading to Algebra*. Coordinates using rational numbers. Graphs from experiments. Place-holders and variables. Linear equations and inequalities; simultaneous equations and linear programming. Comparison of the algebra of numbers with the algebra of sets and switches.

Logic and Sets. Classification and sorting. All, some, none, etc. Venn diagrams. Dienes' logical attribute blocks. Negation. If ... then ... Logical puzzles and paradoxes.

Computers and Young Children. Computers in society. Flowcharts; a computer game; programming. Report of some work in schools with an Olivetti-101 desk computer.

Logic and Computers. Simple electrical circuits. Logic gates. The binary adder.

A Guide to the Guides. To supersede *The Story So Far*, published May 1969. A brief summary of, and index to, all the Guides published to date, together with a number of schemes of work produced by different schools taking part in the Project.

Problems (Purple Set). A set of problems intermediate in difficulty between the Green and Red Sets, already published, also with a Teachers' Book.

Checking up II and III. Continuation of *Checking up I*, already published.

Mathematics in Practice. Project book for upper juniors, using mathematics with other subjects.

APPENDIX F. FILMS AND APPARATUS PRODUCED BY, OR IN CONJUNCTION WITH, THE PROJECT

Films

I Do and I Understand. A 15-minute film taken in a Junior school in Blackpool in 1964. Available on free hire from British Petroleum Film Bureau, 4 Brook Street, London, W. 1. Available for purchase, at £12·50, from Sound Services Ltd., Wilton Crescent, Merton Park, London S.W. 19.

Children and Mathematics. Five 30-minute B.B.C. Television programmes made in 1967. Available for hire at £1·50 a film from Concord Film Council, Nacton, Ipswich, and for purchase at £35 a film from B.B.C. Television Enterprises, B.B.C. Television Centre, White City, London, W. 12.

Into Secondary School. A 15-minute film taken in two Secondary schools. Available for hire and purchase as *I Do and I Understand*. Purchase price £16.

Apparatus

Multiboard. Available from E. Marshall Smith (School Utilities Ltd.), 5–9 Church Lane, Romford, Essex. Price £6·50.

Pandora Box. Available from S.T.A., Colquhoun House, 27 Broadwich Street, London, W. 1. Price £3·50.

APPENDIX G. PUBLICATIONS WITH FURTHER INFORMATION ABOUT THE PROJECT

The Schools Council Field Report No. 1: *New Developments in Mathematics Teaching*. Published by H.M.S.O., May 1966. A first progress report on the Nuffield Mathematics Project, up to September 1965.

The Schools Council Field Report No. 4: *Progress in Primary Mathematics*. Published by H.M.S.O., July 1967. A sequel to Field Report No. 1, reporting progress up to the end of 1966.

Where, Supplement 13: published by The Advisory Centre for Education (ACE) Ltd., Spring 1968. A guide for parents to the Nuffield Mathematics and Junior Science Projects.

Mathematics Projects in British Secondary Schools. A pamphlet prepared for the Mathematical Association. Published by G. Bell & Sons Ltd., 1968.

International Clearing House Report on Science and Mathematics Curricula Developments. Obtainable from J. David Lockard, Science Teaching Centre, University of Maryland, U.S.A. An annual publication, containing information about the Nuffield Mathematics Project in issues from 1966 onwards.